Focus on Electrochemistry Research

FOCUS ON ELECTROCHEMISTRY RESEARCH

MAGDALENA NUNEZ
EDITOR

Nova Science
New York

For permission to use material from this book please contact us:
Telephone 631-231-7269; Fax 631-231-8175
Web Site: http://www.novapublishers.com

NOTICE TO THE READER

LIBRARY OF CONGRESS CATALOGING-IN-PUBLICATION DATA:
Available Upon Request

ISBN 1-59454-545-6

Published by Nova Science Publishers, Inc. ✦ New York

CONTENTS

Preface **vii**

Chapter 1 The Effects of Permanent Magnets on Oxygen Reduction Reaction and their Application in Polymer Electrolyte Fuel Cells (Pefc) **1**
N. I. Wakayama, T. Okada and L. B. Wang

Chapter 2 Electrodeposition and Characterization of Mo–Se Thin Film Layers **37**
N. Dukstiene

Chapter 3 Potassium Tellurocyanate as a Tellurating Agent **67**
Ali Z. Al-Rubaie

Chapter 4 Copper Electrochemical Polishing for ULSI Fabrications **91**
J. Huo, R. Solanki and J. McAndrew

Chapter 5 Electrocatalytic Hydrogenation of Unsaturated Organic Molecules: Some Features and Preparative Opportunities **145**
L. M. Korotaeva and T. Ya. Rubinskaya

Chapter 6 Development of CO Tolerant Electrocatalysts for Reformate Hydrogen and Methanol Oxidations on Proton Exchange Membrane Fuel Cells **167**
Elisabete I. Santiago, William H. Lizcano-Valbuena, Edson A. Ticianelli and Ernesto R. Gonzalez

Index **213**

PREFACE

Electrochemistry is the branch of chemistry that deals with the chemical action of electricity and the production of electricity by chemical reactions. In a world short of energy sources yet long on energy use, electrochemistry is a critical component of the mix necessary to keep the world economies growing. Electrochemistry is involved with such important applications as batteries, fuel cells, corrosion studies, hydrogen energy conversion, bioelectricity. Research on electrolytes, cells, and electrodes is within the scope of this old but extremely dynamic field.

In: Focus on Electrochemistry Research
Editor: Magdalena Nunez, pp. 1-35

ISBN: 1-59454-545-6

Chapter 1

THE EFFECTS OF PERMANENT MAGNETS ON OXYGEN REDUCTION REACTION AND THEIR APPLICATION IN POLYMER ELECTROLYTE FUEL CELLS (PEFC)

***N. I. Wakayama*[a]*, T. Okada*[b] *and L. B. Wang*[c]**

[a]National Institute for Materials Science, Tsukuba, Ibaraki 305-0003, Japan
[b]National Institute of Advanced Industrial Science and Technology, Tsukuba, Ibaraki, 305-8565, Japan
[c]Lanzhou Jiaotong University, Lanzhou, Gansu 730070, China

ABSTRACT

In a polymer electrolyte fuel cell (PEFC), the cathode is the performance-limiting component, and the performance of the cathode depends primarily on the oxygen transport rate, which is strongly affected by the liquid water which are caused by liquid water generated by the reaction and electro-osmic drag, accumulating in the porous catalyst and gas diffusion layers. Cathode flooding caused by excessive liquid water in the cathode is generally recognized as the primary reason for poor fuel cell performance. This chapter describes a new method of using permanent magnets to manage the flow of liquid water and O_2 gas in a porous cathode. Permanent magnets attract iron, and thus attract paramagnetic O_2 gas, while rejecting diamagnetic water, nitrogen gas, etc. This kind of magnetic force is called Kelvin force. Recently, the effect of permanent magnets on electrochemical oxygen reduction was studied for possible application to cathode gas reactions in PEFCs. In electrochemical and rotating electrode experiments, when a permanent magnet was placed in contact behind a Pt foil or Pt-dispersed carbon paper cathode and O_2 was supplied toward a magnet, i.e., a stronger magnetic field, electrochemical flux was enhanced and the current increased with increasing magnetic field. These magnetic effects can be explained by the magnetic repulsive force acting on water and by the attractive force toward O_2 gas. Furthermore, when small magnet particles were inserted in the cathode catalyst layer of a PEFC and then magnetized, the performance of a single PEFC was superior to that of a PEFC not magnetized; the performance was particularly enhanced in the limiting current region. Because these magnetized particles worked as permanent magnets, the use of magnet particles will provide a method to manage liquid water in the cathode of PEFC and thus improve the fuel cell performance. To clarify the mechanism by which the fuel cell performance is thus improved, the numerical simulation was conducted, using a two-dimensional, two-phase model. The numerical results indicate that magnetic repulsive force acting on liquid water mainly improves the fuel cell performance by decreasing the overall level of liquid water saturation near the interface of the catalyst/diffusion layers and thus increasing the pore space for O_2 to transport to and contact with the catalyst layer. The results also show that fuel-cell performance was particularly enhanced in the limiting current region in the presence of magnetic force, agreeing with experimental results. The use of permanent magnets will be especially useful for portable fuel cell in which there is no power to supply air.

1. Introduction

Polymer electrolyte fuel cells (PEFC) are considered as one of the most promising energy-conversion devices of the 21st century due to their high-energy efficiency, high energy and power density, and environmental friendly characteristics (i.e., low pollutant emissions and absence of radiation). However, to be cost competitive with current energy-conversion devices, the performance of PEFC needs to be further improved [1,2]. A fuel cell is an electrochemical energy conversion device that shares similarities with a battery, yet is distinctly different. Whereas a battery stores energy in the chemicals contained in it, a fuel cell acts as a converter where the reactants and products are merely in transit within the cell; battery electrodes are consumed, wheres fuel cell electrodes are not. A typical PEFC with a hydrogen fuel and air as oxidant is illustrated in Fig. 1. At the anode, a fuel (H_2) is oxidized, whereas, at the cathode, an oxidant (O_2 in air, O_2,) is reduced via Pt catalyst ($1/2O_2 + 2H^+ + 2e \rightarrow H_2O$). In this spontaneous process, electrons that are generated at one electrode and then consumed at the other circulate in an external circuit that can drive, for example, an electric motor.

A typical polarization curve of a PEFC (Fig. 2) shows that the loss in cell potential occurs due to activation resistance, ohmic losses, and mass transfer limitations [1,7]. The sharp voltage drop of the first region associated with the activation resistance is attributed to the type of catalyst and the catalyst surface area that is contact with the electrolyte and its accessible to the reaction gasses. The gradual drop in voltage of the second region, known as the ohmic voltage loss, is directly related to the protonic conductivity of the membrane and the electronic conductivity of the current collectors.

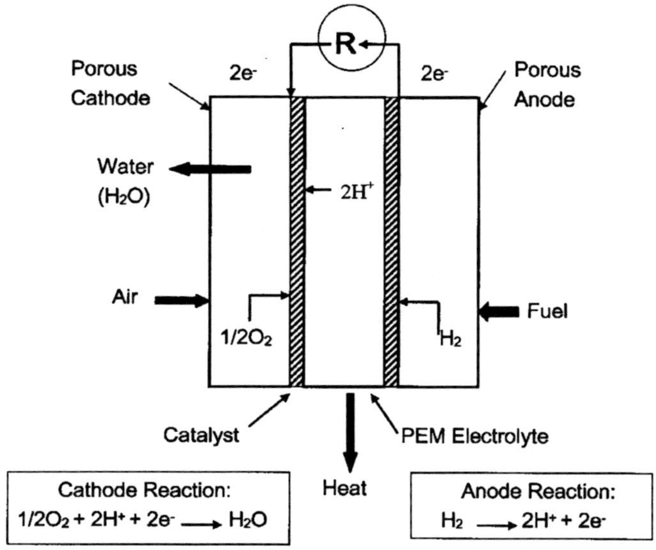

Figure 1. Polymer electrolyte fuel cell (PEFC)

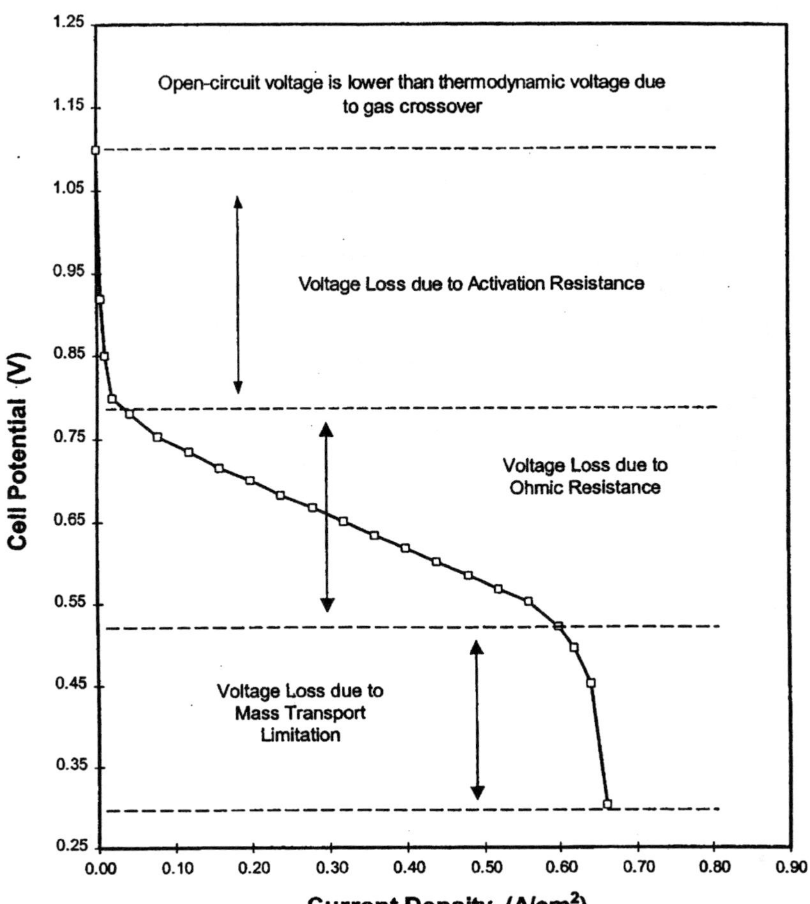

Figure 2. Typical polarization curve of a polymer electrolyte fuel cell (PEFC)

The sharp voltage drop of the third region, known as the voltage loss associated with the mass-transfer limitations, is attributed to concentration polarization, which occurs as a result of the depletion of the reactant at the reaction interface because its transport to the reaction sites fails to keep up with the reaction rate [3]. This phenomenon is especially severe at the cathode of PEFC where O_2 gas is the reactant due to the presence of liquid water within the porous structure of the electrode and on the catalyst/membrane surface (see Fig. 1). This liquid water, which is the product of the cathodic reaction and proton transport from the anode, acts as an additional barrier to the transport of O_2 to the reaction sites. Minimizing this resistance will allow the ohmic region to be extended, resulting in much higher power density.

It is a widely accepted fact that the cathode is the performance-limiting component due to slow kinetic rate of the oxygen reduction reaction and to the mass-transfer limitations caused by liquid water generated by the reaction and electro-osmic drag, accumulating in the porous catalyst and gas diffusion layers. The performance of the cathode depends primarily on the oxygen transport rate, which is strongly affected by the liquid water. Cathode flooding caused by excessive liquid water in the cathode is generally recognized as the primary reason for poor fuel-cell performance.

Several methods are available to minimize the quantity of liquid water in the cathode, and include modifications to reactor design, electrode design, and operating conditions [4-7]. A common method is to maintain a higher pressure near the cathode than the anode [4]. This pressure difference partially counteracts the electro-osmotic flow of water through the membrane, which proceeds with the protons from the anode moving to the cathode side of the membrane. Another method is to use hydrophobic electrodes [4,5]. The gas-diffusion layers are often waterproofed with polyfluoroethylene (PTFE), which increases the resistance to liquid water flow and ensures that the pores in the carbon cloth (or carbon paper) do not become clogged with water. Another method is to use an interdigitated flow-field design [6,7]. The cell improvement by this method was attributed to the dead-end flow channels converting the transport of the reactants and products in the porous electrode from diffusion to a forced-convection mechanism. For other liquid-water mitigation methods, the electrode porosity can be changed to accommodate changes occurring along the flow field length, and operating conditions can be manipulated by allowing the temperature to increase near the outlet to evaporate the water produced and/or by lowering the temperature in a dry region to condense the water.

Recently, the effects of permanent magnets on oxygen reduction reaction via Pt catalysts and PEFC have been studied, and a method for improving the fuel-cell performance has been demonstrated [8-10]. Permanent magnets attract iron, and thus attract paramagnetic oxygen gas, while rejecting diamagnetic water. If the liquid water flow and O_2 gas flow in the cathode gas electrode can be promoted by this kind of magnetic force, the performance of the PEFC will be improved. The aim of this chapter is to describe research on the use of permanent magnets to enhance the cathode gas reaction, and to show a possible application of permanent magnets in the cathode of PEFC system. Section 2 of this chapter gives an overview of magnetic force. Sections 3 and 4 describe the effects of permanent magnets on the electrochemical oxygen reduction with relevance to the cathode gas reactions in PEFC. Section 5 describes the experimental research on the use of small magnets particles in PEFCs, and Section 6 describes numerical simulation of such use.

2. What is Magnetic Force?

Lorentz force acts on charged particles when they cross a magnetic field. There exists another kind of magnetic force. In the 19th century, In the 19th century, Faraday carried out experiments about the effect of magnetic force on many materials, and categorized the materials as either 'attractive' or 'repulsive'. He found that most of the materials were repulsed by a magnet, whereas O_2 gas was attracted by a magnet [11]. This kind of magnetic force, called Kelvin force or magnetization force, is a body force and acts on every material, even materials that are not ferromagnetic.

This magnetic force per unit volume can be expressed as $\mathbf{f}_m = \mu_0(\mathbf{M}\cdot\nabla)\mathbf{H}$, where μ_0 ($4\pi \times 10^{-7}$ H m^{-1}) is the absolute magnetic permeability of vacuum, **M** is the magnetization (i.e., magnetic moment per unit volume), and **H** is the local magnetic field. $\mathbf{M} = \chi\,\mathbf{H}$, where χ is the magnetic susceptibility per unit volume.

Generally, a unit volume of material in a magnetic field gradient experiences magnetic force [12] expressed as

$$\mathbf{F}_m = \mu_0\chi(\mathbf{H}\cdot\nabla)\mathbf{H} = \mu_0\chi_g\rho\nabla H^2/2 \approx (\chi_g\rho/\mu_0)\nabla B^2/2 \quad (1)$$

where ∇H^2 is the gradient of the square of *H*, χ is defined as the product of material density (ρ) and magnetic susceptibility per unit weight (χ_g). Here the magnetic flux density $\mathbf{B} = \mu_0(\mathbf{M}+\mathbf{H}) = \mu_0(\chi\mathbf{H}+\mathbf{H}) \approx \mu_0\mathbf{H}$ because χ is small (<<1).

It is characteristic that a magnetic force is induced under a magnetic field gradient and the resultant force along site-coordinate *z*-axis, is proportional to the value of dB^2/dz. Table 1 shows χ, χ_g and ρ of several diamagnetic (D) and paramagnetic (P) materials at room temperature. Most materials, such as water and a majority of organic compounds, are diamagnetic ($\chi_g < 0$) and thus experience a weak repulsive force along the steepest gradient of the magnetic field strength (*H*). An attractive force acts on paramagnetic materials ($\chi_g > 0$). Oxygen gas is a unique gas that has paramagnetic properties because of its parallel spins in the electron configuration of a molecule [13], and the absolute value of χ_g for O_2 gas is much larger than that for diamagnetic materials.

Table 1. Volume magnetic susceptibility (χ), mass magnetic susceptibility (χ_g), and density (ρ) of diamagnetic (D) and paramagnetic materials (P) at room temperature [14]

		χ	χ_g [kg$^{-1}\cdot$m^3]	ρ [kg·m^{-3}]
Pure water	D	-9.00x10-6	-9.00x10-9	1.0x103
Acetone	D	-5.82x10-6	-7.35x10-9	0.785x103
Benzene	D	-6.51x10-6	-8.82x10-9	0.874x103
Ethanol	D	-7.75x10-6	-9.02x10-9	0.786x103
N2 gas	D	-6.79x10-9	-5.40x10-9	1.25
O2 gas	P	+1.91x10-6	+1.34x10-6	1.43

In the 20th century, Pauling et al. [15] invented equipment for measuring the concentration of O_2 gas, utilizing the magnetic attractive force toward O_2 gas. This magnetic attractive force acting on O_2 gas has recently been shown to induce gas flows [16,17] and to affect chemical reactions involving O_2 gas, for example, combustion in diffusion flames [18-20], and catalytic combustion [21]. This research field was called 'magnetoaerodynamics'. In diffusion flames, combustion reaction occurs at the surface of a flame, i.e. the reaction zone. Combustion in a diffusion flame is promoted when the magnetic force transports O_2 gas to the reaction zone and then removes combustion gases such as diamagnetic water and CO_2 from the reaction zone [18,20].

A promising application of magnetoaerodynamics is promotion of the oxygen reduction reaction in PEFC; this reaction is the major rate-limiting process, and thus causes high overpotential and loss of efficiency of the system. If water flow and O_2 gas flow in the cathode gas electrode are controlled by a magnetic force, the performance of PEFC can be improved.

3. Effects of Permanent Magnets on O_2 Reduction Reaction Revealed by Three- Electrode Electrochemical Experiments

The effects of permanent magnets on the electrochemical oxygen reduction were studied based on three-electrode electrochemical experiments, with relevance to cathode gas reactions in PEFC where O_2 gas is reacted with H^+ to produce water on a platinum (Pt) catalyst [8,9].

3.1. Experimental Procedures

3.1.1. Experimental Apparatus and Procedure

In the electrochemical experiments, Pt foil (12 μm thick) was used as the working electrode in a three-electrode glass cell. The reference electrode was Ag/AgCl (0.588 V vs. RHE, the hydrogen electrode in the same solution), and the counter electrode was a Pt plate. One side of the Pt foil was open (diameter 16 mm) to the glass cell, which was filled with 0.05 mol dm^{-3} H_2SO_4 solution. The other side of the Pt foil was in contact with the magnetic pole of a rectangular permanent magnet (Nd–Fe–B, NEOMAX, B_r=1.42 T, Sumitomo Special Metals Co. Ltd.). As shown in the plot in Fig. 3, the magnetic flux density B decreased with increasing distance z from the magnet. A steep magnetic field gradient applied to the Pt foil yielded a magnetic force (F_m) that attracted the O_2 gas towards the cathode electrode (arrow pointing left in the figure) and a magnetic force that repulsed water in the reverse direction (arrow pointing right). The B and dB^2/dz at the surface of the magnet were varied by using magnets of different thickness (l×20×30 mm, l=10, 20, 30 mm, see section 3.1.2).

O_2 gas was bubbled into the solution, and electrochemical measurements were carried out using a potentio-galvanostat (model 263A, EG&G Instruments) at 25 °C. Potential was scanned at 0.1 V s^{-1} for cyclic scanning, and at 0.01 V s^{-1} in the cathodic direction for the polarization measurements. The Pt foil was used in either its bare state or covered with

Nafion® polymer film. The covered Pt foil was adopted as a model of the catalyst layer existing in the gas diffusion electrodes of PEFCs. For the preparation of the covered Pt foil, 20 μl of 5% Nafion® solution (Aldrich Chemicals) was poured on the surface of Pt foil, and allowed to dry. This process was repeated until the desired amount of Nafion® was deposited. In the same manner, carbon supported Pt deposited on 0.25mm-thick carbon paper (20 mass% Pt/C, 1 mg cm^{-2} Pt/Vulcan XC-72 laminated to a hydrophobic backing graphite filter paper, Electrochem. Inc.) was also tested as the working electrode with various amounts of Nafion® polymer impregnation.

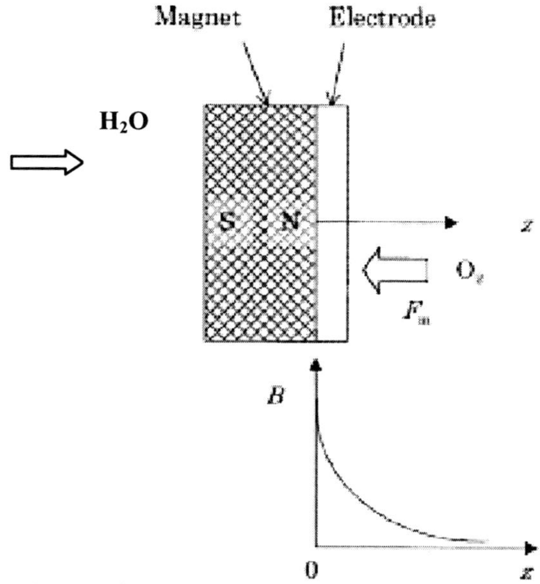

Figure 3. Experimental setup of the cathode and permanent magnet block, and corresponding spatial distribution of magnetic flux density (B).

3.1.2. Spatial Distribution of Magnetic Flux Density B

For three-electrode electrochemical cell measurements, rectangular magnets were used. According to Eq.(1), the magnetic force (F_m) is proportional to dB^2/dz. Accurate measurement of the spatial distribution of B near the permanent magnet is extremely difficult because of the large size of the sensor required for the measurement, i.e. Hall probe (0.4 mm thick) of a Gauss meter (5080, F.W. Bell). Therefore, the spatial distribution of B and dB^2/dz of a rectangular bar magnet was calculated, using the following equation [22]:

$$B_z(z) = \frac{B_r}{\pi}\left(\tan^{-1}\frac{ab}{2z\sqrt{a^2+b^2+4z^2}} - \tan^{-1}\frac{ab}{2(z+l)\sqrt{a^2+b^2+4(z+l)^2}} \right) \quad (2)$$

where $a \times b$ is a cross section of the pole of the magnet, l is the thickness of the magnet, B_r is the residual magnetic flux density of a permanent magnet, an angle is expressed in radian, and z is the distance from the center of the magnetic pole, as shown in Fig. 3.

Fig. 4 (a) shows the calculated spatial distribution of B and dB^2/dz along the z-axis near a rectangular magnet of thickness l when B_r=1.42 T, a=20 mm and b=30 mm. The calculated B

decreases drastically with increasing z and a steep magnetic field gradient exists in the region where z is below several mm. Hereafter, we will use dB^2/dz at z=0.1 mm, i.e. $(dB^2/dz)_0$ as a parameter to indicate the magnetic field gradient. When l=10, 20 and 30 mm, $(dB^2/dz)_0 = -167$, -337 and -489 T^2 m^{-1}, respectively, and B_0 = 0.572, 0.644 and 0.657 T, respectively.

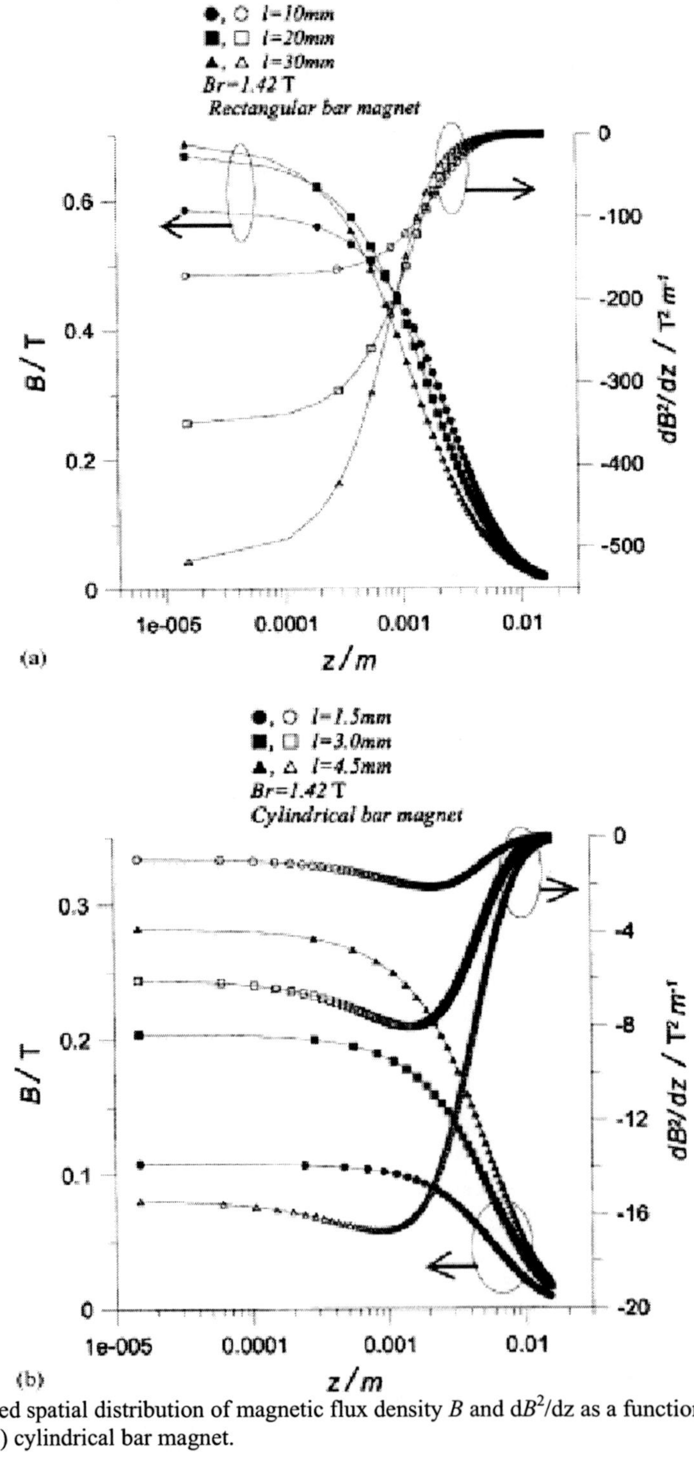

Figure 4. Calculated spatial distribution of magnetic flux density B and dB^2/dz as a function of z near a (a) rectangular and (b) cylindrical bar magnet.

3.2. Effects of Permanent Magnets on O_2 Reduction Reaction Revealed by Three Electrode Electrochemical Experiments

Fig. 5 shows the potential-current density curves measured on bare Pt foil when O_2 gas was bubbled in 0.05 mol dm^{-3} H_2SO_4 solution, either with or without a rectangular permanent magnet of 30 mm $[(dB^2/dz)_0 = -489\ T^2\ m^{-1}]$ in the three electrode electrochemical measurements. The potential was cycled at a scan rate of 0.1 V s^{-1}. The cathodic current that begins at around 0.8 V versus RHE and then increases at negative potential, corresponds to the reduction current of dissolved O_2 in the solution. Although the amount of dissolved O_2 in 0.05 mol dm^{-3} H_2SO_4 solution, ca. 1.4×10^{-6} mol dm^{-3}, is very small with respect to the cathode reaction of PTFC, a clear promotion effect of the magnetic field is observed, and the limiting current density of oxygen reduction increases over 80% at 0.1 V s^{-1} and 50% at 0.01 V s^{-1} scan rates. This enhancement effect was the same whether the N-pole or S-pole of the permanent magnet was placed behind the Pt foil. Furthermore, the circulation of solution by Lorentz force was negligible because the direction of the magnetic field was perpendicular to the surface of the electrode. Therefore, the magnetic attractive force toward O_2 gas and repulsive force acting on water near the cathode are considered responsible for this enhancement effect.

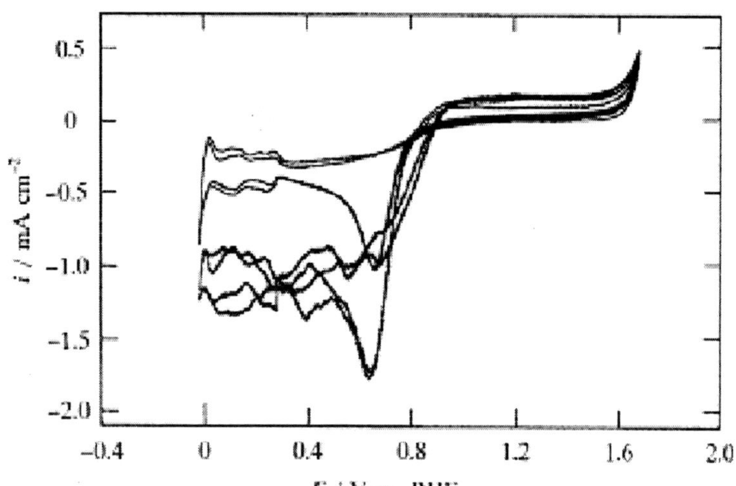

Figure 5. Cyclic voltammograms of Pt foil at potential scan rate of 0.1 V s^{-1} for magnetic field $(dB^2/dz)_0 = 0\ T^2\ m^{-1}$ (top curve) and $-489\ T^2\ m^{-1}$ (bottom) in 0.05 mol dm^{-3} H_2SO_4 solution saturated with O_2 gas with stirring at 25 °C.

The absolute value of dB^2/dz, which is proportional to a magnetic force, decreases with increasing *z,* as shown in the calculation results in Fig. 4(a). Therefore, the magnitude of a magnetic force decreases drastically with increasing *z,* and a strong magnetic force acts on O_2 gas and water only near the electrode. When $dB^2/dz = -489\ T^2\ m^{-1}$ at $z = 0.1$ mm, the magnetic forces acting on O_2 gas and on liquid water are 372 N m^{-3} and 1750 N m^{-3}, respectively, calculated using Eq.(1). The repulsive magnetic force acting on water is about 18 % of gravitational force, whereas the attractive magnetic force toward O_2 gas is about 27 times stronger than the gravitational force (1 atom, 273 K). So far, we considered the magnetic force acting on O_2 gas as small bubbles in the solution. However, if dissolved O_2 gas exists as independent molecules, then the magnetic attractive force acting on a molecule is $F_m = 2\mu_B \, grad \, H$, where μ_B is Bohr magneton and H is magnetic field strength.

Figure 6 shows the dependence of $(dB^2/dz)_0$ on the limiting current density of oxygen reduction on the bare Pt. The limiting current tended to increase with increasing $(dB^2/dz)_0$, and this dependence increased when the solution was stirred (i.e., O_2 gas feeding condition in the solution was improved). Sweep rate dependence on the current was not large compared with its dependence on the stirring condition. This suggests that oxygen transport determines the net rate of a cathode reaction, and that the magnetic effect occurs in the diffusion layer that develops in the solution near the electrode surface.

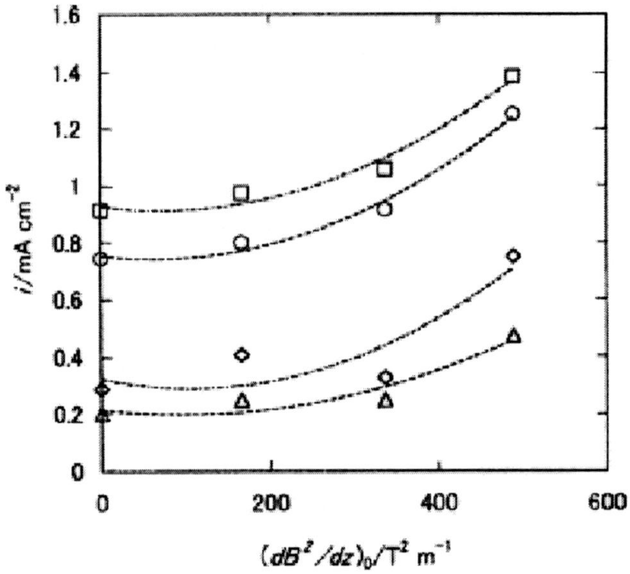

Figure 6. Oxygen reduction current at 0.59 V vs. RHE on Pt foil measured at various $(dB^2/dz)_0$ in 0.05 mol dm^{-3} H_2SO_4 solution saturated with O_2 gas with stirring (top curve) or without stirring (bottom) at 25 °C at a potential scan rate of 0.01 V s^{-1} (○, △) and 0.1 V s^{-1} (□, ◇).

When the Pt foil covered with various amounts of Nafion® was used as an electrode, the effect of diffusion limitation of O_2 gas in Nafion® film appeared, and the magnetic-enhanced increase in current at high Nafion® loading was not as large as that observed for the bare Pt foil.

A similar measurement was done for Pt dispersed carbon paper, the thickness of which was 0.25 mm. Figure 7 shows the dependence of $(dB^2/dz)_0$ and Nafion® amount on the oxygen reduction current at 0.59 V. The magnetic field effectively increased the current at low Nafion® loading. At high loading of the Nafion® film, the magnetic promotion is no more apparent, indicating diffusion limitation of oxygen in the film. Based on these results, an optimum amount of Nafion® polymer film exists to attain the maximum current density.

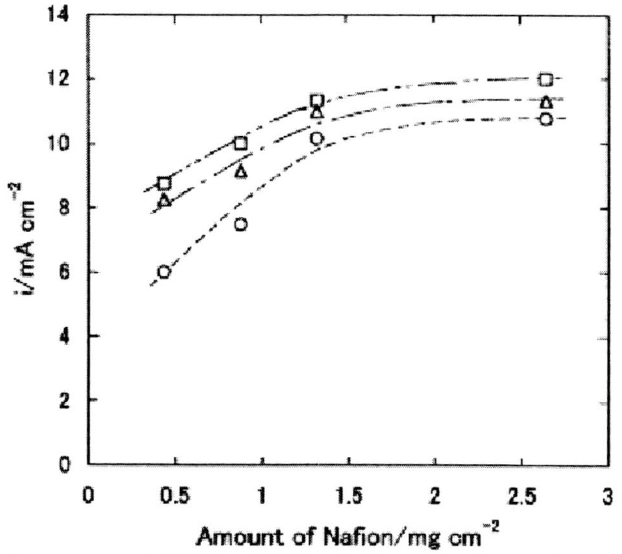

Figure 7. Oxygen reduction current at 0.59 V vs. RHE on Pt/C supported on carbon paper at various amounts of Nafion® coatings in a 0.05 mol dm^{-3} H_2SO_4 solution saturated with O_2 gas with stirring at 25 °C in a magnetic field gradient $(dB^2/dz)_0$=0 T^2 m^{-1} (○), −167 T^2 m^{-1} (△) and −337 T^2 m^{-1} (□).

When Pt dispersed carbon paper was used as the electrode, the current per apparent area was about 20 times higher than that for Pt foil. Assuming the roughness factor of 5, the amount of Pt atoms on the outermost surface of the foil, the calculated W_{Pt}(foil) is ca. 2×10^{-3} mg cm^{-2} for the apparent area. The calculated amount of Pt atoms available for Pt/C on carbon paper (1 mg Pt cm^{-2}) is then $W_{Pt}(Pt/C)=8.4\times10^{-8}/2r$ (mg cm^{-2}), where $2r$ (cm) is the particle size of Pt dispersed on C. Assuming $2r=5\times10^{-7}$ cm gives $W_{Pt}(Pt/C)=1.7\times10^{-1}$ mg cm^{-2}, which is significantly larger than W_{Pt}(foil), in accordance with the experimental results.

However, the enhancement effect by the magnetic field was enfeebled for the Pt dispersed carbon paper (current density of oxygen reduction increased only by about 25% at 0.1 V s^{-1}, and by 20% at 0.01 V s^{-1} when $(dB^2/dz)_0$=−489 T^2 m^{-1}). This limitation might come from the experimental condition, where dissolved O_2 in the solution was used as the reactant source, and where diffusion of O_2 was hindered in the porous media. Another factor might be the difference in thickness used for the electrode (12 μm for platinum foil, while 0.25 mm for Pt dispersed carbon paper). Thus, the magnetic effect would strongly depend on the electrode fine structure, on the electrode thickness, and on the spatial distribution of (dB^2/dz) where O_2 transport occurs.

4. Magnetic Effects on O_2 Reduction Reaction Revealed by Rotating Vertical Electrode Experiments

In Section 3, the effects of permanent magnets on the electrochemical oxygen reduction were discussed based on three-electrodes electrochemical experiments. When a permanent magnet was placed in contact behind a cathode, the current of O_2 reduction increased with increasing $(dB^2/dz)_0$ and with increasing magnetic force. In this section, the effects of permanent magnets on the electrochemical oxygen reduction are discussed based on experiments using a rotating electrode [8,9].

4.1.1. Experimental Apparatus and Procedure

A partially immersed rotating vertical electrode in Fig. 8 [23] was used as a testing apparatus modeling the gas diffusion electrode to study the magnetic effect. A test electrode was mounted in a Teflon® holder with 11 mmϕ opening (area 1 cm^2), which was half immersed (vertically) in 0.05 mol dm^{-3} H_2SO_4 solution and revolved at slow rotation speed (typically 0.81 rpm). The upper half of the electrode surface was in contact with the air, and while revolving, polarization measurements were conducted (using potentio-galvanostat) for the lower half of the surface exposed to the solution at 25 °C. Test specimen was Pt foil or carbon-supported Pt deposited on carbon paper, impregnated with various amounts of Nafion®. A cylindrical permanent magnet (Nd-B-Fe, NEOMAX, B_r=1.1 T, Sumitomo Special Metal Co. Ltd.) (D=15 mmϕ, l=1.5, 3, 4.5mm) was placed in contact behind the electrode. The magnetic flux density and its gradient were varied by using different magnets of different thickness (l). The reference electrode was SCE (0.316 V vs. RHE), and the counter electrode was a Pt sheet. Electrochemical measurements were performed as described previously. Hereafter, all the potentials are referred to RHE.

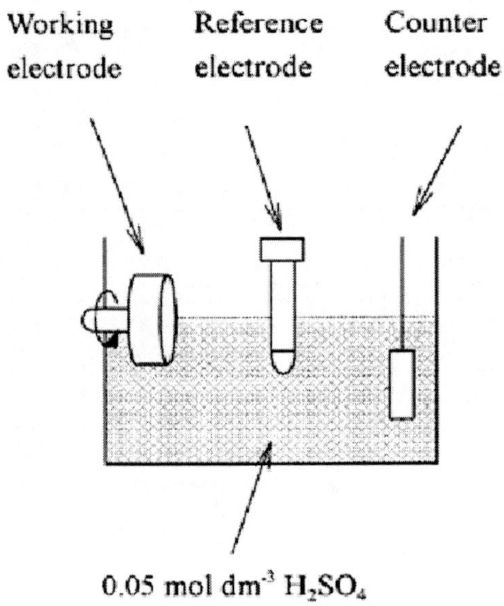

Figure 8. Schematic of a rotating electrode apparatus.

4.1.2. Spatial Distribution of Magnetic Flux Density B

For the rotating vertical electrode measurements, cylindrical magnets were used. The spatial distribution of B and dB^2/dz near a cylindrical bar magnet was also calculated, using the following equation [22]:

$$B_z(z) = B_r\left(\frac{z+l}{\sqrt{4(z+l)^2+D^2}} - \frac{z}{\sqrt{4z^2+D^2}}\right) \tag{3}$$

Fig. 4(b) shows the calculated spatial distribution of B and dB^2/dz along a central z-axis near the magnet for various thickness (l) when D=15 mm ϕ and B_r=1.42 T. The point z=0 mm corresponds to the center of the magnetic pole as shown in Fig. 3. A magnetic field gradient also exists near the magnet, and both B and dB^2/dz calculated for the cylindrical bar magnet are much smaller than those for the rectangular block magnets (Fig. 4a). When l=1.5, 3 and 4.5mm, $(dB^2/dz)_0$ at z=0.1mm are −1.03, −6.29 and −15.7 T^2 m^{-1}, respectively, and B_0 are 0.107, 0.203 and 0.28 T, respectively.

4.2. Effects of Permanent Magnets on O_2 Reduction Reaction Revealed by Rotating Vertical Electrode Experiments

Fig. 9 shows the cyclic voltammograms of oxygen reduction on Pt foil covered with 1.6 mg cm^{-2} Nafion® film, measured using a partially immersed rotating electrode, either with or without a cylindrical magnet of 3 mm in thickness [$(dB^2/dz)_0$=−6.29 T^2 m^{-1}]. When a magnetic field gradient was applied, the current increased by about 100%. Thus, although $(dB^2/dz)_0$ for the cylindrical bar magnet was much smaller than that when a rectangular bar magnet was used (Section 3.2.), the application of a magnetic field gradient drastically increased the current, and the oxygen reduction current increased relatively linearly with the potential. The elimination of the oxygen diffusion barrier in the solution in this experimental configuration enhanced the magnetic promotion more than in the three electrode electrochemical cell.

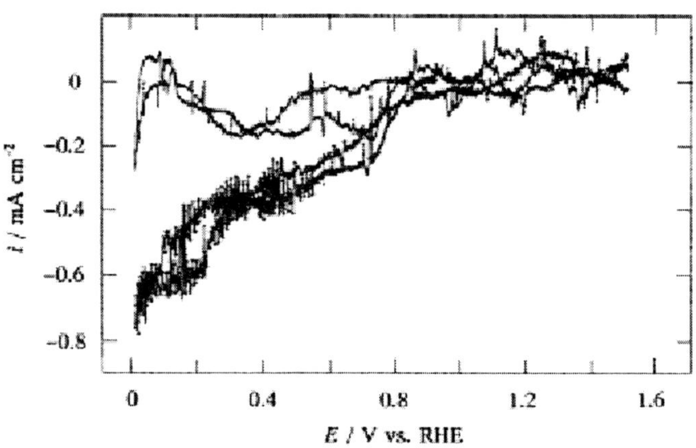

Figure 9. Polarization curves of oxygen reduction on Pt foil covered with 1.6 mg cm^{-2} Nafion® measured using a partially immersed rotating vertical electrode apparatus at 25 °C in 0.05 mol dm^{-3} H_2SO_4 solution in a magnetic field gradient $(dB^2/dz)_0$=0 T^2 m^{-1} (top curve) and −6.29 T^2 m^{-1}(bottom).

The oxygen reduction current for the apparent area of the Pt-dispersed carbon paper was much higher than that for Pt foil. The largest current was obtained when Pt-dispersed carbon paper was tested using the partially immersed rotation electrode (Fig. 10). In the absence of the magnetic field gradient (a), limiting current behavior showed hysteresis between forward and backward potential scans, whereas in the presence of the magnetic field gradient (b-d), the reduction current increased relatively linearly with the potential, and the current was enhanced by increasing $(dB^2/dz)_0$. For example, when a 1.5 mm thick magnet was placed behind the electrode and $(dB^2/dz)_0=-1.03$ T^2 m^{-1} (b), the current at maximum increased more than 50%. The magnetic promotion effect for Pt- dispersed carbon paper was smaller than that for Pt foil, possibly due to the difference in electrode structure (specially dimensions) in which the magnetic field acts. Experiments also showed the existence of an optimum amount of Nafion® polymer for magnetic enhancement effect on oxygen reduction; this optimum was 1.6 mg cm^{-2}.

Thus, the magnetic effect on the current increase was stronger for a rotating vertical electrode cell than for a three-electrode cell experiments although the magnitude of $(dB^2/dz)_0$ of the former (1–16 T^2 m^{-1}) was much smaller than that of the latter (167–489 T^2 m^{-1}). This stronger effect is probably due to the better O_2 supply in the former experimental arrangement.

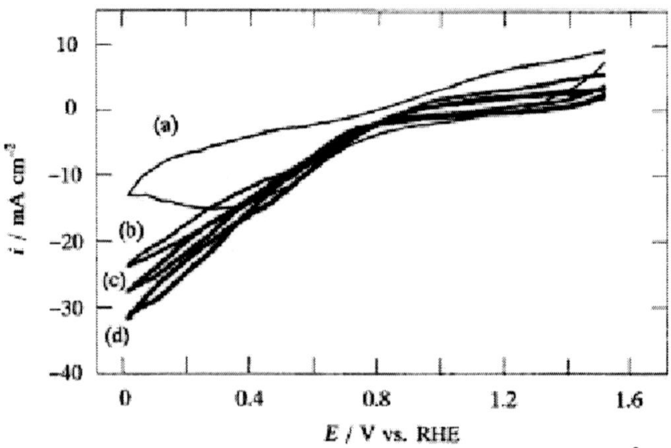

Figure 10. Polarization curves on Pt-dispersed carbon paper (Pt loading of 1 mg cm^{-2}) impregnated with 1.3 mg cm^{-2} Nafion measured using a partially immersed rotating vertical electrode at the potential scan rate 0.01 V s^{-1} in 0.05 mol dm^{-3} H_2SO_4 solution when (a) $(dB^2/dz)_0=0$ T^2 m^{-1}, (b) −1.03 T^2 m^{-1}, (c) −6.29 T^2 m^{-1}, (b) −15.7 T^2 m^{-1}.

5. Incorporation of Magnet Particles into the Catalyst Layer of the Cathode and Single Fuel Cell Experiments

In Sections 3 and 4, the effect of permanent magnets on oxygen reduction reaction was discussed with possible application to the cathode gas reactions in PEFCs where O_2 gas is reacted with H^+ to produce water on Pt catalyst. The experimental results clearly show that a permanent magnet placed in contact behind a cathode, i.e. application of a magnetic field gradient on the cathode, can effectively increase the O_2 reduction current. This magnetic effect can be explained by enhanced supply of O_2 gas to the catalyst layer and enhanced removal of liquid water from the electrode, caused by the magnetic forces acting on O_2 gas and liquid water, respectively.

In fuel cell operations, as shown in Fig. 1, oxygen (or air) is supplied through the porous carbon backing of the gas diffusion electrode. The O_2 gas is then dissolved in the polymer electrolyte layer, diffuses through it onto the active surface of Pt catalyst sites that are dispersed on fine carbon particles. The physical structure of PEFC makes it impossible to place a block of a permanent magnet behind the cathode inside a single fuel cell. Therefore, a new design was proposed for the cathode catalyst layer of fuel cells such that the magnet particles of small size were incorporated together with platinum loaded carbon powder [9].

5.1. Experimental Apparatus and Procedures

The magnet particles were Nd–Fe–B powder (Magnetoquench Co. Ltd., MQPC-powder, B_r=0.78 T). To protect the surface of the particles against corrosion, the particles were covered with polyimide in a vacuum [24] by first simultaneously depositing pyromellitic anhydride acid and oxydianiline (heated at 200–240 and 180–200 °C, respectively) in a vacuum and polymerizing on the surface of Nd–Fe–B particles. The resulting polyamide was heated at 300 °C, and dehydrated to produce polyimide film. The film thickness was ca. 1 μm.

For the cathode catalyst layer of a membrane electrode assembly (MEA), first Pt catalyst on carbon powder (20 wt.% Pt/Vulcan XC-72, Tanaka Kikinzoku Kogyo) was mixed with the magnet particles by weight ratio of 9:28 (volume ratio 3:2) together with 20 wt.% Nafion® solution and 1-buthanol. Then, the mixture was pasted on carbon paper (SGL Carbon) that was pretreated by FEP (tetrafluoroethylene-hexafluoropropylene copolymer, Daikin Kogyo) dispersion with Vulcan XC-72R.

For the anode catalyst layer, Pt catalyst on carbon powder was likewise pasted on carbon paper. Pt loading was 0.5 mg cm^{-2} for both the anode and the cathode catalyst layers. The polymer electrolyte membrane was Nafion® 117 (DuPont, Wilmington, DE, thickness 183 μm). The MEA was made by hot-pressing the catalyst-loaded carbon papers (5 cm×5 cm) on both sides of Nafion® 117 at 140 °C and at 5 MPa for 120 s. A cathode electrode or MEA was either magnetized in 4 T, or used as prepared. After the magnet particles were thus magnetized, they could supply a magnetic field. The direction of magnetization was perpendicular to the electrode surface.

A single fuel cell was assembled, consisting of the MEA, bipolar plates with gas channels, current collectors, and fastener bolts. Operating condition of the fuel cell utilizing pure hydrogen and air was cell temperature of 80 °C, humidifier temperature of 75 °C, hydrogen gas utilization of 70% and oxygen gas utilization of 40%. Polarization measurements were taken after 8 h operation at these conditions. Furthermore, oxygen utilization was varied between 10 and 90% at 400 mA cm^{-2}, and the effect of oxygen utilization on cell voltage was tested with and without a magnetic field (i.e. the magnetization of magnet particles).

5.2. Magnet Particles Incorporated into the Catalyst Layer of the Cathode and a Rotating Vertical Electrode Measurement

Two sets of cathode electrode in which Nd–Fe–B magnet particles were mixed with Pt catalyst on carbon powder were prepared, and one set was magnetized by applying a magnetic field of 4 T, and the other was not magnetized. After magnetization, magnetic particles in the cathode could supply a magnetic field. The performance of these two cathodes was examined by using the same rotating vertical electrode apparatus as described in Section 4.1.2. For this measurement here, however, the cylindrical permanent magnet was not used.

Fig. 11(a) shows the polarization curves of O_2 reduction in 0.05 mol dm^{-3} H_2SO_4 solution on these cathodes with 0.88 mg cm^{-2} Nafion impregnation with and without magnetization of magnet particles. The current of oxygen reduction for the magnetized cathode (bottom curve) was larger than that for the nonmagnetized cathode (top). When E=0 V, the current of the magnetized cathode was about 27% larger than that of the nonmagnetized cathode. These results show that the magnetized magnet particles are effective to increase the current of oxygen reduction.

Fig. 11(b) shows the dependence of the amount of Nafion on O_2 reduction current with and without cathode magnetization. There exists an optimum amount of Nafion for the current, and at every amount of Nafion, the O_2 reduction current was higher for the magnetized cathode than for the nonmagnetized cathode. The efficiency of magnetic increase of the current was largest at the highest Nafion loading.

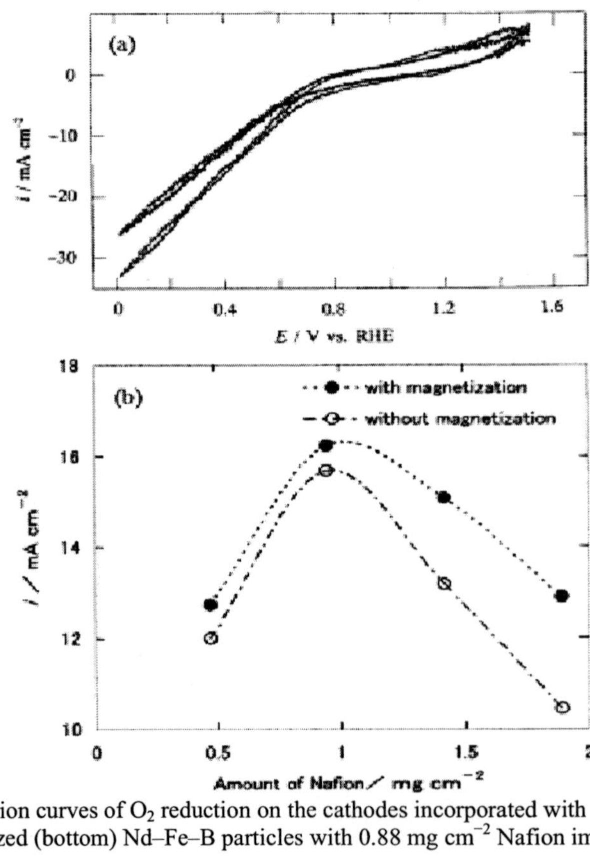

Figure 11. (a) Polarization curves of O_2 reduction on the cathodes incorporated with nonmagnetized (top curve) and 4T magnetized (bottom) Nd–Fe–B particles with 0.88 mg cm^{-2} Nafion impregnation, measured

using a partially immersed rotating vertical electrode apparatus in 0.05 mol dm^{-3} H_2SO_4 solution at 25 °C. (b) Effect of Nafion impregnation on O_2 reduction current on the cathodes incorporated with Nd–Fe–B magnetic particles with and without magnetization.

5.3. Magnet Particles Incorporated into the Catalyst Layer of Cathode and the Performance of a Single Fuel Cell

Figure 12 shows the polarization curves of a single fuel cell including Nd–Fe–B particles in the cathode catalyst layer with and without magnetization of magnet particles. The performance of the magnetized fuel cell (•) is superior to that of the nonmagnetized cell (○), especially in the limiting current region. This fact suggests that the magnetic force manages the flow of liquid water and O_2 and thus contributes to the improvement in fuel cell performance.

Figure 13 shows the cell voltage measured at a fixed current density of 400 mA cm^{-2} for various level of oxygen gas utilization. The magnetized fuel cell shows the superior cell-voltage behavior at all levels of oxygen gas utilization. Impedance measurements of the catalysts and membrane across the MEA during the fuel cell operation revealed smaller semi-circles in the Cole–Cole plots for the magnetized cell than for the nonmagnetized cell. Based on these single fuel cell experiments, magnetized particles improve cell performance. Unfortunately, the net performance of the cell was not good as compared with reported data [25, 26]. This would be due to the use of insulating material as the protective layer of magnet particles and the large thickness of a membrane. Further research toward these prospects is needed. This method will be especially useful for a portable fuel cell in which there is no power to supply the air required for fuel cell operation.

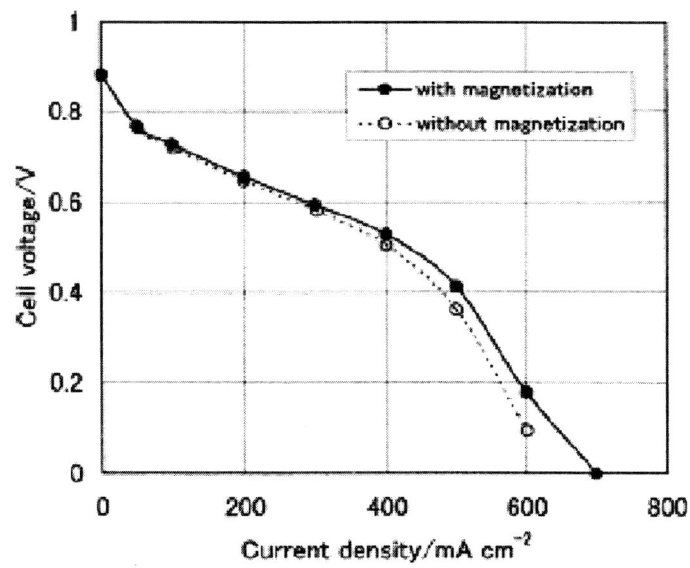

Figure 12. Polarization curves for H_2/air single fuel cell, including nonmagnetized (○) and 4T magnetized (•) Nd–Fe–B particles in the cathode catalyst layer. Operation condition is: cell temperature of 80 °C, humidifier temperature of 75 °C, hydrogen gas utilization of 70%, and oxygen gas utilization of 40%.

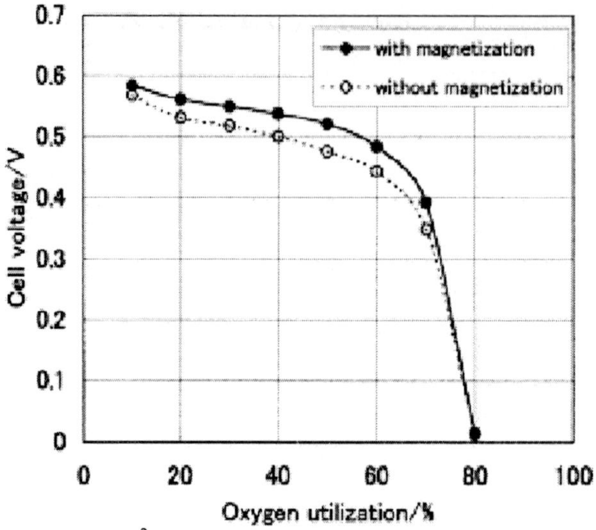

Figure 13. Cell voltage at 400 mA cm^{-2} at various oxygen utilization for H_2/air single fuel cell, including nonmagnetized (○) and 4T magnetized (●) Nd–Fe–B particles in the cathode catalyst layer.

6. Numerical Simulation of Water and O_2 Transport in the Cathode of PEFC by Using Permanent Magnet Particles Deposited in the Cathode-Side Catalyst Layer

6.1. Introduction

In Section 5, instead of a block of a permanent magnet, Nd-B-Fe magnet particles were deposited in the catalyst layer of a cathode, and the performance of a single fuel cell with and without magnetization was evaluated. As shown in Figs. 12 and 13, when the magnet particles were magnetized, the performance of a single fuel cell was superior to that of a nonmagnetized cell, especially in the limiting current region. Thus, the use of magnet particles is effective to manage liquid water in the cathode of fuel cells, and it will supply a new method to improve the performance of low temperature fuel cells such as PEFCs.

The main mechanism of the magnet particle method is the use of magnetic force to promote the transfer of chemical species in the cathode of a fuel cell system. To clarify quantitatively how the magnet particles improve fuel cell performance, the numerical simulations were conducted, considering magnetic forces acting on liquid water and O_2 gas.

When a fluid is placed in a static magnetic field, **H**, and the fluid is electrically insulating, the force per unit volume, **f**, can be expressed as [27]

$$\mathbf{f} = \nabla p + \mu_0 \mathbf{M} \cdot \nabla \mathbf{H} \tag{4}$$

where p is the pressure, $\mathbf{M} = \chi\mathbf{H}$, $\mathbf{B} = \mu_0\mathbf{H}$, μ_0 is the magnetic permeability of vacuum, χ is the magnetic susceptibility of fluid per unit volume, and **B** is the magnetic flux density.

The second term $\mu_0 \mathbf{M} \cdot \nabla \mathbf{H} = (\chi/\mu_0) \mathbf{B} \cdot \nabla \mathbf{B}$ in Eq. (4) is the magnetic force shown by Eq. (1). This magnetic force tends to move diamagnetic fluids such as water toward regions of lower magnetic fields and move paramagnetic fluids such as oxygen gas toward higher magnetic fields. In a fuel cell system, steep concentration gradients of both liquid water and oxygen gas exist near the interface between the catalyst layer and gas diffusion layer at the cathode side. Therefore, if a steep magnet-field gradient is applied to this region, the non-uniform magnetic force will push diamagnetic liquid water from this region and thus will make more pore space available for paramagnetic oxygen gas.

6.2. Governing Equations for the Numerical Model

A numerical model was developed to simulate the effects of magnet particles on liquid-water motion in cathodes [10]. Figure 14 shows the computational domain, which is the cross section of a cathode electrode. The modeled region consists of a porous layer (gas diffusion layer) in contact with an interdigitated gas distributor, which has an inlet channel, a shoulder, and an outlet channel to allow gas to flow through the electrode (see Fig. 14). The catalyst layer (long black rectangle in figure) was treated as a line in this model. The width of the electrode, L, was 0.2 cm, and that of the inlet, shoulder, and outlet were 0.05, 0.1, and 0.05 cm, respectively. The height of the porous layer, h, was 0.025 cm.

In typical operation of PEFC, humidified air is drawn into the inlet channel as the O_2 gas source. Water is generated as a product of the oxygen reduction reaction at the catalyst layer. Depending on the local partial pressure of water and its saturation vapor pressure at the fuel-cell operating temperature, water can either evaporate or condense. If the water generation rate exceeds the removal rate of water vapor, liquid water collects in the cathode electrode. The presence of liquid water causes electrode flooding, which blocks active sites on the catalysts and reduces the gas volume in the porous electrode. In the simulation, the electrode layer was treated as a homogeneous porous medium with uniform morphological properties, such as porosity, tortuosity, and permeability. The gas and liquid phases within the electrode were assumed continuous, and thus, Darcy's law applies. The gas mixtures were assumed to be perfect gases.

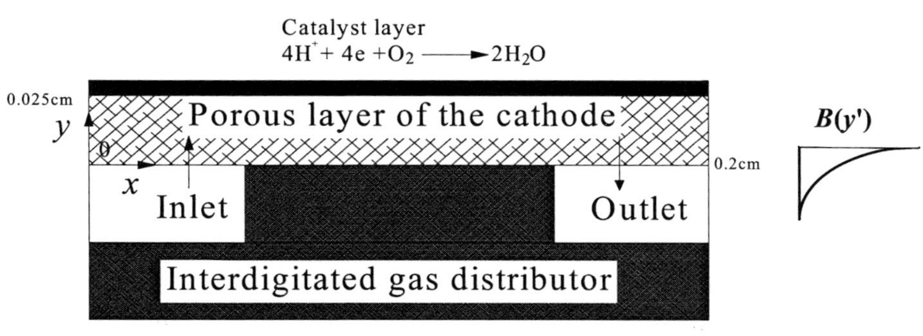

Figure 14. Computational domain.

6.2.1. Mass Transfer between Gas and Liquid Phases

The interfacial mass transfer rate of water by either condensation or vaporization (in kg/s), r_w, can be expressed as:

$$r_w = k_c \frac{M_w \varepsilon^g x_w}{RT}(x_w p - p_w^{sat})(1-q) + k_v \varepsilon^0 s \rho_w (x_w p - p_w^{sat}) q \tag{5}$$

where k_c and k_v are the rate constants for condensation (in s^{-1}) and vaporization (in $(s\cdot atm)^{-1}$), respectively, x_w is the mass fraction of water in the gas phase, p_w is the water partial pressure, p is the total gas pressure, p_w^{sat} is the saturation pressure of water at the operating temperature, M_w is the molecular weight of water, ε^0 is the dry porosity of the electrode, s is the saturation of liquid water, (defined as the fraction of the void volume occupied by liquid), ε^g is the gas porosity defined as $\varepsilon^g = \varepsilon^0(1\text{-}s)$, and ρ_w is the density of liquid water. The first term on the right-hand side of Eq. (5) represents the condensation rate, whereas the second term represents the evaporation rate. q is a switching function, where q = 0 for $px_w=p_w > p^{sat}$ and $q = 1$ otherwise, and can be expressed as [2,28]:

$$q = \frac{1 + \left| p_w^{sat} - x_w p \right| / (p_w^{sat} - x_w p)}{2} \tag{6}$$

6.2.2. Gas-Phase Model

The continuity equation for the gas phase is

$$\nabla \cdot (\rho \mathbf{u}^g) - r_w = 0 \tag{7}$$

where $\mathbf{u}^g$ is the superficial gas velocity and ρ is the gas density.

Because the magnetic force is a body force similar to the gravitational force, $\mathbf{u}^g$ can be expressed as:

$$\mathbf{u}^g = -\frac{\kappa^g}{\mu^g}\nabla p + \frac{\kappa^g}{\mu^g}\frac{\chi_{om}}{2\mu_0}\rho \varepsilon^g x_o \nabla \mathbf{B}^2 \tag{8}$$

where κ^g is the gas permeability and $\kappa^g = \kappa^0(1\text{-}s)$, μ^g is the gas viscosity, χ_{om} is the mass magnetic susceptibility of O_2 gas (m^3/kg), and x_o is the mass fraction of O_2. The last term of Eq. (8) is the magnetic attractive force acting on the oxygen. The magnetic repulsive force acting on nitrogen and on water vapor is negligible, because the absolute value of their magnetic susceptibilities is much smaller than that of oxygen (see Table 1).

Using the mass conservation equation and assuming that **B** changes only in the y-direction (see Section 6.2.5), the following equations for the gas velocity and pressure are derived as

$$u^{g} = -\frac{\kappa^{g}}{\mu^{g}}\frac{\partial p}{\partial x}, \quad v^{g} = -\frac{\kappa^{g}}{\mu^{g}}\{\frac{\partial p}{\partial y} - \frac{\chi_{om}}{2\mu_0}\frac{dB^2}{dy}\rho\varepsilon^{g}x_{o}\} \tag{9}$$

$$\frac{\partial}{\partial x}(\rho\frac{\kappa^{g}}{\mu^{g}}\frac{\partial p}{\partial x}) + \frac{\partial}{\partial y}(\rho\frac{\kappa^{g}}{\mu^{g}}\frac{\partial p}{\partial y}) - \frac{\partial}{\partial y}\left\{\rho\frac{\kappa^{g}}{\mu^{g}}\frac{\chi_{om}}{2\mu_0}\frac{dB^2}{dy}\rho\varepsilon^{g}x_{o}\right\} + r_{w} = 0 \tag{10}$$

where u^{g} and v^{g} are the gas velocities along x- and y-directions, respectively. The governing equations of O_2 gas and water vapor mass transfer are

$$\frac{\partial}{\partial x}(\rho u^{g}x_{o}) + \frac{\partial}{\partial y}(\rho v^{g}x_{o}) = \frac{\partial}{\partial x}(\rho D_{o}^{eff}\frac{\partial x_{o}}{\partial x}) + \frac{\partial}{\partial y}(\rho D_{o}^{eff}\frac{\partial x_{o}}{\partial y}) \tag{11}$$

$$\frac{\partial}{\partial x}(\rho u^{g}x_{w}) + \frac{\partial}{\partial y}(\rho v^{g}x_{w}) = \frac{\partial}{\partial x}(\rho D_{w}^{eff}\frac{\partial x_{w}}{\partial x}) + \frac{\partial}{\partial y}(\rho D_{w}^{eff}\frac{\partial x_{w}}{\partial y}) + r_{w} \tag{12}$$

where D_{o}^{eff} and D_{w}^{eff} are the effective diffusivities of oxygen gas and water vapor, respectively, and are related to their binary diffusivities as $D_{i}^{eff} = D_{i}(\varepsilon^{g})^{1.5}$. The pressure dependence of D_{o}^{eff} and D_{w}^{eff} and the effect of dispersion are neglected in this model due to small pressure difference between inlet and outlet. The temperature dependence of the binary diffusion coefficients is $D_{i} = D_{i}^{ref}(T/T^{ref})^{1.75}$ [29].

6.2.3. Liquid-Phase Model

The mass conservation for liquid water can be expressed as

$$\nabla\cdot(\rho_{w}\mathbf{u}^{lw}) + r_{w} = 0 \tag{13}$$

$$\mathbf{u}^{lw} = -\frac{\kappa^{lw}}{\mu^{lw}}\nabla p^{lw} + \frac{\kappa^{lw}}{\mu^{lw}}\frac{\chi_{lwm}}{2\mu_0}\rho_{w}\varepsilon^{0}s\nabla\mathbf{B}^2 \tag{14}$$

where $\mathbf{u}^{lw}$ is the superficial velocity, χ_{lwm} is the mass magnetic susceptibility of liquid water, κ^{lw} is the liquid water permeability of the electrode, and μ^{lw} is the liquid water viscosity. From Ref. [27], we have

$$\frac{\kappa^{lw}}{\mu^{lw}}\nabla p^{lw} = \frac{\kappa^{lw}}{\mu^{lw}}\cdot\frac{\mu^{g}}{\kappa^{g}}\cdot\frac{\kappa^{g}}{\mu^{g}}\nabla p - \frac{\kappa^{lw}}{\mu^{lw}}\cdot\frac{dp^{c}}{ds}\nabla s \tag{15}$$

where p^{c} is capillary pressure.

The interfacial drag coefficient, *f*, can be expressed as [2, 28]

$$f = (\kappa^{lw}\mu^{g})/(\kappa^{g}\mu^{lw}) = \mu^{g}\kappa^{lw}(s+0.01)/\{\kappa^{0}(1-s)\mu^{lw}\} \tag{16}$$

and the capillary diffusion coefficient, D_c, can be expressed as

$$D_c = -\frac{\kappa^{lw}}{\mu^{lw}}\frac{dp^c}{ds} = \kappa^{lw}\rho_w g\, AE(s+0.01)(e^{-A(s-C)} + e^{A(s-C)})/\mu^{lw} \quad (17)$$

where A, C, and E are constants.
From Eqs.(14) to (17), $\mathbf{u}^{lw}$ can be expressed as

$$\mathbf{u}^{lw} = f\mathbf{u}^g - D_c\nabla s + \frac{\kappa^g}{\mu^g}\frac{\chi_{lwm}}{2\mu_0}\rho_w\varepsilon^0 s\nabla\mathbf{B}^2 \quad (18)$$

From Eqs. (13) and (14), a differential equation for s can be expressed as

$$\frac{\partial}{\partial x}(D_c\rho_w\frac{\partial s}{\partial x}) + \frac{\partial}{\partial y}(D_c\rho_w\frac{\partial s}{\partial y}) - \frac{\partial}{\partial x}(f\rho_w u^g) - \frac{\partial}{\partial y}(f\rho_w v^g) - \frac{\partial}{\partial y}\left\{f\rho_w\frac{\kappa^g}{\mu^g}\frac{\chi_{lwm}}{2\mu_0}\frac{dB^2}{dy}\rho_w\varepsilon^0 s\right\} - r_w = 0 \quad (19)$$

6.2.4. Boundary Conditions

At the inlet, the boundary conditions for p and for each gas species are

$$p = p^{in}, \quad v^g x_i - D_i^{eff}\frac{\partial x_i}{\partial y}\bigg|_{y=0^+} = v^g x_i^{in}, \quad v^{lw} = 0 \quad (20)$$

where *in* and *i* indicate the inlet and the species, respectively.
For v^{lw}=0 there are two possible conditions:

$$\text{if} \quad s_2 < \frac{fv_{in}^g}{D_c}\delta y, \quad s_1 = 0 \quad (21)$$

$$\text{if} \quad s_2 > \frac{fv_{in}^g}{D_c}\delta y, \quad s_1 = s_2 - fv_{in}^g\delta y / D_c \quad (22)$$

where the subscripts 1 and 2 represent the inlet boundary point and the internal grid point next to inlet, respectively, and δy is the distance between these two points.

At the interface in contact with the shoulder of the gas distributor, the boundary conditions are

$$\frac{\partial p}{\partial y} = 0, \quad \frac{\partial x_i}{\partial y} = 0, \quad \frac{\partial s}{\partial y} = 0 \quad (23)$$

At the outlet interface, the boundary conditions are

$$p = p^{\text{out}}, \quad \frac{\partial x_i}{\partial y} = 0, \quad \frac{\partial s}{\partial y} = 0 \tag{24}$$

At the boundaries corresponding to either x = 0 cm or x =0.2 cm, the following symmetry conditions are used:

$$\frac{\partial p}{\partial x} = 0, \quad \frac{\partial x_i}{\partial x} = 0, \quad \frac{\partial s}{\partial x} = 0 \tag{25}$$

At the interface of the catalyst/diffusion layers, the boundary conditions are:

$$\left.\frac{\partial p}{\partial y}\right|_{y=h^-} = \frac{\chi_{\text{om}}}{2\mu_0}\frac{dB^2}{dy}\rho\varepsilon^{g}x_{\text{o}}, \quad \rho(v^{g}x_{\text{o}} - D_{\text{o}}^{\text{eff}}\left.\frac{\partial x_{\text{o}}}{\partial y}\right|_{y=h^-}) = \frac{IM_{\text{o}}}{4\text{F}}, \quad v^{g}x_{\text{w}} - D_{\text{w}}^{\text{eff}}\left.\frac{\partial x_{\text{w}}}{\partial y}\right|_{y=h^-} = 0 \tag{26}$$

$$f v^{g} - D_{\text{c}}\frac{\partial s}{\partial y} + f\frac{\kappa^{g}}{\mu^{g}}\frac{\chi_{\text{lwm}}}{2\mu_0}\frac{dB^2}{dy^2}\rho_{\text{w}}\varepsilon^{0}s = -(0.5+\alpha)\frac{IM_{\text{w}}}{\text{F}\rho_{\text{w}}} \tag{27}$$

where M_o is the molecular weight of O_2, I is the local current density, α is the net water transfer coefficient of the membrane, and F is Faraday constant.

I is defined by the Tafel equation as

$$I = I_0\frac{\varepsilon^{g}}{\varepsilon^{0}}\frac{x_{\text{o}}}{x_{\text{o}}^{\text{ref}}}\exp(\frac{\text{kF}}{\text{R}T}\eta) = I_0(1-s)\frac{x_{\text{o}}}{x_{\text{o}}^{\text{ref}}}\exp\left(\frac{\text{kF}}{\text{R}T}\eta\right) \tag{28}$$

where I_0 is the exchange current density, k is the transfer coefficient for the oxygen reduction reaction, η is the electrode overpotential, T is temperature and R is the gas constant.

The effect of liquid water on the surface availability of the reaction site is accounted for by the term $\varepsilon^g/\varepsilon^0$ in the Tafel equation. Generally, the magnetic field affects the movement of charged carriers, i.e., the proton flux toward the cathode. In our model, I_0 reflects the effect of the magnetic field on the proton flux. However, we ignored this magnetic effect because the apparent direction of the magnetic field is parallel to the direction of the proton movement [9] and thus the Lorentz force becomes zero.

6.2.5. Magnetic-Field Model

Generally, a steep magnetic field gradient exists in the vicinity of a small magnet particle. The spatial distribution of B and dB^2/dx near a small magnet particle was calculated, using Eq.(2). For example, at a distance 1 μ m from the surface of a 5- μ m-cube magnet of Nd-Fe-B and B_r=1.42 T, $B(dB/dx')$ = 1.5x10^4 T^2/m, and the calculated magnetic repulsive force acting on liquid water along x'-direction is about 1.08×10^8 N/m^3. This high magnetic force can push liquid water.

In the experiment study [9], magnet particles were deposited in the cathode-side catalyst layer, which is indicated by the upper long black rectangle in Fig. 14. Therefore, in the porous layer (gas diffusion layer), B decreases rapidly with increasing the distance from the interface of catalyst/diffusion layers (y'), and a steep magnetic field gradient is generated in the y-direction near the interface. To simulate a magnetic field equivalent to the actual magnetic field in these experiments, we used the distribution of magnetic field along the central axis perpendicular to the pole surface of a cubic magnet [Eq.(2)], and modified it by multiplying by the volume fraction of magnet particles deposited in the catalyst layer, 0.28:

$$B(y') = 0.28\frac{B_r}{\pi}\left(\tan^{-1}\frac{a^2}{(y'+c)\sqrt{2a^2+(y'+c)^2}} - \tan^{-1}\frac{a^2}{(y'+c+2a)\sqrt{2a^2+(y'+c+2a)^2}}\right) \quad (29)$$

where B_r is the residual flux density of a permanent magnet, $2a$ is the size of magnet particles (i.e., 5 μm), c is the thickness of the film covering the magnet (0.1 μm), and y' =0.025 cm -y.

6.3. Methods and Validation of Numerical Model

The governing differential equations were discretized by using a finite-difference method. The primary independent variables were p, x_o, x_w, and s. After discretization, a system of algebraic equations was solved by using a general-purpose computational fluid dynamics (CFD) code. The grid was refined until a grid-independent solution was obtained. The final grid was 81×61. The coupled set of equations was solved simultaneously, and the solution was considered converged when mass conservation was satisfied to within 1% of a mass scale with 10^{-5}kg/s. The parameters used in the simulations were taken from References [2, 28]. Unless indicated otherwise, the parameters used in the simulations are those listed in Table 2.

To validate our numerical model, the simulation results were compared with a set of experimental data [28]. The experimental parameters were used in our comparative simulations, namely, a membrane thickness of 125 μm, catalyst layer thickness of 70 μm, membrane conductivity of 11 S/m, and catalyst layer conductivity of 53 S/m. The same treatment of anode overpotential loss as reported in [28] was used. To simulate the actual airflow rate, an inlet pressure of 1.0133 atm was used, and the relative humidity of the inlet air was 0%. The open-circuit potential of the cell was 1.0V. The cell voltage was expressed as

$$V = 1.0 - \eta - 1.7857\times10^{-5} I - 1.3207\times10^{-6} I \quad (30)$$

The simulated polarization curve agrees with the measured curve (Fig. 15), confirming the validity of our numerical model.

Table 2. Base Case Data

Electrode permeability (^{0})	10^{-12} m^2
Electrode porosity (^{0})	0.3
Inlet O_2 mole fraction	0.21
Inlet stream relative humidity	100%
Inlet N_2 mole fraction	0.79
Inlet channel pressure	1.007 atm
Outlet channel pressure	1.000atm
Temperature	60˚C
Liquid water permeability (^{lw})	0.7×10^{-14} m^2
Gas viscosity ( )	20.3×10^{-6}kg/m·s
Viscosity of liquid water at 60°C	5.0×10^{-4} kg/m·s
Density of liquid water at 60°C	983.2 kg/m^3
Constants used in capillary head expression *A, C,* and *E*	3.7, 0.494,and 1.73×10^{-4} m, respectively
Exchange current density at 60°C (I_0)	100 A/m^2
Condensation rate constant (k_c)	100/s
Vaporization rate constant (k_v)	100/atm·s
Transfer coefficient of O_2 reduction reaction (k)	0.5
Overpotential ( )	0.55 V
Net water transfer coefficient of the membrane ( )	0.1
Mass magnetic susceptibility of liquid water (χ_{lwm}) $\chi_{lw} = \chi_{lwm}\rho_w$.	-9.05×10^{-9} m^3/kg,
Mass magnetic susceptibility of O_2 (χ_{om}) $\chi_o = \chi_{om}\rho_o$.	1.35×10^{-6} m^3/kg,

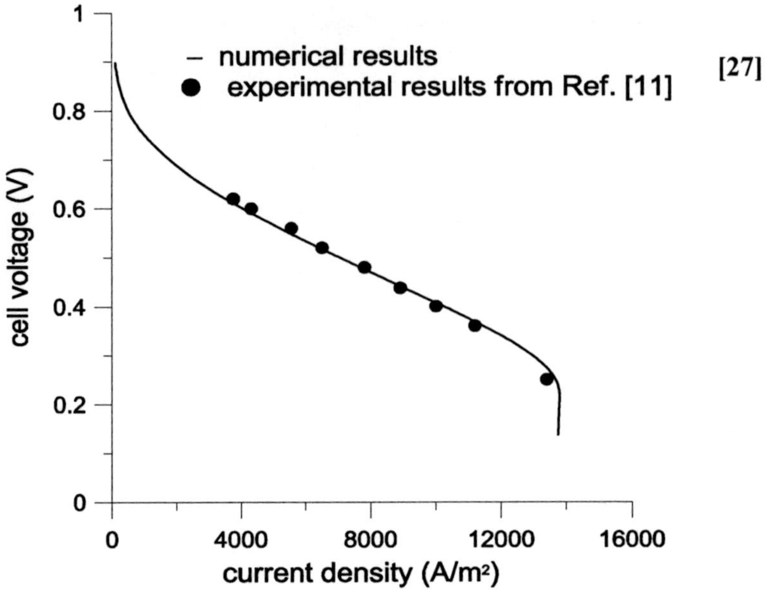

Figure 15. Measured and simulated cell voltage (V) vs. current density (I).

6.4. Numerical Results

6.4.1. Effect of Permanent Magnet Particles on Polarization Characteristics

In the calculation of the cell voltage, the conductivity of the catalyst layer was assumed to decrease by 40% after the magnet particles were deposited in this layer. The cell voltage was calculated as:

$$V = 1.0 - \eta - 1.7857 \times 10^{-5} I - 1.4 \times 1.3207 \times 10^{-6} I \tag{31}$$

Fig. 16 shows the simulated cell voltage-current (V vs. I) characteristics for B_r = 0, 0.5, 1, and 1.5 T. The magnetized magnet particles in the catalyst layer induce magnetic force, which is proportional to B_r (Eq. 2). With increasing magnetic force acting on liquid water and O_2 gas, the fuel cell performance improves when $I > 2000$ A/m^2, especially in the limiting current region. When $V = 0.4$ V, I for B_r = 0.5, 1.0, and 1.5 T, increases about 0.8, 3, and 6.5%, respectively, compared to I (B_r = 0T). These trends are similar to those observed in previously reported experimental results (see Fig. 12) [9], even though the simulated gas diffusion layer is different from that used in the experimental system (interdigitated vs. conversional). The power density is also an important parameter for PEFC. Fig. 16 also shows the power density-current characteristics. The enhancement in fuel cell performance increases with increasing B_r, especially in $I > 2500$ A/m^2, and the maximum power density occurs near $I = 4500$ A/m^2. Compared with the case when B_r = 0T, the maximum power density increases by about 0.3% when B_r =0.5T, 2.3% when B_r =1.0T, and 5.4% when B_r =1.5T.

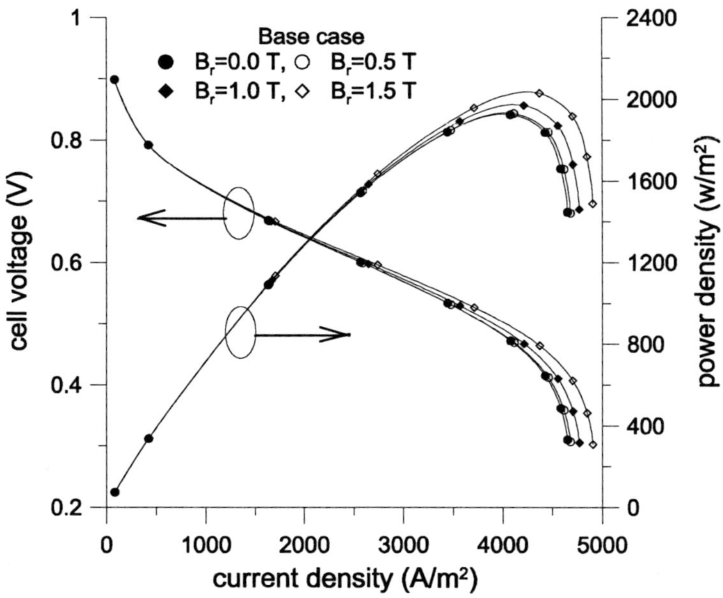

Figure 16. Calculated cell voltage (V) and power density vs. current density (I) for B_r = 0, 0.5, 1.0, and 1.5 T.

Recently, improvement of PEFC performance was simulated for the following two cases; when magnetic force acts on only liquid water and when it acts on both liquid water and oxygen gas [30]. Improvement in performance was nearly equal for both cases, although performance for magnetic force acting on both liquid water and O_2 gas was slightly superior to that for magnetic force acting only on liquid water. This result suggests the magnetic force acting on liquid water manages the flow of water and mainly contributes to the improvement in fuel cell performance. The magnetic force per unit volume rather than the force per unit weight seems to determine the magnetic effects.

6.4.2. Effects of Permanent Magnet Particles on the Distribution of Mole Fraction of Oxygen Gas (X_o), Liquid Water Saturation (S), Current (I), Velocity of Liquid Water (U_{lw}, V_{lw}) along the Interface of the Catalyst Layer and Diffusion Layer

Water is generated at the catalyst layer by an oxygen reduction reaction. The presence of liquid water causes electrode flooding, which blocks active sites on the catalyst and thus reduces the gas volume in the porous electrode. This leads to a smaller active area available for reaction and a reduced volume for O_2 gas transport. Fig. 17 shows the simulated s and x_o along the interface of the catalyst layer and diffusion layer for various B_r and η = 0.55 V. When the magnet particles deposited in the catalyst layer are not magnetized (B_r = 0T), I = 4585 A/m^2 and V = 0.355 V. On the other hand, when the magnet particles are magnetized; i.e., the magnetic force acts on liquid water and oxygen gas (B_r > 0T), both s and x_o decrease with increasing B_r, decreasing more rapidly with increasing B_r. The maximum decrease in s of about 13% (compared with that without magnetization, B_r = 0T) occurs at x=0.2cm when B_r=1.5T. Thus, the magnetic force caused by the magnetized magnet particles decrease in s at the interface of the catalyst layer and, thus more space is provided for the gas phase to transport to the reaction sites. The decrease in s helps improve the fuel-cell performance, while, at first sight, the decrease in x_o seems to degrade the fuel-cell performance.

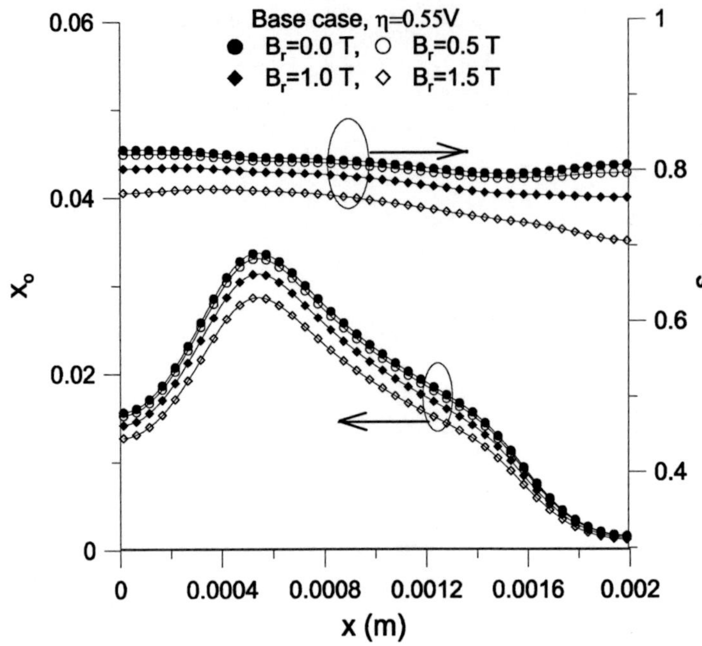

Figure 17. Simulated mass fraction of oxygen (x_o) and saturation of liquid water (s) vs. position along the interface of the catalyst/diffusion layer (x) at a height of $y = 0.025$ cm, for various B_r.

The total amount of O_2 gas at the interface of catalyst/diffusion layers with and without magnetic forces generated by magnetized magnet particles was also evaluated. The contribution of the reactant oxygen gas to the fuel-cell reaction can be expressed by the combined parameter $x_o\varepsilon^0(1\text{-}s)$ contained in the Tafel equation (see Eq. (28). Fig. 18 shows the simulated distributions of $x_o\varepsilon^0(1\text{-}s)$ and I along the interface of the catalyst/diffusion layers when $\eta = 0.55$V. Both parameters show maxima near x = 0.06 cm, and increase with increasing B_r. This indicates that the magnetic forces improve the fuel-cell performance by decreasing s, increasing the pore space, and increasing the mass of O_2 that contacts the catalyst sites. The decrease in x_o in Fig. 17 suggests that more O_2 is consumed as a result of the improvement in fuel-cell performance.

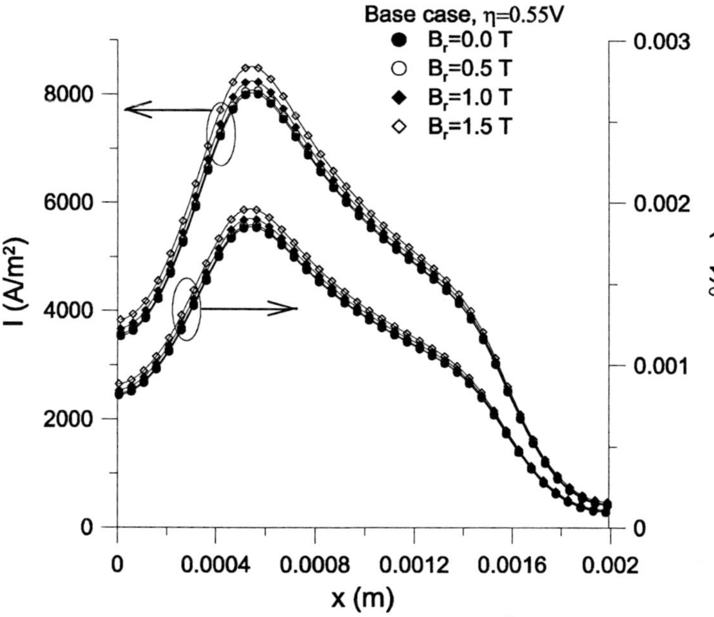

Figure 18. Simulated current density, I, and combined parameter, $x_o\varepsilon^0(1\text{-}s)$, vs. position along the interface of the catalyst/diffusion layers, x for various B_r.

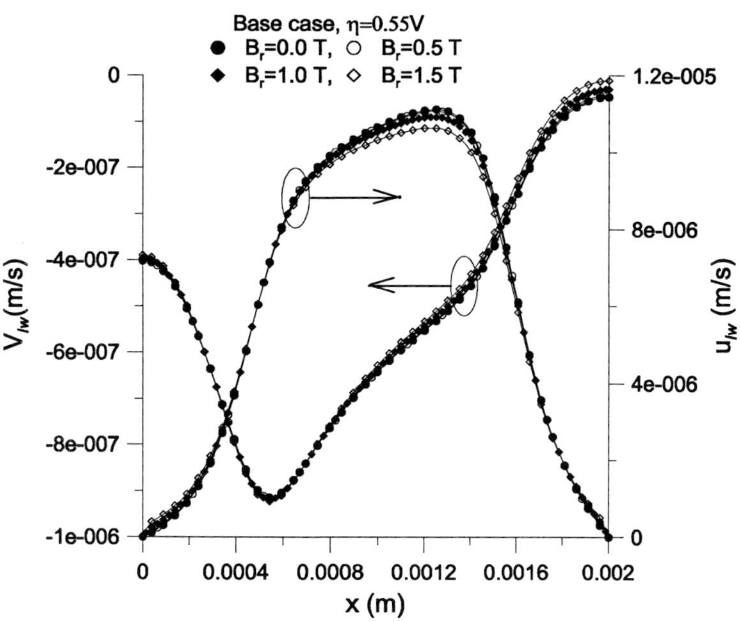

Figure 19. Simulated liquid-water velocity components, u_{lw} and v_{lw}, vs. position along the interface of the catalyst/diffusion layers, x for various B_r.

Fig. 19 shows the distribution of liquid water velocity in the *x*- and *y*-directions, u_{lw} and v_{lw}, along the interface of the catalyst/diffusion layers. The v_{lw} is negative because the liquid water flows out from the catalyst layer. The magnitudes of u_{lw} and v_{lw} are small, u_{lw} is less than 11 μm/s, and v_{lw} is less than 1 μm/s. The magnetic force does not change the velocity distribution in most of the interface. On the other hand, the magnetic force decreases the velocities both u_{lw} and v_{lw}; u_{lw} decreases in the central area, whereas v_{lw} decreases near the outlet. The decrease in $|v_{lw}|$ near the outlet region is caused by the decrease in the gradient of s at the interface of the catalyst layer and diffusion layer because u_{lw} and v_{lw} partially depend on the capillary force caused by the gradient of s.

6.4.3. Effect of Permanent Magnet Particles on Distribution of Liquid Water Saturation (S), Velocity of Liquid Water (V_{lw}), and Mole Fraction of Oxygen Gas (X_o), along the Central Line of the Cathode

The velocity of liquid water v_{lw} partly depends on the capillary force caused by the gradient of s. Fig. 20 shows the distribution of s and v_{lw} along the central line perpendicular to the catalyst/diffusion layers (x = 0.1 cm). When the magnet particles are not magnetized (B_r = 0T), s decreases with decreasing y. This is reasonable because water is generated at the catalyst layer. On the other hand, when the particles are magnetized (B_r > 0T), s at the interface of the catalyst/diffusion layers decreases with increasing B_r. For B_r =0 and 1.5 T, s = 0.81 and 0.76, respectively. This difference in s means that the magnetic repulsive force does move liquid water from the interface. The effective region of the magnetic repulsive force is sufficiently small that liquid water accumulates in the region about 10 μm from the interface. As B_r increases, the negative gradient of s at the interface of the catalyst/diffusion layers increases and the capillary force almost completely counteracts the acceleration caused by the magnetic force. Therefore, v_{lw} at the interface is only weakly affected by the magnetic field. However, near the interface, v_{lw} increases with increasing B_r, and reaches a maximum about 10 μm from the interface of the catalyst/diffusion layers. For B_r=1.5 T, the maximum velocity is v_{lw} = 4.2 μm/s, whereas for B_r=0T, v_{lw} = 0.2 μm/s, and the magnetic force increases v_{lw} by a factor of about 20. The reason for this increase is that near the interface, repulsive magnetic force accelerates the liquid water. As the liquid water moves away from the catalyst layer, the magnetic force becomes sufficiently weak that the resistance force finally changes the direction of acceleration. Therefore, more water is carried out from the region near the catalyst layer by the magnetic force. This leads to an increased active area available for reaction and an increased area and volume for O_2 transport.

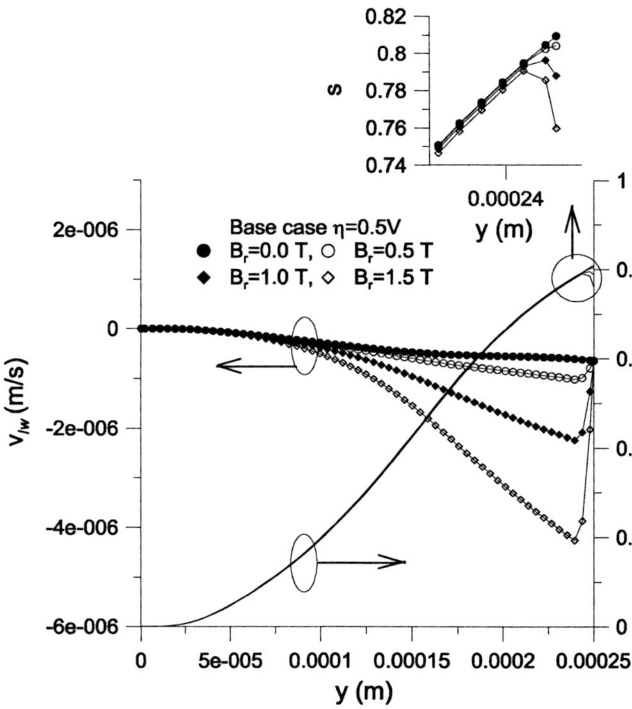

Figure 20. Simulated liquid water saturation, s, and y-direction liquid-water velocity component, v_{lw}, vs. vertical position along the centerline of the cathode electrode (i.e., x = 0.1 cm), for various B_r.

6.4.4. Effect of Permanent Magnet Particles on Field Characteristics in the Electrode

Fig. 21 shows the contour plots of x_o for the base-case conditions (η = 0.55V), for B_r=0 and 1.5 T, and indicates that x_o decreases slightly when the magnetic force acts on liquid water and O_2 gas. This decrease occurs primarily because in the presence of magnetic force, I increases, more oxygen is consumed in the fuel-cell reaction, and the back diffusion of O_2 gas at the inlet causes a slight decrease in x_o.

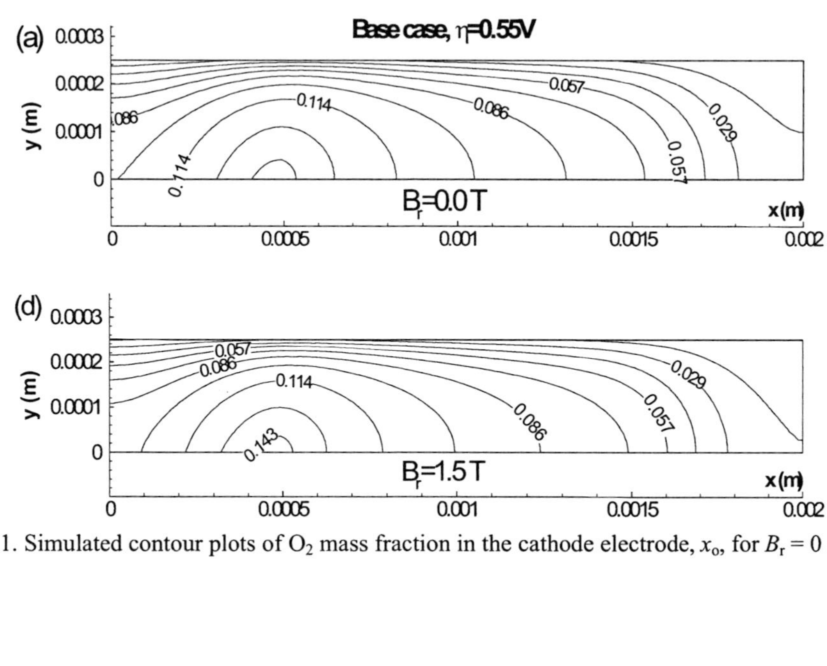

Figure 21. Simulated contour plots of O_2 mass fraction in the cathode electrode, x_o, for B_r = 0 and 1.5 T.

Under the base-case conditions ($\eta = 0.55$V), most of the improvement in fuel-cell performance comes from the decrease in s near the catalyst layer due to the magnetic repulsive force. Fig. 22 shows the contour plots of s in the gas diffusion layer for B_r=0 and 1.5 T. In both cases, near the shoulder of the current collector (x=0.05cm), s is small because the shear force caused by the gas flow removes nearly all of the liquid water from this region. In other regions, the fraction of the pore space clogged by liquid water is large, even at the gas inlet region, thus preventing the transport of O_2 gas to the reaction sites. When the magnetic force acts (B_r=1.5T), s decreases at the region near the catalyst layer, especially near the outlet region. Note that the comparisons are carried out at relatively large cathode overpotential condition and the gas diffusion layer is already flooded to some extent. Under large η, most of the improvement in cell performance comes from the decrease in s near the catalyst layer due to the magnetic force.

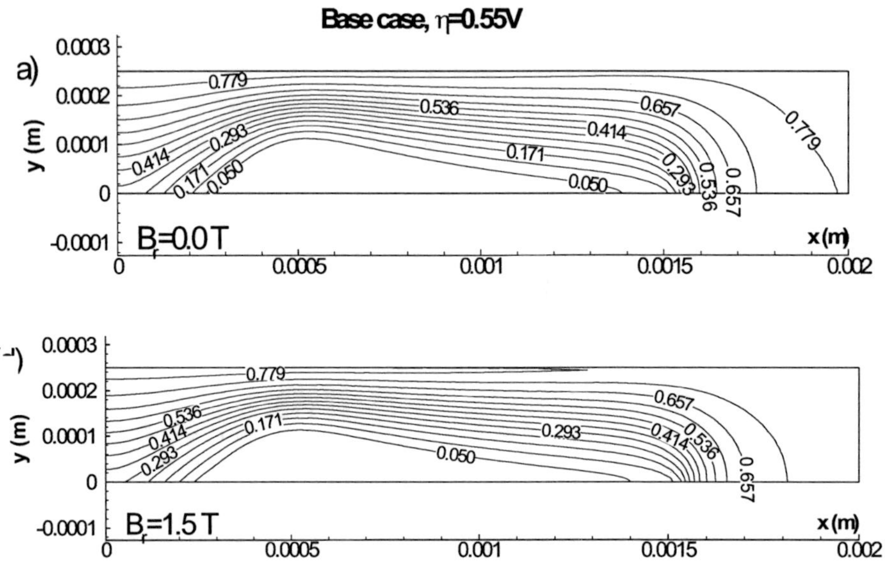

Figure 22. Simulated contour plots of liquid water saturation in the cathode electrode, s, for B_r = 0, and 1.5 T.

The gas velocity distribution in the gas diffusion layer is not presented here because no significant difference was observed in the distribution with and without magnetic force. Fig. 23 shows the distribution of the liquid water velocity in the absence (B_r = 0T) and presence (B_r =1.5 T) of magnetic force. Due to the relatively high gas velocity, the liquid water flow is primarily in the gas flow direction, especially at the outlet region. Without the magnetic force (B_r = 0), in the upstream region near the interface, the main driving force is the capillary force, which pushes liquid water away from the interface. In the middle of this interface, the shear force created by the gas flow transports the liquid water downstream and pushes the water out of the backing layer. However, in the presence of magnetic force (B_r = 1.5T), the velocity field of liquid water changes significantly. In the region near the interface, the y-direction velocity component, v_{lw} increases with increasing B_r, and more liquid water is carried out from this region. Because more pore space is thus provided for O_2 to contact the catalyst layer, the fuel-cell performance is improved. With the magnetic force, the velocity of liquid water at the outlet also increases slightly.

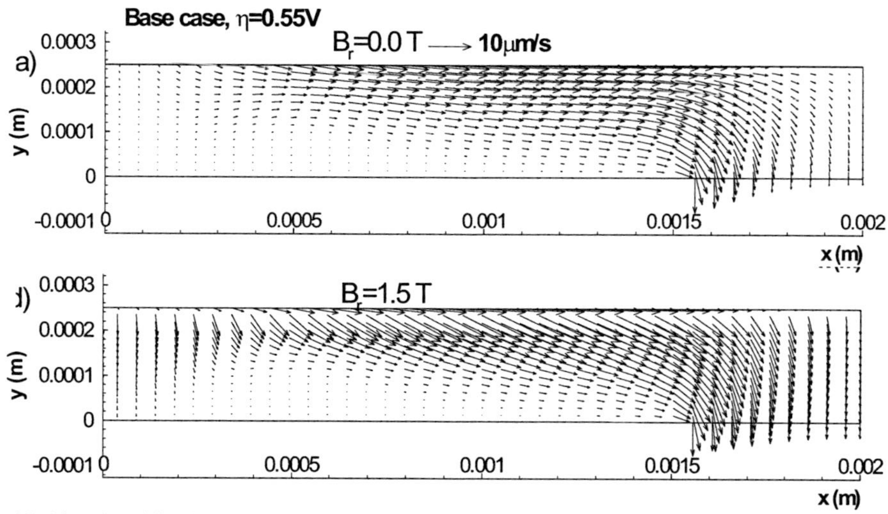

Figure 23. Simulated liquid-water velocity vectors in the cathode electrode for $B_r = 0$, and 1.5 T.

6.4.5. Conclusions from Numerical Simulations

The numerical simulations yield two main conclusions. First, magnetic repulsive force caused by the magnetized magnet particles improves the fuel cell performance by decreasing the overall level of liquid water saturation near the interface of the catalyst/diffusion layers, *s* and by increasing the pore space near the interface of the catalyst/diffusion layers for O_2 to transport to and contact with the catalyst layer. Second, the cell performance is improved specially in the limiting current region in the presence of magnetic force. This tendency agrees with experimental results [9].

7. Summary

This chapter describes a new method to manage liquid water in cathode for improving fuel cell performance. The key points of this chapter are as follows:

1) The effect of permanent magnets on electrochemical oxygen reduction was studied for possible application to cathode gas reactions in PEFC. In electrochemical and rotating electrode experiments, when a permanent magnet was placed in contact behind a Pt foil or Pt-dispersed carbon paper cathode and O_2 was supplied toward a stronger magnetic field, electrochemical flux was enhanced and the current increased with increasing magnetic field.

2) When small magnet particles were inserted in the cathode catalyst layer of a PEFC and then magnetized, the performance of a single PEFC was superior to that of a PEFC not magnetized; the performance was particularly enhanced in the limiting current region.

3) These magnetic effects are explained qualitatively by the magnetic repulsive force acting on water and by the attractive force toward O_2 gas. The numerical simulation indicates that magnetic repulsive force acting on liquid water mainly improves the fuel cell performance by decreasing the overall level of liquid water saturation near the interface of the catalyst/diffusion layers and thus increasing the pore space for O_2 to transport to and contact with the catalyst layer.

4) This method will be especially useful for a portable fuel cell in which there is no power to supply the air required for fuel cell operation.

REFERENCES

[1] K. Kordesh and G. Simmader, Fuel Cells and Their Applications, VCH, Weinheim (1996).
[2] D. Natarajan and T. Van Nguyen, J. Electrochem. Soc., 148(2001) A1324.
[3] T.E. Springer, M.S. Wilson, and S. Gottesfeld, J. Electrochem. Soc., 140 (1993) 3513.
[4] D. M. Bernardi and M. W. Verbrugge, J. Electrochem. Soc. 139 (1992) 2477.
[5] D. M. Bernardi and M. W. Verbrugge, AIChE J. 37 (1991) 1151.
[6] T. V. Nguyen, J. Electrochem. Soc. 143 (1996) L103.
[7] D. III, Wood, S. J. Yi, and T. V. Nguyen, Electrochimica Acta 43 (1998) 3795.
[8] N. I. Wakayama, T. Okada, J. Okano and T. Ozawa, Jpn. J. Appl. Phys. 40 (2001) L269.
[9] T. Okada, N. I. Wakayama, L. B. Wang, H. Shingu, J. I. Okano, and T. Ozawa, Electrochimica Acta 48 (2003) 531.
[10] L. B. Wang, N. I. Wakayama, T. Okada, Electrochemistry Communications 4 (2002) 584.
[11] M. Faraday, Philos. Mag. S3 (1847) 401.
[12] McGraw-Hill Encyclopedia of Science and Technology, vol. 10, 7th ed., MacGraw-Hill, New York, 1992, p. 315; J. Huang, D.D. Gray, Phys. Rev. E, 58 (1998)5164.
[13] G. Herzberg, Molecular Spectra and Molecular Structure 1. Spectra of Diatomic Molecules, Van Nostrand Reinhold, New York (1939), p. 366, 462.
[14] Chemical Society of Japan (Ed.), Kagaku Binran (Tables of Physical and Chemical Data), 2nd ed., Maruzen, Tokyo, 1975, p. 1253.
[15] L. Pauling, R.E. Wood and J.H. Sturdivant, J. Am. Chem. Soc. 68 (1946) 795.
[16] N.I. Wakayama, J. Appl. Phys. 69 (1990) 2734.
[17] B. Bai, A. Yabe, J. Qi and N.I. Wakayama, AIAA J. 37 (1999) 1538.
[18] N.I. Wakayama, Combust Flame 93 (1993) 207.
[19] N.I. Wakayama, H. Ito, Y. Kuroda, O. Fujita and K. Ito, Combust Flame 107 (1996) 187.
[20] S. Kinoshita, T. Takagi, H. Kotera, N.I. Wakayama, IEEE Transactions for Applied Superconductivity, 14 (2004)1685.
[21] N.I. Wakayama, Jpn. J. Appl. Phys. 39 (2000) L436.
[22] K. Yamakawa, K. Okawa, and K. Miyamoto, Design of the magnetic flux loop with permanent magnet and its application (1997), Sogodensi Publishing Co. Ltd. pp. 89-96
[23] T. Okada, T. Higashino, M. Yuasa and I. Sekine, Electrochemistry 69 (2001) 758.
[24] Y. Takahashi, K. Iida and M. Iijima, J. Vac. Soc. Jpn. 42 (1999) 992.
[25] M.S. Wilson and S. Gottesfeld, J. Electrochem. Soc. 139 (1992) L28.
[26] M. Uchida, Y. Fukuoka, Y. Sugawara, H. Ohara, and A. Ohta, J. Electrochem. Soc. 145 (1998) 3708.
[27] L. D. Landau and E. M. Lifhitz, Electrodynamics of Continuous Media (2nd Edition), Butterworth Heineman, Chap.4, p.127.
[28] W. S. He, J. S. Yi, and T. V. Nguyen, AIChE J. 46 (2000) 2053.

[29] R. B. Bird, W. E. Stewart, E. N. Lightfoot, Transport Phenomena (1960), Wiley, New York.

[30] L.B. Wang, N.I. Wakayama, T. Okada, to be submitted.

In: Focus on Electrochemistry Research
Editor: Magdalena Nunez, pp. 37-67
ISBN: 1-59454-545-6

I wish you to review English for style, grammar, spelling and scientific English.

Chapter 2

ELECTRODEPOSITION AND CHARACTERIZATION OF MO–SE THIN FILM LAYERS

N. Dukstiene

Department of Physical Chemistry, Kaunas University of Technology Radvilenu pl.19, Kaunas, Lithuania, LT- 50299. E-mail: nijole.dukstiene@ktu.lt

ABSTRACT

The present paper concerns the electrodeposition of Mo–Se thin films on a conducting tin oxide (SnO_2) coated glass substrates from a sulfamatic solution containing Na_2MoO_4 and H_2SeO_3. A discussion on Mo–Se cathodic electrodeposition in terms of thermodynamics in association with an investigation on the voltammetric behavior of SnO_2/glass electrode in Na_2MoO_4 + H_2SeO_3 system is presented. The formation of Mo–Se thin film has been determined to be of a multistage character. An assumption has been made that the reduction of elemental selenium to HSe^- is a sluggish and deposition-controlled reaction.

The influence of electrodeposition conditions on the structure, composition, optical properties and morphology of Mo–Se deposits correlated with the appearance and the growth of layer is discussed.

The deposition potential varied from –0.8 V to –1.1 V (vs SHE). Deposition temperature T = 20 °C – 40 °C. pH 6.5. X-ray diffraction analysis revealed that the overall composition of the films deposited is consistent with the formation of MoO_2 and $MoSe_2$. The lattice parameters of “as-deposited” $MoSe_2$ are a = b = 3.2340 A°, c = 13.2859 A° which fits into a hexagonal structure.

The Mo–Se thin film layers are indirect band-gap semiconductors. The values of E_{og} are in the range of 1.68 and 2.2 eV.

The data presented in this paper serve as an effective basis for studying the electrodeposition mechanism of transition metal selenide thin film layers.

1. Introduction

Molybdenum diselenide absorbs in the visible and near-IR regions and exhibits an inherent resistive nature to photocorrosion [1]. $MoSe_2$ has layered sandwich structure and the layers of Se-Mo-Se interact with each other by Van der Waals forces, which allow adjacent layers to slide. It is easy for ions such as Li^+ or Na^+ to intercalate in the interlayer gaps, due to a large d – spacing between layered sandwich structures. All these properties cause layered $MoSe_2$ to have special applications in precursors of intercalation superconductor [2], materials for photoelectrochemical cell [3, 4] and solid – state lubricants.

Different growth techniques such as intercalation-exfoliation [5], self-propagating high temperature synthesis (SHS) [6], solvothermal synthesis [7], direct vapour transport technique [8, 9] are currently used for Mo–Se thin film deposition.

However, these high temperature methods suffer from the difficulty of producing large amounts of materials and are expensive.

The high production cost of such materials requires the search for a cheaper method. These techniques include electrodeposition and chemical deposition.

The preparation of transition metal chalcogenide thin films by electrosynthesis method is currently attracting considerable attention [10-13]. Electrodeposition appears to offer the greatest potential in terms of low costs and compatibility with applications requiring large photoactive areas.

It is commonly known that it is impossible to deposit pure molybdenum by the electrolysis of aqueous solutions. However, it can be readily deposited as an alloy with Co, In and Fe [14-16]. The co-deposition mechanism of molybdenum with the iron group metals has been proposed [17-19]. A kind of the induced co-deposition mechanism has also been suggested for many metal selenides: WSe_2 [13], PbSe [20] and EuSe [21].

The co-deposition of Se and Mo with simultaneous formation of a chemical compound, $MoSe_2$, can be achieved by cathodic electroreduction from a solution containing H_2SeO_3 (or SeO_2) and MoO_4^{2-} ion. An electrochemical model accounting for the preparation method relies on the deposit–electrolyte thermodynamics and the rate components transfer in solution, as proposed by Kroger and co–workers [22]. According to this description, a semiconductive binary compound of well-defined stoichiometry can be prepared by under potential deposition (upd) of less noble element, induced by the free energy gained by compound's formation. The above model provides valid predictions as to the equilibrium potentials of electrodeposited compounds and has been the starting point for efficient thin film (such as CdSe [23], ZnSe [24]) deposition inasmuch as proper experimental procedures are followed.

Sanjeeviraja [10] et. al have reported on the electrodeposition of highly textured $MoSe_2$ film with a polycrystalline nature from an ammonaical solution of H_2MoO_4 and SeO_2

In this paper, an attempt is made to study the electroreduction of H_2SeO_3 and Na_2MoO_4 onto SnO_2/glass electrode in order to determine the potential range of $MoSe_2$ formation. The influence of experimental conditions (temperature, potential applied, bath composition) on the structure, composition, optical properties and morphology of Mo–Se deposits correlated with an appearance and the growth of the layer is also discussed.

2. EXPERIMENTAL

The electrodeposition mechanism was studied employing cyclic voltammetry. Cyclic voltammetry (CV) experiments were conducted in a thermostat-controlled conventional three-electrode JSE-2 cell with a PI-I-50 potentiostat coupled to a PR-8 programmer and an XY potentiometer recorder (Russia, "ZIP"). Cyclic voltammograms were obtained at a sweep rate of 50 $mV\cdot s^{-1}$.

The working electrodes were glass sheets covered with conductive and transparent tin oxide (SnO_2). The geometrical area of the electrode was 5.0 cm^2 and the thickness of SnO_2 film was 0.5 μm, resistivity: $8.3\cdot 10^{-4}$ Ω·cm. The conducting SnO_2/glass electrodes were cleaned with acetone and then electrochemically in 0.1 M H_2SO_4 rinsed with distilled water and dried in an alcoholic vapour [21]. The auxiliary electrode was a platinum wire. A standard Ag,AgCl|$KCl_{(sat)}$ electrode was used as the reference electrode. The tip of the reference electrode was placed very close to the cathode surface so that the exact potential at the surface was monitored, unaffected by the solution resistance. All reported values of electrode potentials are expressed relative to the standard hydrogen electrode (SHE).

The experiments were performed in an electrochemical aqueous bath containing different amounts of Na_2MoO_4, and H_2SeO_3. To prepare electrolyte solutions having relative concentrations of the components, basic solutions of 1M Na_2MoO_4 and 1 M H_2SeO_3 were first prepared. The different volumes of basic solutions were mixed to give electrolyte solutions. A 0.1 M $NH_4SO_3NH_2$ solution was used as a supporting electrolyte. The bath pH was maintained at 6.5–6.8. Purified water and chemicals of minimum 98% purity were used.

X-ray diffractometry was carried out under a Brag Brendan circuit on a difractometer (Dron - 3.0 Russia, Bourevestnic Inc.) using CuK_α (λ=0.154178 nm) radiation, 30 kV voltage and 20 mA current. The 2θ scanning range was from 0 to 90. The scanning speed was 1 ^{0}/min. The results were registered in "in situ" mode by a computer and then processed by the "Searechmath" and X-Fit[1] programs.

X-ray fluorescence analysis was carried out with a VRA-20 X-ray spectrometer (Germany) by direct radiation of the thin layer so as not to destroy the Mo–Se layers. The conditions of specimens investigated by the VRA-20 device were the following: a Rh tube as anode, 35 kV voltage and 26 mA current. The specimens were revolved in the horizontal plane in order to improve the X-ray signal stability. The calibration curve was formed in the interval from 0.25 to 1 μg, using for the calibration "thin" layers formed from the standard solutions on the filter paper. Distribution of Mo and Se on the standard specimen area was monitored by the UV fluorescence mark method. The "thin" layers criteria were kept for concentrations up to 40 $\mu g.mm^{-2}$. Various measurements were carried out on different regions of the deposited films and the average composition was determined. The calculated analysis error was less than 5%.

Optical microscopy was carried out on a computer microscope (Intel QX3, Intel USA) at a magnification of 60x. Some samples were removed in a reflection mode and some in a trough mode at additional illumination by an intense heat lamp. The program Jmage J[2] processed the results.

---

[1] (URL: http://www.ccp14.ac.uk/tutorial/xfit-95/xfit.htm).

[2] URL: http://rsb.info.nih.gov/ij/download/zips/ij128.zip

An atomic force microscope (Nanotop– *206* Microtestmashines Belarus) operating in a contact–state mode was used to investigate the morphology of the surface. It used a silicon cantilever, with a force constant 5.0 N/m. The image processing and analysis of the scanning probe microscopy data were performed using a Windows-based program (*Surface View version 1.0)*.

The optical parameters of Mo–Se films were deduced from the UV, Vis-spectrophotometry measurements carried out in the spectral region of 225–700 nm, using a UV/VISIBLE spectrophotometer (Spectronic Genesis). The optical absorption spectrum was taken for Mo–Se thin film using an identical transparent tin oxide-coated glass plate as a reference, and the values of optical absorption coefficient (α) were not corrected for the reflectance from the film surface.

3. RESULTS AND DISCUSSION

3.1. Thermodynamic Aspects and Background

3.1.1. An Examination of Standard Reduction Potentials

The voltammetric behavior of selenium compounds, exhibiting electrochemical activity in only two different valent states, Se^{2-} and Se^{4+}, is determined by the composition and pH value of the working solution through variations in the composition of the various species present (i.e. H_2Se, HSe^-, Se^{2-} SeO_3^{2-}, $HSeO_3^-$, H_2SeO_3) and the structure of the electrical double layer as well. [25]. Because of its multiple oxidation states, selenium can engage in numerous redox couples. Table 1 lists the species and the thermodynamic data culled from literature sources.

It should be noted that while the reduction of H_2SeO_3 has a fairly high (thermodynamic) potential 0.74 V, the rather sluggish kinetics coupled with the passivating influence of the incipient Se^0 on the electrode surface, shifts the reduction effectively to more negative potentials. Furthermore, different reduction potentials of H_2SeO_3 to selenium and Se (0) reduction to Se (-II) have been reported. [32, 34-43]. These differences may be attributed to different substrates and conditions of experiments such as stirring solution, pH, temperature and concentration.

Figure 1 illustrates the Pourbaix diagram for the Mo–H_2O system at 25 ^0C [44]. Candidate reactions and thermodynamic data for this diagram are shown in Table 2.This table lists the standard potential values for the (-II), (+IV) and (+VI) oxidation states available in the literature sources [10, 33, 44].

Table 1. Standard potentials of selenium redox couples at 25^0 and $1.01314 \cdot 10^5$ Pa pressure (vs SHE)

Electrode reaction	E^0,V	Literature source
Se(4+)/Se(0) couple		
$H_2SeO_{3(aq)} + 4H^+ + 4e^-$ $Se_{(s)} + 3H_2O$	0.740	27
		29
		26
		31
		30
	0.744	28
$HSeO_3^-{}_{(aq)} + 5H^+ + 4e^-$ $Se_{(s)} + 3H_2O$	0.778	23
$SeO_3^{2-} + 3H_2O + 4e^-$ $Se + 6OH^-$	-0.336	40
Se(0)/Se(-2) couple		
$Se(s) + 2H^+$ $2e^-$ $H_2Se_{(aq)}$	-0.400	31
		26
	-0.115	25, 30
$Se(s) + 2H^+$ $2e^-$ $H_2Se_{(g)}$	-0.082	44
$Se(s) + H^+$ $2e^-$ HSe^-	-0.512	26
	-0.227	36
$Se(s) + 2e^-$ Se^{2-}	-0.927	26
	-0.670	44
	-0.641	30
	-0.680	32
$2Se + 4e^-$ Se_2^{2-}	-0.592	44
Se(4+)/Se(-2) couple		
$H_2SeO_{3(aq)} + 4H^+ + 6e^-$ $Se^{2-} + 3H_2O$	0.269	32
$HSeO_3^-{}_{(aq)} + 5H^+ + 6e^-$ $Se^{2-} + 3H_2O$	0.292	32
$H_2SeO_{3(aq)} + 6H^+ + 6e^-$ $H_2Se + 3H_2O$	0.360	23
$HSeO_3^-{}_{(aq)} + 7H^+ + 6e^-$ $H_2Se + 3H_2O$	0.386	23

The analyses of standard potentials and Pourbaix diagram predict the following for the pH6.5–6.8 range pertinent to this study: the stable MoO_4^{2-} species reduces to MoO_2 by the following reaction:

$$MoO_4^{2-} + 2H_2O + 2\bar{e} \rightarrow MoO_2 + 4OH^- \qquad (1)$$

On further cathodic poliarization of electrode potential, MoO_2 should form Mo via reaction:

$$MoO_2 + 4H^+ + 4\bar{e} \rightarrow Mo + 2H_2O \qquad (2)$$

Rationalization of experimental data on real system in the light of the above thermodynamic treatment has to include complications arising from great difference in the deposition potentials of Mo and Se.

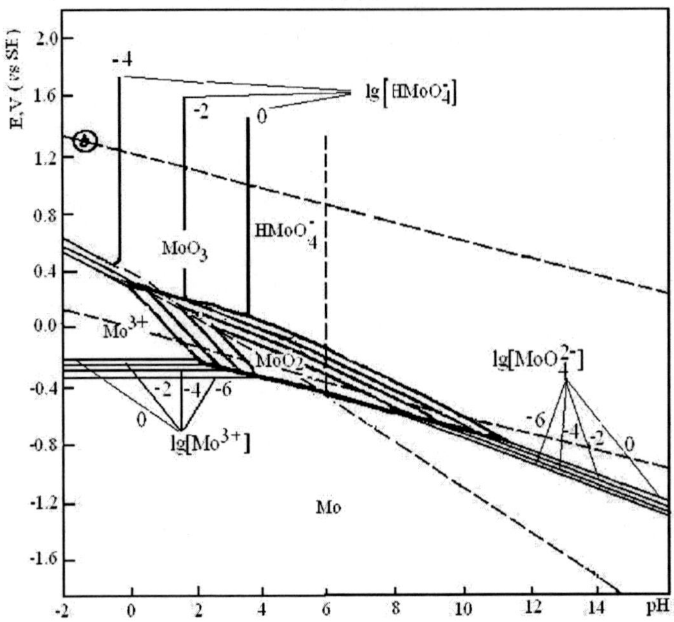

Figure 1. Pourbaix diagram for the Mo-H_2O system at 25 ^{o}C [44].

Table 2. Standard potentials of molybdenum redox couples at 25^{0} and 1.01314·10^{5} Pa pressure (vs SHE)

Electrode reaction	E^{0},V	source
Mo(6+)/Mo couple		
$MoO_4^{2-} + 6H^+ + 6e\ \ Mo^0 + 4H_2O$	0.0 V	10
	0.155 V	33
$MoO_4^{2-} + 4H_2O + 6e\ \ Mo^0 + 8OH^-$	-1.05 V	10, 44
Mo(6+)/Mo(4+) couple		
$MoO_4^{2-} + 4H^+ + 2e\ \ MoO_2 + 2H_2O$	0.606	44
	-0.772	33
$HMoO_4^- + 3H^+ + 2e\ \ MoO_2 + 2H_2O$	0.429	44
$MoO_3 + 2H^+ + 2e\ MoO_2 + H_2O$	0.320	44
	0.537	33
Mo(6+)/Mo(3+) couple		
$HMoO_4^- + 7H^+ + 3e\ Mo^{3+} + 4H_2O$	0.390	44
$MoO_4^{2-} + 8H^+ + 3e\ \ Mo^{3+} + 4H_2O$	0.508	44
$MoO_3 + 6H^+ + 3e\ \ Mo^{3+} + 2H_2O$	0.317	44
Mo(4+)/Mo(3+) couple		
$MoO_2 + 4H^+ + e\ \ Mo^{3+} + 2H_2O$	0.311	44
Mo(4+)/Mo(0) couple		
$MoO_2 + 4H^+ + 4e\ Mo + 2H_2O$	-0.072	
	-0.983	33
Mo(3+)/Mo(0) couple		
$Mo^{3+} + 3e\ Mo$	-0.200	33

Another problem that arises with the electrodeposition of Mo–Se film layers is that the free energy of formation of the anion MoO_4^{2-} is -838 kJ/mol [33], which indicates its high thermodynamical stability. The molybdate ion is stabilized by its inner orbital hybridized with the p-orbital. The energy required for breaking such hybridization exceeds the energy necessary for the cathodic discharge of hydrogen ion.

3.1.2. Mose₂ Electrodeposition – Prediction of Equilibrium Potentials

For the sake of simplicity let us consider that $MoSe_2$ compound electrodeposition from solutions close to neutral (pH 6.5) involves the following reactions at the cathode:

$$HSeO_3^- \ 5H^+ + 4e \Leftrightarrow Se_{(s)} \ 3H_2O \ E^0 = 0.780 \text{ V vs. SHE} \quad (3)$$

$$MoO_4^{2-} + 4H_2O + 6e \Leftrightarrow Mo^0 + 8OH^-, \ E^0 = -1.05 \text{ V vs. SHE} \quad (4)$$

$$Mo + 2Se \Leftrightarrow MoSe_2 \quad (5) \ [45]$$

$\Delta G^0_{f,298}(MoSe_2) = -218.94$ kJ/mol [46].

The activities of Mo and Se components in a $MoSe_2$ deposit are interrelated as determined by equilibrium constant of Eg (5):

$$K = \frac{a_{MoSe_2}}{a_{Mo}a^2_{Se}} = \exp\left(\frac{-\Delta G^0_{f,298},(MoSe_2)}{RT}\right) \xrightarrow{a_{MoSe_2}} \quad (6)$$

$$a_{Mo}a^2_{Se} = \exp\left(\frac{\Delta G^0_{f,298}(MoSe_2)}{RT}\right) \quad (7)$$

Deviations from stoichiometry practically signify the existence of excessive Mo or Se atoms located in lattice or interstitial positions as well as the formation of distinct elemental phases. In a simple case, $MoSe_2$ deposit may present a composition range limited on one hand, by coexistence of the compound with a pure Se phase, (Se + $MoSe_2$) and, at the other hand, by coexistence with pure Mo phase ($MoSe_2$+Mo). The activities of component elements, in coexisting solid phases, correspond to the limiting values of a_{Mo} and a_{Se}. Therefore, Eq.(7) gives:

At Se|$MoSe_2$| phase boundary:

$$a_{Se} = 1 \text{ and } a_{Mo} = \exp\left(\frac{\Delta G^0_{f,298}(MoSe_2)}{RT}\right) \quad (8)$$

At $MoSe_2|Mo$ phase boundary

$$a_{Mo} = 1 \text{ and } a_{Se} = \sqrt{\exp\left(\frac{\Delta G^0_{f,298}(MoSe_2)}{RT}\right)} \tag{9}$$

The variation of a_{Se} and a_{Mo} causes a shift in equilibrium potentials of the particular Se and Mo elements though a homogeneous deposited film of a specific composition, in contact with the working electrolyte, possesses only one equilibrium potential equal to the potentials of the constituent elements. That is $E_{MoSe_2} = E_{MoO_4^{2-}/Mo} = E_{H_2SeO_3/Se}$, an equation that determines the relation between ion activities at the cathode/solution interface and consequently it allows the adjustment of bulk solution concentrations according it mass transport conditions. In particular, a further analysis on the above relationship postulates the requirement of having a large excess of MoO_4^{2-}in electrolyte solution.

The corresponding Nernst potential for Eq. (4) is

$$E_{MoO_4^{2-}/Mo} = E^0_{MoO_4^{2-}/Mo} + \frac{RT}{nF}\ln\left(\frac{a_{MoO_4^{2-}}a^4_{H_2O}}{a_{Mo}a^8_{OH^-}}\right) \tag{10}$$

Since $a_{H_2O} = 1$ and $a_{OH^-} = \frac{K_{H_2O}}{a_{H^+}} = \frac{1\cdot 10^{-14}}{a_{H^+}}$ at T=298 ^{o}K and pH 6.5 one will obtain:

$$E_{MoO_4^{2-}/Mo} = -0.465 + 0.0043\ln\left(\frac{a_{MoO_4^{2-}}}{a_{Mo}}\right) \tag{11}$$

Generally, according to Eq. (11) Mo could be co–deposited with Se at cathodic potentials ranging from $E_{MoO_4^{2-}/Mo} = -0.465 + 0.0043\ln\left(\frac{a_{MoO_4^{2-}}}{a_{Mo}}\right)$ at the $Se|MoSe_2$ phase boundary to $E_{MoO_4^{2-}/Mo} = -0.465 + 0.0043\ln a_{MoO_4^{2-}}$ at the $MoSe_2|Mo$ phase boundary.

Replacing Eq. (8) in Eq.11 and subtracting derived relations one finds that over the whole composition range $E_{MoO_4^{2-}/Mo}$ is changed by $\left(\frac{-\Delta G_{f,298}(MoSe_2)}{6F}\right)$. At T = 298 ^{o}K , the formation of $MoSe_2$ shifts the $E_{MoO_4^{2-}/Mo}$ potential by a +0.378 V. Since $\Delta G^0_{f,298}(MoSe_2) <$ 0, the shift in potential, following the activity reduction is positive. Hence Mo could be deposited at more positive potentials than its equilibrium value. Similarly $E_{H_2SeO_3/Se}$ potential changes by $\left(\frac{-\Delta G_{f,298}(MoSe_2)}{4F} \right)$ and this change is equal +0.567 V.

The equilibrium potentials of E_{MoSe_2} within the potential range that the compound could be deposited are calculated by applying $E_{MoSe_2} = E_{MoO_4^{2-}/Mo}$ and Eq. (10) twice, for $a_{Mo} = \exp\left(\frac{\Delta G^0_{f,298}(MoSe_2)}{RT} \right)$ and $a_{Mo} = 1$ at the boundaries of Se|MoSe$_2$ and MoSe$_2$|Mo phases, respectively. At T = 298 ^{0}K and [MoO_4^{2-}] = 0.2M, it follows that $E_{MoSe_2} = E_{MoO_4^{2-}/Mo} = -0.0935$ V at Se|MoSe$_2$ and $E_{MoSe_2} = E_{MoO_4^{2-}/Mo} = -0.472$ V at MoSe$_2$|Mo.

3.2. Voltammetic Measurements

The cyclic voltammograms obtained from electrolyte solutions can provide the basic information about the electrodeposition process occurring in the real system,

The point to emphasize here is that the formation of $MoSe_2$ can be achieved in solutions containing a small amount of H_2SeO_3 and a sufficient excess of Na_2MoO_4 (chapter 3.1.2.). In the usual electrodeposition of metal selenides the Me/Se ratio in the solution should be much larger than unity (≥ 5) [47]. Therefore the electrolytes with various concentrations of Na_2MoO_4 (0.05 M, 0.10 M, 0.15 M, and 0.20 M) at a concentration constant 0.01 M of H_2SeO_3 were used.

3.2.1. Study of H_2SeO_3 Reduction Features on the SnO_2/Glass Electrode

Addition of a selenite acid in an aqueous media results in the formation of dissolved Se (IV) species establishing an equilibrium, the state of which depends on solution acidity and temperature. Specifically, the H_2SeO_3, $HSeO_3^-$, SeO_3^{2-} species precipitate in the following acid–base equilibrium at 25 ^{0}C [30]:

$$H_2SeO_3 \Leftrightarrow HSeO_3^- + H^+ \qquad (12) \qquad (pK_a = 2.46)$$

$$HSeO_3^- \Leftrightarrow SeO_3^{2-} + H^+. \qquad (13) \qquad (pK_a = 7.31)$$

Hence at 20 ^{0}C and pH adjusted to 6.5–6.8, SeO_3^{2-} and $HSeO_3^-$ comprise about 24 and 76% respectively. The content of H_2SeO_3 being very low, it can be neglected.

On the cyclic voltammogram for the tin oxide electrode in 0.1 M $NH_4SO_3NH_2$ containing H_2SeO_3 (Fig. 2), on the forward cathodic scan a single reduction wave (c_1) can be observed from $E_{st.} = -0.182$ to -0.95 V before the current increases rapidly (c_2).

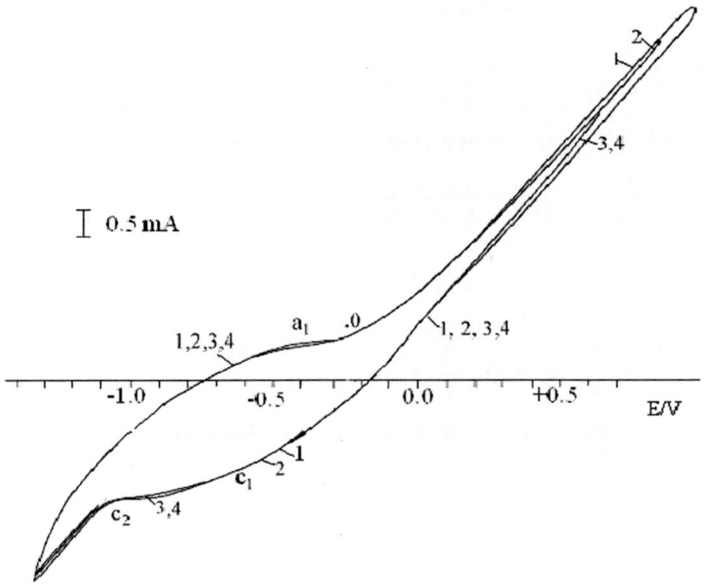

Figure 2. Cyclic voltammograms for the SnO_2 electrode obtained in the electrolyte containing 0.01 M H_2SeO_3 + 0.1 M $NH_4SO_3NH_2$. Temperature 20 °C, pH 6.5. 1–the first scan, 2–the second scan, 3–the third scan 4 – the tenth scan.

The electrolysis performed at the potentials from –0.20 to –0.65 V leads to the formation of a product strongly bound to the electrode surface. X-ray fluorescence analysis revealed the formation of elemental selenium. However, due to the amorphous nature of the deposited selenium film, it was impossible to confirm its formation by XRD. Formation of the amorphous selenium film on the electrode surface can be expressed by the following reaction:

$$HSeO_3^- + 5H^+ + 4\bar{e} \rightarrow Se + 3H_2O \tag{14}$$

At potentials ranging from –0.7 V, the film gradually disappears and the SnO_2 electrode keeps it initial color. The catholyte becomes red-colored. The selenium deposited at the potentials varying from –0.20 to –0.70 V on the surface of the SnO_2 electrode has appeared to be electroactive and is reduced at E ≤ –0.7 V by the reaction

$$\mathrm{Se} + H^+ + 2\bar{\mathrm{e}} \rightarrow H\mathrm{Se}^- \tag{15}$$

Formed HSe^- and reacts with $HseO_3^-$ or SeO_3^{2-} in the solution as follows:

$$2HSe^- + HSeO_3^- + 3H^+ \rightarrow 3Se + 3H_2O \tag{16}$$

$$2HSe^- + SeO_3^{2-} + 4H^+ \rightarrow 3Se + 3H_2O \tag{17}$$

Such a coloration of the solution has been observed previously during the reduction of selenious acid upon various solid electrodes [34-38] and upon a SnO_2 electrode [42].

Cathodic reduction potentials of $HSeO_3^-$ ion to elemental selenium are affected by H^+ ion concentration in the plating solution. Presuming that only $HSeO_3^-$ species are involved in the electrochemical reaction (14), the potential approaching the four electron reduction can be derived by the Nernst formula as follows:

$$E_{HSeO_3^-/Se} = E^0_{HSeO_3^-/Se} + \left(\frac{2.3RT}{4F}\right)\lg\frac{\left[a_{HSeO_3^-}\right]}{\left[a_{Se}\right]} + \left(\frac{2.3RT}{4F}\right)\lg\frac{\left[a^5_{H^+}\right]}{\left[a^3_{h_2o}\right]} \tag{18}$$

Using an appropriate temperature coefficient to correct the standard potential $(\partial E/\partial T)$ = -0.52 mVK^{-1} [27] and since a_{se} and a_{H_2O} = 1, at T = 20 ^{0}C and pH 6.5 we obtain:

$$E_{HSeO_3^-/se} = +0.281 + 0.01476\lg(a_{HSeO_3^-}) \tag{19}$$

The equilibrium (Nerstian) potential values, calculated by Eq. (19) for the fourth electron reduction of $HSeO_3^-$ are E = +0.235 V for c = 1 x 10^{-3} M solution and E = +0.249V for c = 1 x 10^{-2} M solution, respectively. Interestingly, analysis of cyclic voltammograms shows that the cathodic current flow (Fig. 2) is recorded immediately from $E_{st.}$ = –0.182 V. In view of its passive surface a SnO_2/glass electrode constitutes a slow irreversible electrode [48], thus introducing a significant overpotential in $HSeO_3^-$ reduction. Therefore the discharge of $HSeO_3^-$ on a SnO_2/glass electrode starts off at potentials more negative than that approaching the fourth electron reduction equilibrium values.

Thus, the reduction process at wave (c_1) potentials on the SnO_2 electrode (Fig. 2) may represent the overlapping peaks of fourth-electron reduction of $HSeO_3^-$ to elemental Se and its reduction to HSe^-.

The waves attributed to the formation of selenide species are not observed on the voltammogram. Such phenomenon can be related with the properties of selenium. Abrarov and Bigelis [45] have proven that the selenium electrochemically deposited in neutral solutions is less electroactive than Se obtained in an acidic medium and exhibits poor electronic conducting properties. Vedel et al. [42] have demonstrated that the selenium deposited at room temperature on a tin oxide electrode exhibits a passivating effect. The poor conductivity of selenium film is the reason why the wave corresponding to reduction of Se (0) by Eq. (15) does not appear on the curve.

An abrupt increase of the cathodic current at potentials $E \leq -1.0$ V (Fig. 2) may represent the evolution of hydrogen.

3.2.2. Voltammetric Study in Solutions Containing Na_2MoO_4

The cyclic voltammogram of a tin oxide electrode was recorded in 0.1 M $NH_4SO_3NH_2$ solution containing 0.2 M Na_2MoO_4 [Fig. 3]. Note that at potentials negative to –0.4 V cathodic current is clearly found indicating that MoO_4^{2-}-ions can be reduced onto SnO_2/glass substrate. The series of electrodeposition experiments were carried out to demonstrate the effect of cathodic potential on the reduction of the MoO_4^{2-}-ions. Visual observation performed at constant potentials enabled us to establish that thin film layers are only formed at potentials from –0.8 V to –1.2 V. Dark brown films well adherent to the substrate were electrodeposited. In study of molybdenum electrodeposition, important question concerns the chemical nature of the deposited material. Therefore the films were analyzed by x–ray fluorescence and the phases formed were identified by XRD. According to the XRD results presented in the Figure 4, thin MoO_2 films could be successfully electrodeposited on the SnO_2/glass electrode surface at potentials E ≤ –0.8 V. A small diffraction peaks corresponding to MoO_2 deposits are found on XRD diagrams indicating microcrystalline structure of these films. The obtained lattice parameters of hexagonal unit cell a = b =1.4444 A° and c = 22.1111 A° coincide well with reported values of PDF 78-1070.

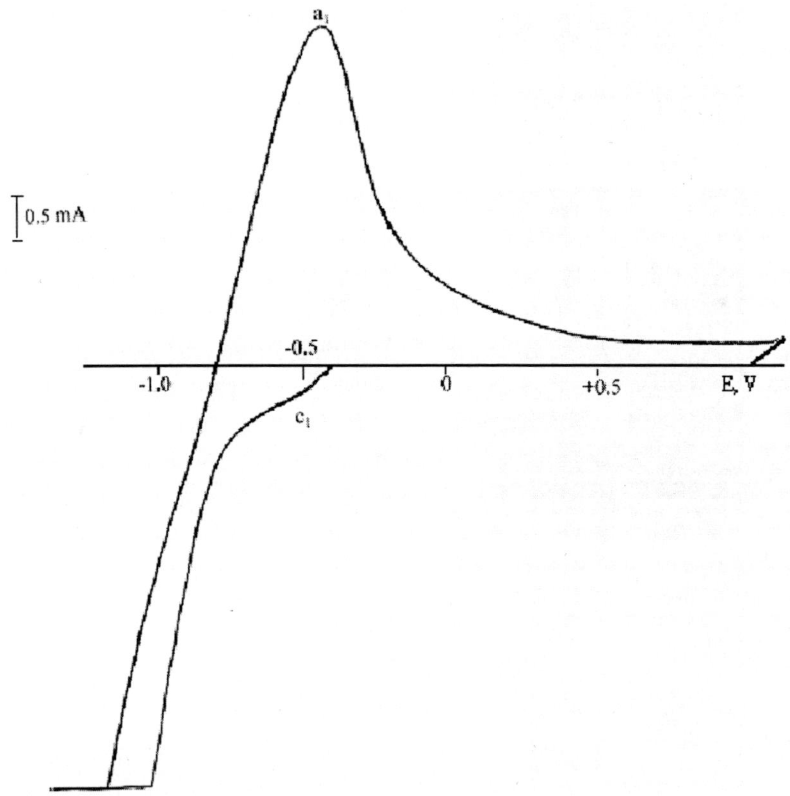

Figure 3. Cyclic voltammograms for the SnO_2 electrode obtained in the electrolyte containing 0.05 M Na_2MoO_4 + 0.1 M $NH_4SO_3NH_2$. Temperature 20 °C, pH 6.5.

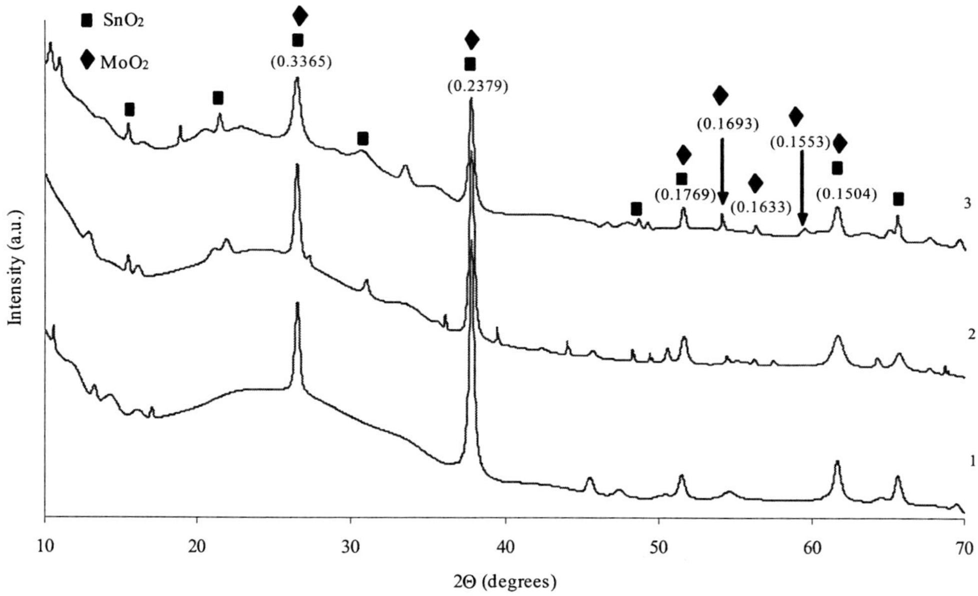

Figure 4. XRD patters of Mo–Se thin film deposited in the electrolyte 0.05 M Na_2MoO_4 + 0.1 M $NH_4SO_3NH_2$. Deposition conditions: T = 20 ^oC; pH 6.5; τ = 1h. Deposition potential (-E, V): 1: 0.8; 2: 0.9; 3: 1.0.

It may be stated that the formation species of MoO_2 onto SnO_2/glass electrode can be expressed by the Eq. (1)

The corresponding equilibrium potential can be derived by the Nernst formula as follows:

$$E_{MoO_4^{2-}/MoO_2} = 0.606 - 0.1182pH + 0.0295 1\text{g}\left[a_{MoO_4^{2-}}\right] \quad (20)$$

At pH 6.5, one has the potential values calculated by Eq (20) for reduction of MoO_4^{2-} E = –0.200 V for 0.05 M and E = –0.188 V for 0.2 M solutions, respectively. The calculated values of electrode potentials are more positive than the observed value of cathodic wave c_1 (Fig.3). Calculations and CV measurements signify that reduction of MoO_4^{2-} initiates at potentials more cathodic than the equilibrium value.

However neither the cathodic deposition nor the anodic oxidation occurs when electrode potentials are located between – 0.4 and – 0.7 V. The formation of wave c_1 was not accompanied by the visible changes of the electrode surface. Moreover, the large anodic current on CV curve at $E_{a,\ p}$ = –0.4 V indicates the more oxidative properties of deposits formed during cathodic scan between – 0.4 and – 0.7 V. An important result of this series of experiments is the conclusion that the anodic process at potentials $E_{a,p}$ = –0.4 V can significantly depress the MoO_2 deposition. This process can presumably be attributed to the simultaneous oxidation of MoO_2 species resulting in deposition of the MoO_2 film at more negative potential values. This assumption is supported by the highly negative standard free energy of MoO_4^{2-}-ions [33]. According to the x-ray analysis, the presence of Mo, previously assumed to be formed by reduction of MoO_4^{2-}-ions (Eqs. 1 and 2, Chapter 3.1.1) did not observed in any deposits. So, the experimental results indicate the formation of MoO_2 film layer at potentials more negative than –0.8 V.

3.2.3. Voltammetric Study in Solutions Containing Na_2MoO_4 and H_2SeO_3

The voltammogram for a SnO_2/glass electrode obtained in 0.1 M $NH_4SO_3NH_2$ solutions containing 0.05 M Na_2MoO_4 and 0.01 M H_2SeO_3 is presented in Fig. 5a. On comparing the successive cyclic voltammograms (Fig. 2, 3 and 5a) it is evident that a simultaneous reduction of MoO_4^{2-} and H_2SeO_3 leading to the formation of Mo-Se thin films is possible in the same range of cathodic potentials. In the presence of Na_2MoO_4 and H_2SeO_3 a current flow is recorded in the early stages of the cathodic scan (Fig. 5a). X-ray fluorescense analysis of a film obtained during the cathodic scan indicated the formation of elemental selenium in the potential region from –0.2 to –0.6 V (c_1). However, due to the amorphous nature of the deposited selenium film, it was impossible to confirm its formation by XRD. An analysis of AFM images [Fig. 6] of the surface shows the formation of amorphous selenium islands. The above results indicate that the reduction of H_2SeO_3 to elemental selenium is the first cathodic process in electrodeposition of Mo–Se thin film layer. An analysis of the cyclic voltammograms exhibits that the cathodic wave (c_1) increases in intensity with increasing the H_2SeO_3 content in the electrolyte. When H_2SeO_3 concentrations were kept constant and Na_2MoO_4 concentrations were changed from 0.05 M to 0.2 M, the current of wave (c_1) decreased and a current flow was recorded at more negative potentials (approximately at E = –0.25 V) (Fig. 5b). The further excursion into the negative potential region is preceded by the process (c_2) (Fig. 5a and 5b). This cathodic current is attributed to the Mo–Se deposition.

3.2.4. Composition of the Solid Phases

According to the diagnostic test on cyclic voltammograms the Mo–Se films can be grown at potentials in the range from –0.8 V to –1.10 V. Mo–Se films were deposited on SnO_2/glass electrode for various electrodeposition conditions. The films composition was analyzed by x–ray fluorescence and the phases formed were identified by XRD. The ratio of $MoSe_2$ and MoO_2 phases in the films was estimated from the respective integrated XRD peaks intensities using the following equation [49]

$$x = \left(1 + 0.8\frac{I_A}{I_B}\right)^{-1} \tag{21}$$

x-is the weight fraction in the sample.
I_A-is the intensity of the $MoSe_2$ diffraction peak.
I_B- is the intensity of the MoO_2 diffraction peak.

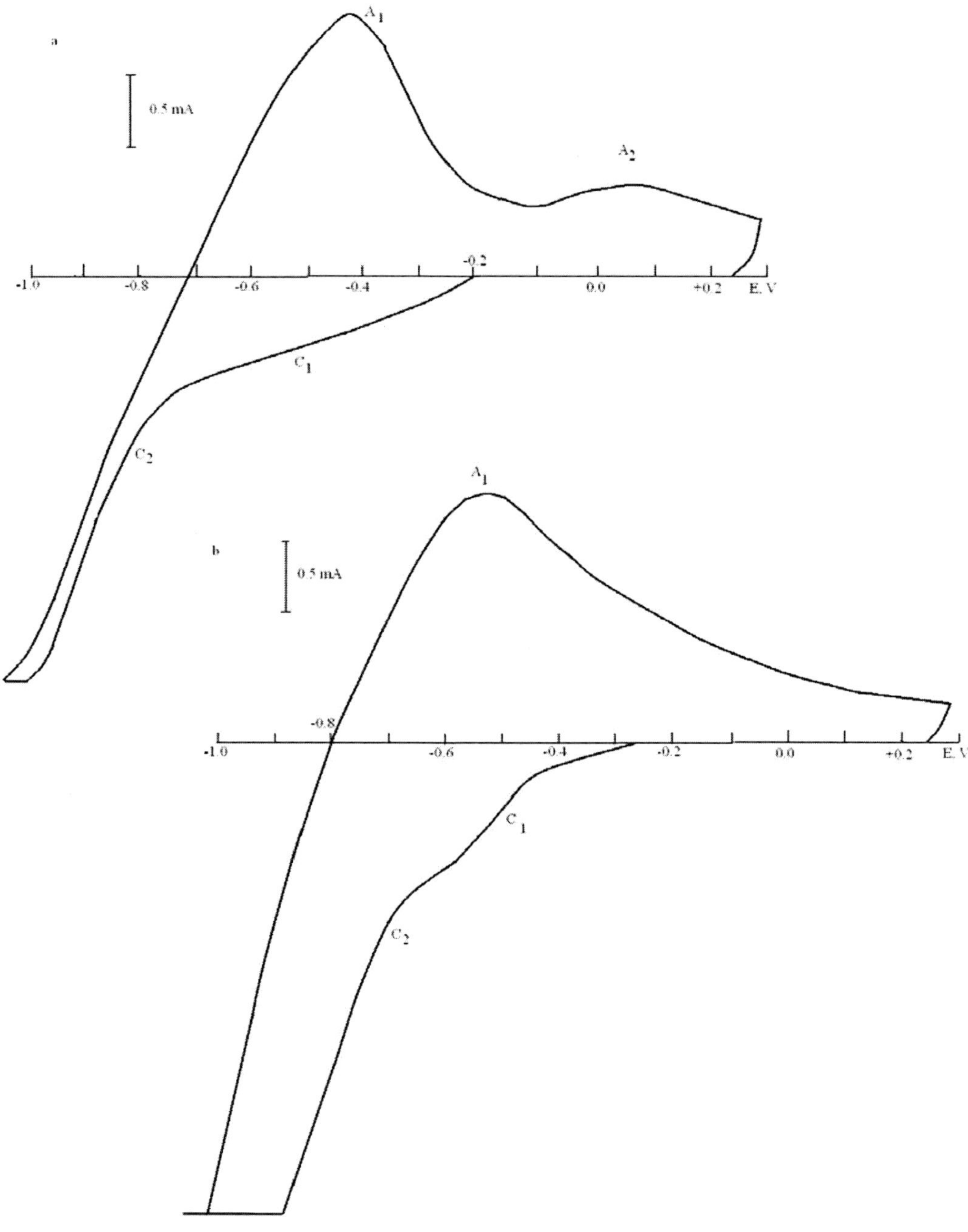

Figure 5. The cyclic voltammogram for the SnO_2 electrode in electrolyte: **a.** 0.05 M Na_2MoO_4 + 0.01 M H_2SeO_3 + 0.1 M $NH_4SO_3NH_2$; **b.** 0.2 M Na_2MoO_4 + 0.01 M H_2SeO_3 + 0.1 M $NH_4SO_3NH_2$. Temperature 20 ^0C, pH 6.5.

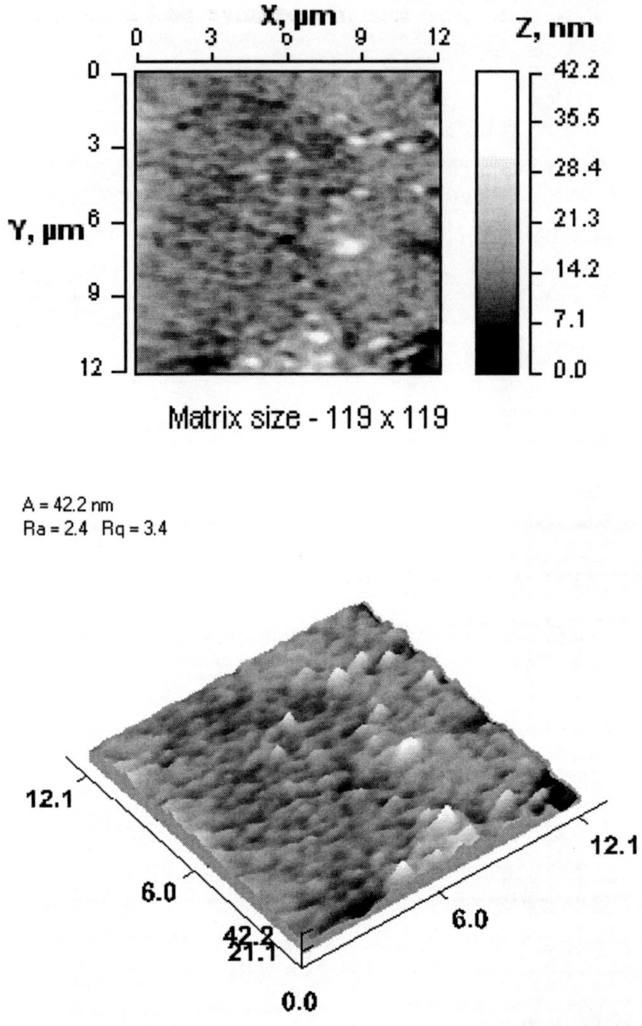

Figure 6. AFM images of the thin film layer deposited on the SnO_2 electrode during cathodic scan from –0.2 to –0.6 V. **a.** 2D topography; **b.** 3D topography.

In order to determine the influence of deposition time on film composition, the electrodeposition was performed by varying deposition time from 10 to 80 min. A comparison of the XRD and x-ray fluorescence analysis data obtained after 15 min, 30 min, 60min and 80 min of deposition (Table N. 3) allows to assuming that the incorporation of Se into Mo–Se film is kinetically sluggish. The films deposited for shorter times are richer in MoO_2 in contrast to the film deposited for 60 min (Table N. 3). The deposits on the SnO_2–coated glass substrate achieve an optimal thickness for the deposition time one hour. It is also found that films are uniform and well adherent to the substrate. It is important to remark that when the thickness exceeded the optimum value the layers started to peel off from substrate.

In the series of experiments the influence of electrodeposition potential on the thin film composition was evaluated. X–ray fluorescence and XRD analyses (Table N. 4) confirms the coexistence of $MoSe_2$ and MoO_2 phases over the whole range of deposition potentials. A shift of electrodeposition potentials to more negative potentials results to an increase of MoO_2 content in thin film layers.

Table 3. Composition and characterization of samples deposited at a different deposition time in the electrolyte containing 0.05 M Na_2MoO_4 + 0.01 M H_2SeO_3 + 0.1 M $NH_4SO_3NH_2$. Electrodeposition conditions: E = –0.8 V; T = 20 ^{0}C; pH6.5

Deposition time, min	Mo/Se in deposits (X-Ray fluorescence analysis)	Identified compound by XRD	$MoSe_2/MoO_2$ in deposits (XRD analysis)	Film characteristic "as deposited"
10	2.2	SnO_2 MoO_2 $MoSe_2$	0.40	Well adherent to substrate, uniform
20	2.0	SnO_2 MoO_2 $MoSe_2$	0.42	Well adherent to substrate, uniform
30	1.13	SnO_2 MoO_2 $MoSe_2$	1.61	Well adherent to substrate, uniform
60	0.62	SnO_2 MoO_2 $MoSe_2$	5.13	Dark brown, well adherent to substrate, uniform
80	0.60	SnO_2 MoO_2 $MoSe_2$	5.2	Dark brown, well poorly adherent to substrate, uniform

Table 4. Composition and characterization of samples deposited at a different deposition potentials in the electrolyte containing 0.05 M Na_2MoO_4 + 0.01 M H_2SeO_3 + 0.1 M $NH_4SO_3NH_2$. τ = 1h. T = 20 ^{0}C; pH6.5

Deposition potential E,V	Mo/Se in deposits (X-Ray fluorescence analysis)	Film thickness, nm	Identified compound by XRD	$MoSe_2/MoO_2$ in deposits (XRD analysis	Film characteristic "as deposited"
-0.8	0.62	94.0	SnO_2 MoO_2 $MoSe_2$	5.13	Brown, well adherent to substrate, uniform
-0.9	0.79	273,3	SnO_2 MoO_2 $MoSe_2$	1.68	Dark brown, well adherent to substrate, uniform
-1.0.	1.18	332.0	SnO_2 MoO_2 $MoSe_2$	1.36	Dark brown, well adherent to substrate, uniform
-1.1		348.0	SnO_2 MoO_2 $MoSe_2$		Black, not uniform

The influence of Na_2MoO_4 and H_2SeO_3 ratio in electrolyte on Mo–Se film composition is given in Table N. 5.

The main points that can be extracted from this experimental part are: (1) the electrodeposition of $MoSe_2$ and MoO_2 phases is observed in the whole potential range; (2) the films electrodeposited at more negative potentials are richer in MoO_2; (3) in a narrow potential range from –0.8 V to –1.10 V the composition of deposits is slightly determined by the ratio of Na_2MoO_4 and H_2SeO_3 in electrolyte.

Table 5. Composition and characterization of samples deposited at different Na_2MoO_4/H_2SeO_3 ratios in the electrolyte. Electrodeposition conditions: T= 40 ^{0}C; E = –1.0 V. τ = 1h

Ratio Na_2MoO_4/H_2SeO_3	Mo/Se in deposits	Film thickness, nm	XRD analysis	
			Identified compound	$MoSe_2/MoO_2$ in deposits
5/1	1.18	332.0	SnO_2 MoO_2 $MoSe_2$	0.74
10/1	1.13	386.7	SnO_2 MoO_2 $MoSe_2$	0.78
15/1	1.42	387.8	SnO_2 MoO_2 $MoSe_2$	0.58
20/1	1.91	403.8	SnO_2 MoO_2 $MoSe_2$	0.45

3.2.5. An Intuitive Reaction Scheme

The current knowledge on the sequence of reactions of electrodeposition of Mo–Se thin film layers is summarized in this chapter. The CV measurements, x-ray fluorescence and XRD analyses allow us to conclude that two parallel electrode processes are responsible at the potentials E ≤ –0.8 V: the formation of MoO_2 and $MoSe_2$. An electrochemical model proposed for preparation of Mo–Se thin film takes consideration of the following elements:

1) The reduction of H_2SeO_3 to elemental selenium (Eq. 14) is first reaction on the electrode surface. On further negative sweep of electrode potential generation of HSe^- is facilitated by the reduction of Se (0) or Se (IV) by following reactions:

$$\mathrm{Se} + H^+ + 2\bar{e} \rightarrow H\mathrm{Se}^-$$

$$HSeO_3^- + 7H^+ + 6\bar{e} \rightarrow H_2Se + 3H_2O \quad (22)$$

2) When the MoO_4^{2-} reduction potential is attained the electrodeposition of MoO_2 according to reaction (1) begins.

3)MoO_4^{2-} and $HSeO_3^-$are reduced independently. After formation of MoO_2 and HSe^- they react to give molybdenum selenide phase as follows:

$$\mathrm{MoO_2} + 2\mathrm{HSe}^- + 2H^+ \rightarrow \mathrm{MoSe_2} + 2\mathrm{H_2O} \tag{23}$$

or

$$\mathrm{MoO_2} + 2\mathrm{H_2Se}+ \rightarrow \mathrm{MoSe_2} + 2\mathrm{H_2O} \tag{24}$$

The point to emphasize here is that the formation of $MoSe_2$ phase is occurring under experimental conditions at about 0.5 V more cathodic potentials than the above prediction of electrodeposition (part 3.1.2), because of the formation of an insulating selenium layer on SnO_2/glass electrode surface.

One of the products of the electrodeposition could be an amorphous Se, which is not seen by XRD analysis. Note that the hydrogen evolution (a main side reaction for the cathodic deposition) and formation of MoO_2 should cause a local increase in pH near the electrode surface. It is commonly known, that hydrogen selenide reacts rapidly in acidic and intermediate pH but more slowly in alkaline solutions with Se (IV) to give Se [23]. Therefore an increase in OH^- concentration should decrease the rate of amorphous selenium formation. When these effects predominate, the already formed HSe^- (or H_2Se) reacts with MoO_2 resulting in Mo–Se deposition. The reduction of elemental selenium to HSe^- (or H_2Se) is a sluggish and fully deposition-controlled process. This statement is also supported by a comparison of data shown in Table N. 3.

The kinetics of electrodeposition process can be controlled by either the slow surface step, or slow chemical reactions in the deposition zone. Unfortunately, the reaction sequence and electrodeposition mechanism of Mo–Se thin film layer is not fully elucidated. The results presented are interesting from the phenomenological point, as they show a great influence of H_2SeO_3 reduction on the formation of Mo–Se thin films. Further studies are necessary to understand the role of Se reduction to HSe^- in the electrodeposition chemistry. Such studies should concentrate on a systematic comparative investigation of Mo–Se film deposition on SnO_2/glass and modified Se/SnO_2/glass electrodes. This type of study will have the potential to reveal relevant details regarding the influence of electrode surface on Mo–Se electrodeposition

Our findings are in essential agreement with those of previous authors who have reported on the molybdenum electrochemistry. Podlaha and Landolt [50–52] have shown that the molybdate ion is first reduced to Mo (IV) oxide, which as an intermediate is subsequently reduced to molybdenum in Ni–Mo alloys. E. Beltowska-Lechman [53] has reported on the formation of mixed nickel and molybdenum oxides in electrodeposition of Ni–Mo coatings from complex citrate solutions. Kuznetsov et al [54] have described the formation of molybdenum compounds of lower valences in electrodeposition of Ni–Mo alloys from ammonia-citrate electrolytes.

3.3. Characterization of Thin Mo–Se Film Layers

3.3.1. XRD Analysis

As already evidenced, the constitutional phases of the electrodeposited films are separate $MoSe_2$ and MoO_2. The x-ray diffraction patterns of the "as-deposited" Mo–Se thin film are shown in Fig. 7. The observed peaks were analyzed to determine the crystal structure of $MoSe_2$ in the film. The sharp peaks reveal the polycrystalline nature of the "as-deposited" Mo–Se film. All the peaks show nearly the same intensity values, confirming the random orientation of the planes in the film. The observed interplaner distance "d" values were compared with the standard values (PDF 77-1715) to determine the crystal structure. The structural features fit into a hexagonal lattice with the parameters a = b = 3.2340 A°, c = 13.2859 A°.

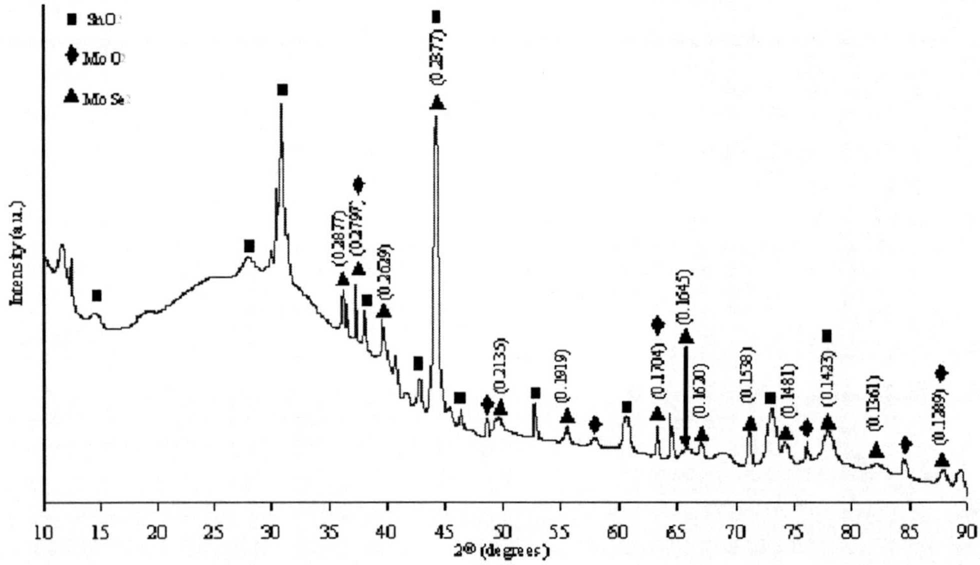

Figure 7. XRD patterns of Mo–Se thin film. Film composition: a) 83.7 % $MoSe_2$ and 16.3 % MoO_2.

3.3.2 .Optical Morphology

The optical micrographs indicate the coexistence of the $MoSe_2$ and MoO_2 phases in the thin film layers. Darker and lighter phases are $MoSe_2$ and MoO_2 phases, respectively. Figures 8, 9 and 10 depict the typical morphology of Mo–Se thin film layers. A comparison of micrographs reveals obvious change in morphologies, indicating that the morphology of Mo–Se thin films is significantly influenced by the composition of film and deposition conditions.

a

b

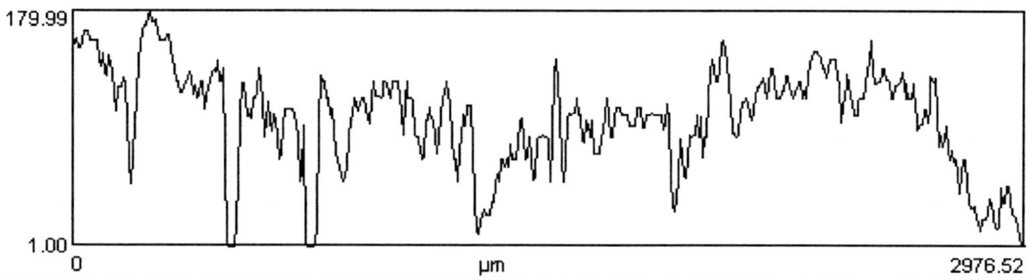

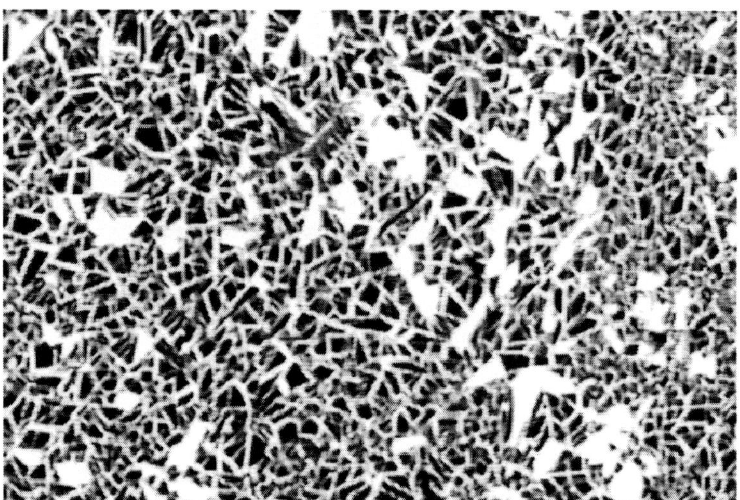

kir29p1

0,1 M NH_2SO_3 NH_4 + 0.15 M Na_2MoO_4+0.01 M H_2SeO_3; E = -0.6 V; Dengimo laikas = 1 val. T = 80 ^{0}C. (po atkatinimo) Fractal D= 1.8528

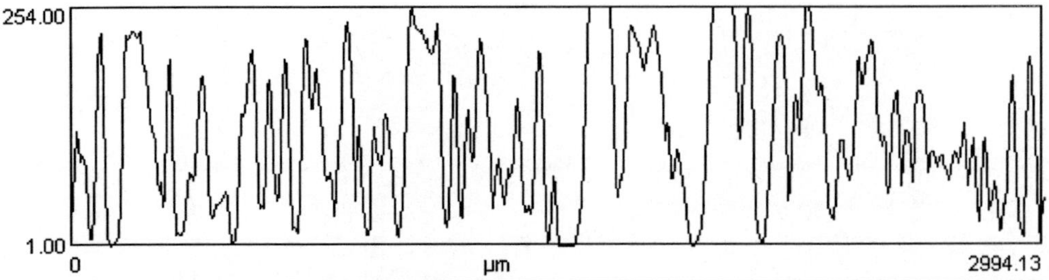

Figure 8. **a**. Optical microscope image of Mo–Se thin films. Film composition: 62.67% $MoSe_2$ and 37.33% MoO_2. **b.** Profile of the fluctuations of sample image optical density along axis x.

a

b

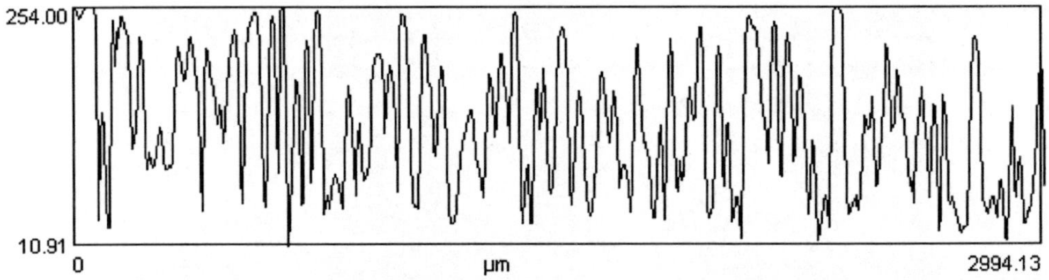

Figure 9. **a**. Optical microscope image of Mo–Se thin films. Film composition 32.80% $MoSe_2$ and 61.20% MoO_2. **b.** Profile of the fluctuations of sample image optical density along axis x.

a

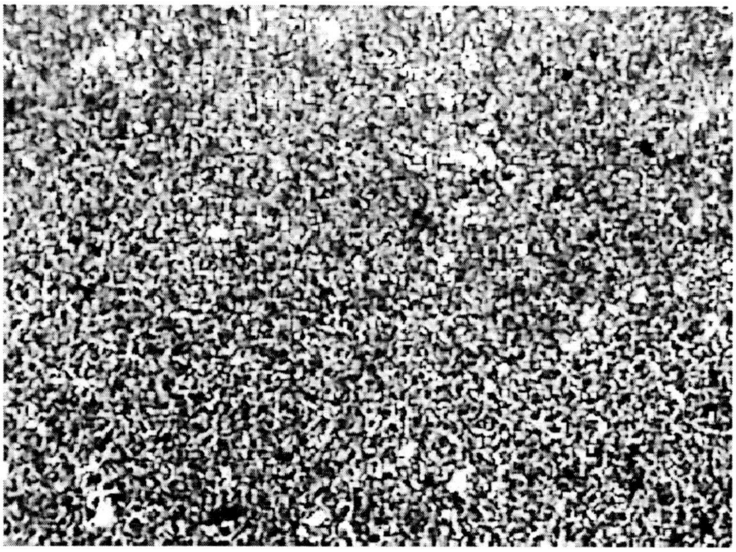

b

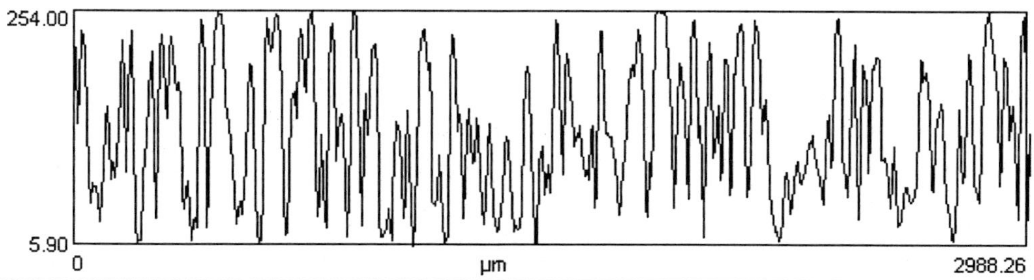

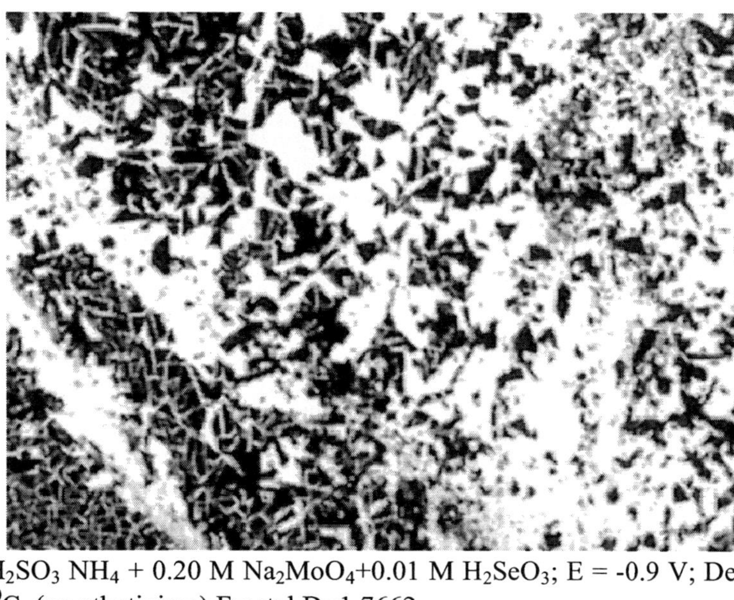

0,1 M NH_2SO_3 NH_4 + 0.20 M Na_2MoO_4+0.01 M H_2SeO_3; E = -0.9 V; Dengimo laikas = 1 val. T = 60 ^{0}C. (po atkatinima) Fractal D=1.7662

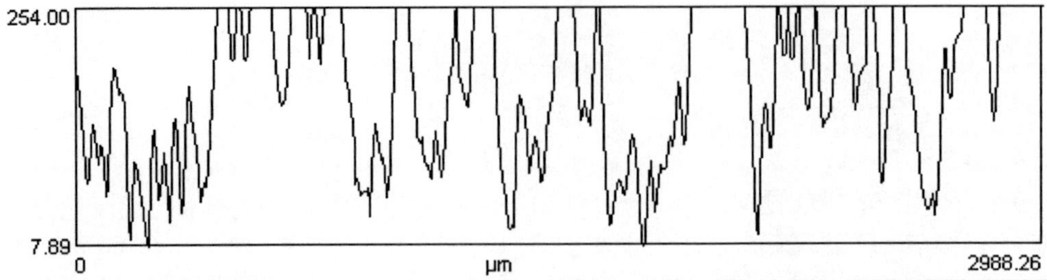

Figure 10. a. Optical microscope image of Mo–Se thin films. Film composition 26.15% MoSe2 and 73.85% MoO2. b. Profile of the fluctuations of sample image optical density along axis X.

The large-scale wires well distributed on the SnO_2 substrate are observed in thin film layers with richer content of $MoSe_2$. The typical top-view image of Mo–Se thin film containing 62.67% $MoSe_2$ and 37.33% MoO_2 is shown in Figure 8. The micrograph of surface “as-deposited” Mo–Se layer shows that film is smooth and homogeneous. The fluctuations of sample image optical density along axis x were in a range from 1.35 up to 164 (at a full scale of optical density from 0 to 255 by software Image J). The fraction dimension D is equal to 1.3034. These results also indicate that closely packed grains are uniformly spread over the substrate without any pinholes.

The shift of the deposition potential to more negative values or increase of the Na_2MoO_4 content in the bath leads to the enrichment of the deposited layer with MoO_2. However these films exhibit different morphology and structure (Fig. 9 and10).

The length of wires will increase with the increase of the Na_2MoO_4 content in the bath. As one can see (Fig. 9) the long wires of MoO_2 are uniformly dispersed in the film. The fluctuations of sample image optical density varied from 11up to 247. The fraction dimension D is equal to 1.8685. It confirms that the thin film layers are heterogeneous and not densely paced on the SnO_2/glass substrate.

The shift of the deposition potential to more negative values leads to the formation of non compact film on the substrate (Fig.10). The fine-grained Mo–Se films were also found in this case, but optical microscopy pictures revealed strongly pronounced borders of grains. The fluctuations of sample image optical density varied from 5.5 up to 254. Based on the results obtained it is possible to draw a conclusion, that grains are located in a distance approximately 2 – 3 microns and do not form the continuous layer. However, these films exhibit the good adhesion to a substrate.

From the results presented above, it is evident that the microstructure of Mo–Se thin layers differs.

Two main conclusions can be drowning from this experimental part. Primarily two types of microstructures were observed in deposits: (1) the fine dispersion of large-scale wires on the electrode surface; (2) the small islands-like phases of $MoSe_2$ and MoO_2 uniformly distributed on the electrode. The different micro structural scale suggests that alternative mechanisms by which Mo–Se film layers are formed might be operative.

It is interestingly to note that in any deposits the presence of the amorphous Se was not observed. Furthermore, the analysis of optical micrographs confirms the existence of $MoSe_2$

and MoO_2 phases over the whole range of deposition, which is in good agreement with XRD results. These observations support the assumptions drown in the chapter 3.2.5.

3.3.3. Optical Studies

The typical absorption spectrum of Mo-Se thin film is shown in Fig. 11. The absorption edge of the film is observed in the range 480–560 nm. A broad absorption maximum between 300–370 nm is observed in all cases. The position of maximum in the absorption peak slightly shifts from 310 nm to 320 nm with increasing MoO_2 content in Mo–Se thin film.

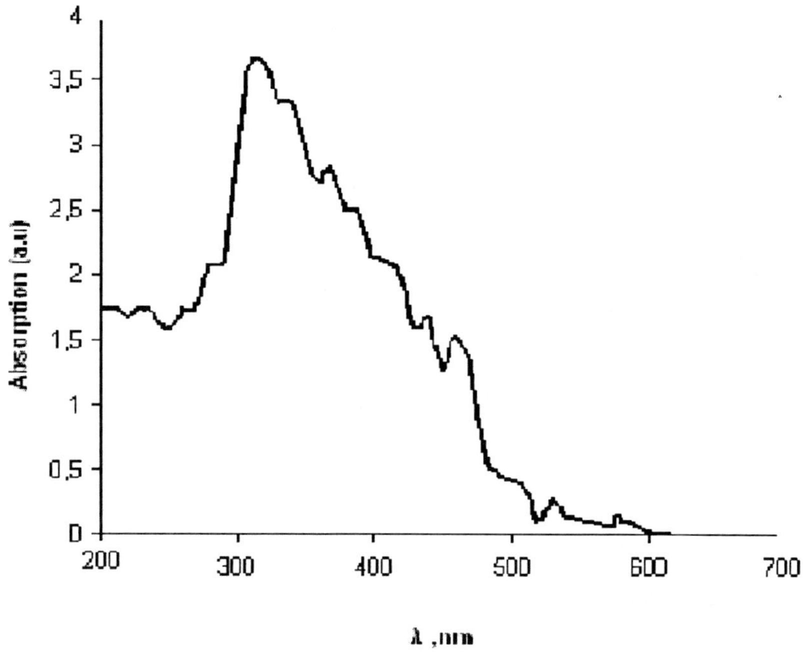

Figure 11. Absorption spectra of electrodeposited Mo–Se thin film.

The optical absorption coefficient is found to have a value of $1.33 \cdot 10^5$ cm^{-1} and $1.85\ 10^5$ cm^{-1} for wavelengths 310 nm and 320 nm, respectively, The value reported by Pathah et al [1] for $MoSe_2$ grown by the direct vapor transport technique is equal to $4.98 \cdot 10^4$ cm^{-1}.

The dependence of the optical transition involved can be found analyzing the dependence of α on hv.

The variation of the optical absorption coefficient near the absorption edge follows the power law [9] in the form of

$$\alpha = \frac{B(E - E_{og})^n}{E}, \tag{25}$$

where E is the photon energy, hν; E_{og} is the band gap energy of the system; B is the function of density of states near the conduction and valence band edges. The value of n is 2 for direct transition and ½ for indirect transition.

Transition metal chalcogenides are usually indirect band-gap semiconductors [10, 12, and 55]. Since the α values remain smaller than 10^5 for wavelengths below the absorption edge, it confirms the existence of an indirect band-gap semiconductor. The graphs of $(\alpha h\nu)^{1/2}$ vs hν are given in Fig. 12. The values of E_{og} obtained from the intercepts of the straight line portion of the curves with the hν axis for the zero absorption are in the range of 1.68 and 2.2 eV. The values of the optical band gap are higher than these reported in literature. The authors of Ref [10] show that the optical band gap energy for polycrystalline $MoSe_2$ film obtained by electrosynthesis is 1.17 eV. Pathak et al [1] have reported the E_{og} value of 1.4 eV for $MoSe_2$ single crystal. The $MoSe_2$ crystals grown by DVT technique [13] are direct band gap semiconductors with the E_{og} value of 1.6 eV.

The observed E_{og} differences could be connected either with the structural changes or different thickness of films. In our experiments, the deposition time was kept constant, and different thicknesses were due to the kinetic reasons. Changes in electrolyte composition and deposition conditions influence the composition of films. Therefore the influence of film composition on the E_{og} value can be regarded as dominant in our experiment.

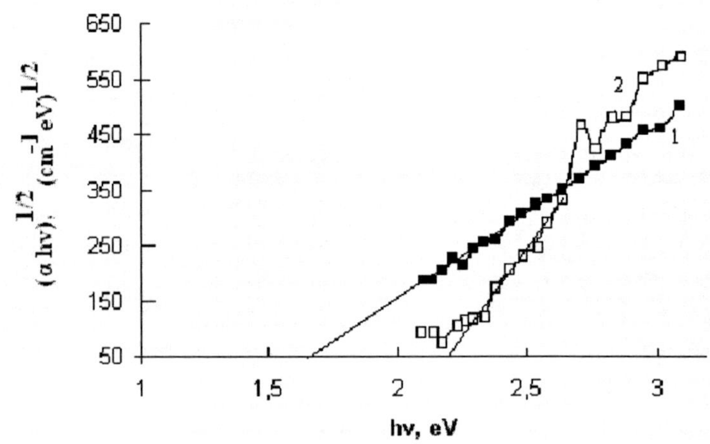

Figure 12. $(\alpha h\nu)^{1/2}$ vs hυ plot obtained by measurement for electrodeposited Mo–Se thin films: 1–film composition 83.7 % $MoSe_2$ and 16,3 % MoO_2; 2–film composition 26,15 % $MoSe_2$ and 73,85 % MoO_2.

4. Conclusions

It has been demonstrated that the electrodeposition of Mo–Se films is possible from sulfamatic solution containing Na_2MoO_4 and H_2SeO_3. However, it is impossible to avoid the parallel formation of the MoO_2 phase. The global composition of the deposited films is consistent with the presence of $MoSe_2$ and MoO_2. The Mo–Se thin film layers are indirect band-gap semiconductors. The values of E_{og} are in the range of 1.68 and 2.2 eV. X-ray measurements performed on the films clearly indicate a hexagonal structure of $MoSe_2$.

On the basis of experimental results obtained by cyclic voltammetry, XRD and x-ray fluorescecse analyses, a reaction sequence on electrode surfacc during the formation of Mo–Se thin films is proposed. The co-deposition of molybdenum and selenium appears as an exception to Kroger's theory, in which the coreduction is an additional electrochemical reaction that is added to the deposition reaction of the constituents. In the present case, it appears that the deposition of selenium does not lead to underpotential deposition of Mo with the subsequent formation of $MoSe_2$. The cathodic decomposition of the initially formed Se appears to be a key to formation of $MoSe_2$. An important feature of model includes two parallel reactions occurring on the electrode surface: the electrodeposition of MoO_2 and formation of HSe^- (or H_2Se). The formation of HSe^- (or H_2Se) results in its reaction with MoO_2 leading to the formation of $MoSe_2$ at potentials $E \leq -0.8$ V (SHE). The fact that Mo–Se layers as synthesized contain the $MoSe_2$ and MoO_2 phases, support this assumption. On the other hand the different micro structural scale of thin film layers suggests that different mechanisms by which Mo–Se film is formed might be operative. Unfortunately, the electrodeposition mechanism of Mo–Se thin film layer is not fully elucidated. The results presented are interesting from the phenomenological point, as they show a great influence of H_2SeO_3 reduction on the formation of Mo–Se thin films. Further work it seems necessary to better understand the selenium assimilation in the solid phase and explain the Mo–Se film formation mechanism.

5. Acknowledgement

The author is thankful to Dr I. Prosycevas for his contribution on the chapter of optical morphology and MS. A. Guobiene for her help with the AFM experiments.

References

[1] Pathak V. M.; Patel K. D.; Pathak R. J.; Srivastava R. Solar Energy Mater Solar Cells 2002, 73, 117-123.
[2] Sampano R. B.; Rembaum A. Phys Rev Lett. 1971, 27, 402-.
[3] Fan R. F.; White H. S.; Wheeler B.; Brad A. J. J Electrichem Soc. 1980, 127, 518-.
[4] Schniemeyer L. F.; Wrighton M.S.; Stacy A.; Siendo M.J. Appl Phys Lett. 1980, 36, 701-.
[5] Castro R. J.; Cabrera C. R. J Electrochem Soc. 1997, 144, 3135-3140.
[6] Bonnean P. R.; Jaris R. F.; Kaner R. B. Nature. 1997, 349, 510-.
[7] Zhan J. H.; Zhang Z. D.; Qian X. F.; Xie Y.; Qian Y. T. Materials Research Bulletin. 1999, 34, 497-501.
[8] Murata H.; Kataoka K.; Koma A. Surface Science, 2001, 478, 131-144.
[9] Deshpande M. P.; Gupta S. K.; Ajay Agarwal.; Adarwal M. K. Synthetic Metals. 2001, 123, 73-81.

[10] Anand T J. S.; Sanjeeviraja C.; Jayachandran M. Vacuum. 2001, 60, 431-435.
[11] Dukstiene N.; Kazancev K.; Prosicevas I.; Guobiene A. J Solid State Electrochemistry. 2004, 8, 330-336.
[12] Devadasan J.J.; Sanjeeviraja C.;Jayachandran M. Materials Chemistry and Physics. 2002, 77,397-401.
[13] Delphine S.M.; Jaychandran M.; Sanjeeviraja C., Materials Chemistry and Physics. 2003, 81, 78-83
[14] Gomez E.; . Pellicer E.; Valles E., J Electroanal Chem. 2001, 517, 109-116.
[15] Zeng Yue.; Zelin Li.; Ming Ma.; Shaomin Zhon. Electrochemistry Comunications 2002, 2, 36-38.
[16] Zhen-Jiang Nin.; Shi-Bing Yao.; Shao-Min Zhon. J Electroanal Chem. 1998, 455, 205-215.
[17] Salvador P.; Chaparro A. M.; Mir A. J Phys Chem-US,1996, 100, 760-768.
[18] Castro R. J.; Cabrera C. R. J Electrochem Soc , 1997, 144, 3135-3140.
[19] Charparo A. M.; Salvador P.; Mir A. J Electroanal Chem, 1997,424, 153-157.
[20] Solinimi H.; Kemell M.; Ritala M.; M. Leskela M. J Mater Chem. 2000, 10, 519-525.
[21] Gaikwad N.S.; Bhosale C.H. Materials Chemistry and Physics. 2002, 76, 198-203.
[22] Panicker P.; Kroger F.A.; Castner M. J Electrochem Soc.1978, 125, 566-.
[23] Bouroushian M.; Kasovic T.; Loizos Z.; Spyrellis N. Electrochem. Commun. 2000, 2, 281-285.
[24] Bouroushian M.; Kasovic T.; Loizos Z.; Spyrellis N. J.Solid State Electrochemistry. DOI:10.1007/s100080100215.
[25] Zhdanov S. I.Encyclopedia of Electrochemistry of the Elements. Bard A. J (Ed) Marcel Dekker, New York, 1974, IV,Ch.7, 93-125 1974, Vol. 4, Ch.7, pp 93-125.
[26] Latimer W.L. The Oxidation States of the Elements and Their Potentials in Aqueous Solutions. 2nd edn. Prentice-Hall, Englewood Cliffs, N.J;1952.
[27] Nevsskii O. B.; Gerasimov A. D; D'yachkova N. N. Elektrochimiya .1968, 4, 624-629.
[28] Osman-Zade S. D.; Vagramyan A.T. Elektrochimiya. 1966, 2, 76-77.
[29] Schott H. F.; Swift E.; Yost D. M. J. Am. Chem. Soc. 1928, 50, 721-727.
[30] Sebey F.; Potin-Gautier M.; Giffaut E.; Borge G.; Donard O. F. X. Chemical geology. 2001, 171, 173-194.
[31] Van Muydler J.; Pourbaix M. Selenium, Atlas d'Equilibres Electrochimiques a 25 ^{0}C. Paris, 1964.
[32] Mishra K. K.; Rajeshwar K. J Electroanal Chim. 1989, 271, 279-294.
[33] Lidin P.A.; Andrejeva L. L.; Molotshko V.A. Spravotshnik po neorganitsheskoj chimij. Cimija, Moscow,1987
[34] Jarzabek G.; Kublik Z. J Electroanal Chem. 1980, 114, 165-177.
[35] Jarzabek G.; Kublik Z. J Electroanal Chem. 1982, 137, 247-259.
[36] Rao G. S. R.; Reddy S. Acta Chim Hungarika 1985, 118, 241-248.
[37] Madolo R.; Traore M.; Vittori V. Electrochim Acta. 1986, 31, 856-867.
[38] Zelic M.; Sipos L.; Branica. Croat Chim Acta. 1985, 58, 43-55.
[39] Mickevicius D.L.; Dukstiene N. V. Chimia Lithuanian Academy of Science 1990, 4, 23-32 (in Russian).
[40] Thouin L.; Rouquette – Sanchez S.; Vedel J. Electrochim Acta 1993, 38, 2387-2394.
[41] Massaccen S.; Sanchez S.; Vedel J. (1996) J Electroanal Chem. 1996, 412, 95-101.

[42] Massaccen S.; Sanchez S.; Vedel J. J Electrochem Soc. 1993, 140, 2540-2546.
[43] Kemell M.; Saloniemi H.; Ritala M.; Leskela M. Electrochim Acta 2000, 45,3737-3748.
[44] Spravotshnik chimika; Nikolski B.P. (Ed.) Moscow, 1964; Vol.3, pp 791-792.
[45] Abrarov O.A.; Bigelis V.M. Elektroosazdenije selena i telura, FAN, Tashkent, 1976 pp32-48. (in Russian)
[46] http://www.Factstage.com
[47] Ichimura M.; Nakamura A.; Takenchi K.; Arai E. Phys Stat Sol (a). 2002, 193,132-138.
[48] Moustafild T.El.; Cachet H.; B. Tribollet B.; D. Festy D. Electrochim Acta, 2002, 47, 1209-1212.
[49] Spurr R.A.; Mayers H. Quantitative Analysis of Elements.X-ray Diffration. 2nd edn. Addison-Wesley, Reading, 1978; p. 94.
[50] Podlaha E.J.; Landolt D. J Electrochem. Soc. 1996, 143, 885-892.
[51] Podlaha E.J.; Landolt D. J Electrochem Soc. 1996, 143, 893-899.
[52] Podlaha E.J.; Landolt D J Electrochem Soc. 1997, 144 1672-1680.
[53] Beltowska-Lehman E. Surface and Coatings Technology. 2002, 14, 440-443.
[54] Kuznetsov V.V.; Pavlov M.R.; K.V. Kuznetsov K.V.; V.N. Kudryavtsev V.N. Russian Journal Electrochemistry.2003, 39 1338-1341.
[55] Patil R. S. Thin Solid Films, 1999, 340, 11-12.

In: Focus on Electrochemistry Research
Editor: Magdalena Nunez, pp. 69-91 ISBN: 1-59454-545-6

Chapter 3

POTASSIUM TELLUROCYANATE AS A TELLURATING AGENT

Ali Z. Al-Rubaie*
Department of Chemistry, College of Science, University of Basrah, Basrah, Iraq.

ABSTRACT

The preparation and the use of potassium tellurocyanate (KTeCN) as a tellurating as well as a cyclization agent to produce various organic tellurocyanates (RTeCN) are reviewed and discussed. Particular attention is given to the use of organic tellurocyanates for the preparation of various classes of organic tellurium compounds focusing on the reactivity of organic tellurocyanates toward oxygen, nitrogen, sulfur and other nucleophiles.

Furthermore, this chapter gives brief survey of our study in this area and contains some recently obtained results in this field. Previously undescribed tellurium-containing pyrazole moieties and their derivatives were prepared by the reaction of pyrazole-1-methanol with phenacyl tellurocyanate in presence of tributylphosphine. Organic isotellurocyanates and the formation of transition metal complexes with tellurocyanate anion are also reviewed and discussed.

INTRODUCTION

Although the first organic tellurium compound was prepared in 1840[1], this branch of organic chemistry has not been systematically studied until the middle of last century. Recent advances in the area of organochalcogen chemistry have been driven by potential applications of selenium/tellurium compounds in modern organic synthesis, ligand chemistry, precursors for metal-organic chemical vapor deposition (MOCVD) and biochemistry. Thus, several reviews and books on organic selenium/tellurium have been published [2-10].

---

* Present address: Department of Chemistry, College of Science, Gerian University, Gerian, P.O. Box 64 200, Libya. E-mail: alrubaie49@yahoo.com

Organic selenocyanates and organic tellurocyanates have been known since many years ago; however, their synthesis and their chemical properties are becoming actively only very recently. Organic tellurocyanates are poorly explored area in contrast to well-established organic selenocyanate chemistry.

The first tellurocyanate anion, a DMF complex of tetraethylammonium tellurocyanate, was prepared in 1968[11]. The chemistry of tellurocyanates has been reviewed by Toshimitsu and Uemura [12], Skopencho and Ivanovg [13] and Sadekov [14], the literature coverage being up to 1995. Since then, this field has grown steadily and much progress has been made in the use of organic selenocyanates and tellurocyanates.

This chapter does not intend to be a review but rather an attempt to give a brief and comprehensive view of synthetic organic tellurocyanate chemistry, focusing attention upon using potassium tellurocyanate to prepare organic tellurocyanates. Furthermore, useful synthetic transformations promoted or suffered by organic tellurocyanates will also be discussed. Emphasis will be given to recent contributions of our laboratory where we explore old reactions trying to establish their generality or making transformations into organic tellurium compounds, which can lead to useful methodologies. Some of our new results in this area will be reported and discussed for the first time in this chapter. In addition we reported the transition metal complexes of tellurocyanate anion.

Synthesis of Organic Tellurocyanates

The syntheses of various organic tellurocyanates have been claimed in many patents [15,16], but none of them has been fully characterized.

1-Azulenyl tellurocyanate (**1**) was the first organic tellurocyanate to be prepared and isolated in extremely low yield (7%) in 1968 by Nefedav [17]. This compound (**1**), Scheme 1, was prepared by refluxing azulene with copper(II) acetate and KTeCN in acetonitrile. Potassium tellurocyanate was prepared by heating for 3 h of a finely ground mixture of tellurium powder and potassium cyanide at temperature gradually increased from 110 to 250°C. The overall reaction is shown in Scheme 1.

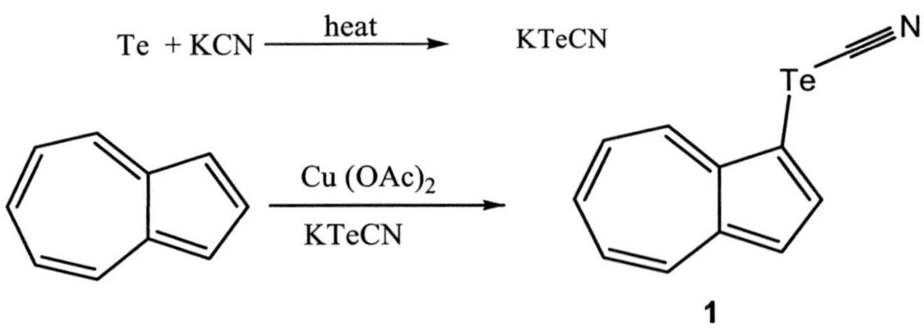

Scheme 1

The proposed mechanism of the above reaction depends on the formation of copper(II) tellurocyanate, $[Cu(TeCN)_2]$, and a direct substitution of hydrogen in azulene by $TeCN^-$

species. However, this substitution reaction could not be applied to other aromatic compounds [12,14].

In general, there are only few methods for preparing organic tellurocyanates and some of them are represented by a few examples only.

The first method based on reaction of aryltellurium(II) halides with silver cyanide [18-20] or with potassium or sodium cyanide [21,22] to produce the corresponding organic tellurocyanates (*i.e.* **2** and **3**), equations 1 and 2.

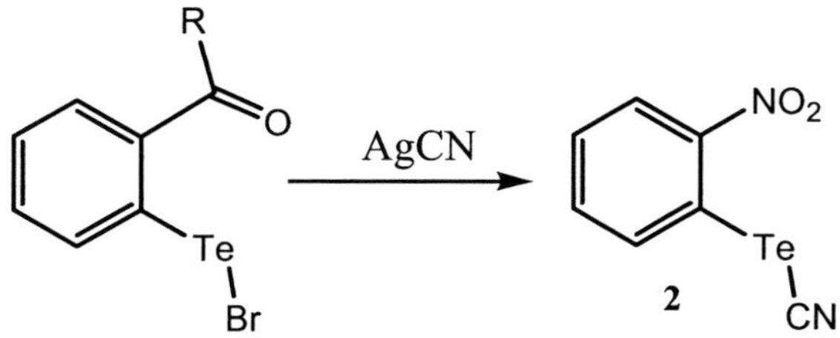

(1)

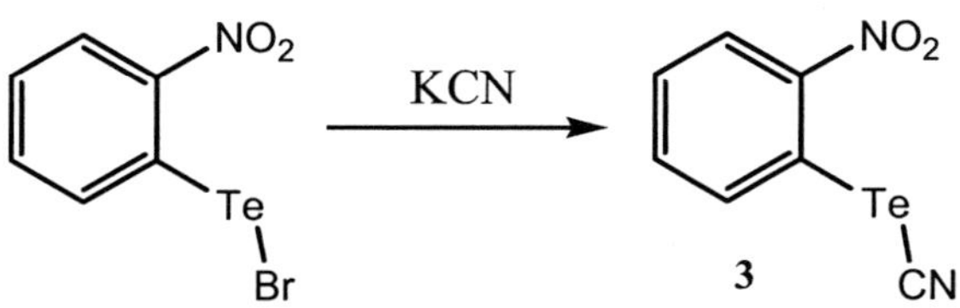

(2)

The application of the above method has been restricted by difficulties in synthetic methodologies of unstable organotellurium(II) halides. The organotellurium(II) derivatives are usually stabilized by attaching chelating groups in close proximity to tellurium atom. Thus, a CHO, COR, COOR, CH=N, CH_2NR_2 or NO_2 group was necessary at position *ortho* to tellurium atom to obtain a stable aryl tellurocyanate due to the interaction of the tellurium with the carbonyl or nitro oxygen atom [23,24,25]. Organyltellurium(II) halides can also be stabilized when they contain a bulky group to prevent substrate attack at the tellurium atom and yet can be converted to tellurocyanate [18, 19, 22, 26], equations 3. We also succeed in preparing 2-tellurocyanatonaphthalene in 56% yield [27] as showing in equation 4.

$$(CH_3)_3Si)_3C\text{-TeI} \xrightarrow{AgCN} (CH_3)_3Si)_3C\text{-TeCN} + AgI \quad (3)$$

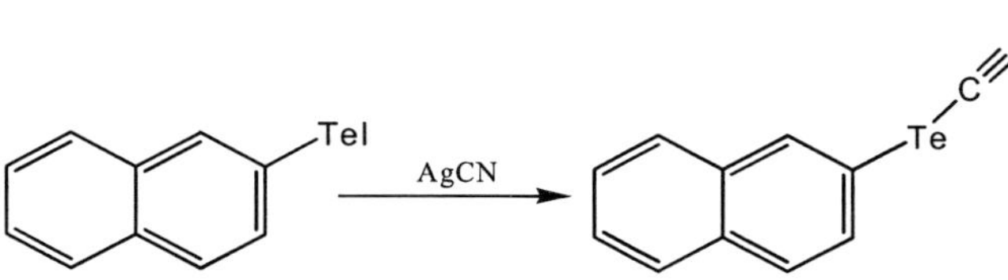

+ AgI (4)

Other method to prepare aryl tellurocyanates is based on cleavage of tellurium-carbon bond of 3-oxo-2, 3-dihydrobenzotellurophene by hydrogen cyanide [28], equation 5.

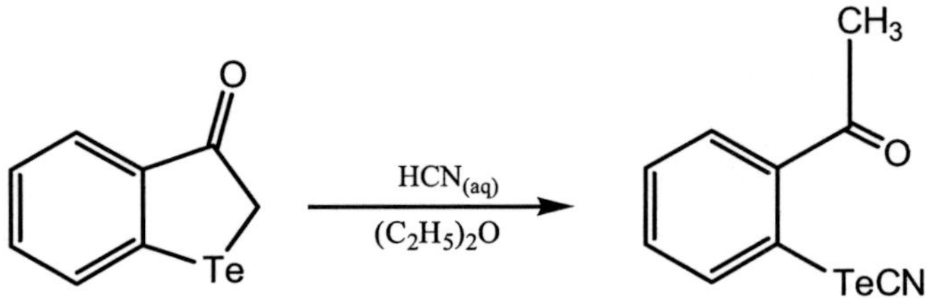

4 **5** (5)

However, reaction in equation 5 is of a limited applicability due to limited availability of tellurophene (compound **4**).

Falcon and Cava [21] proved that *ortho*-substituents (such as COR, CO_2R, NO_2 or CH=N-) are not necessarily required for the preparation of aryl tellurocyanates. Such compounds were prepared either by nucleophilic attack by the CN^- ion on one of tellurium atoms of diaryl ditellurides (**6**), equation 6, or by partial reduction of aryltellurium(IV) trihalides with aqueous sodium metabisulphite to aryltellurium(II) halides, followed by the addition of potassium cyanide [21], equation 7. By this method phenyl tellurocyanates and 4-methoxyphenyl tellurocyanates were prepared in 86% and 93% yields respectively, equation 7.

$$\underset{\mathbf{6}}{\text{Ar-Te-Te-Ar}} + CN^- \longrightarrow \underset{\mathbf{7}}{\text{Ar-TeCN}} + ArTe^- \xrightarrow{O_2} \underset{\mathbf{6}}{\text{Ar-Te-Te-Ar}} \quad (6)$$

$$ArTeX_3 \longrightarrow ArTe(O)X \xrightarrow{Na_2S_2O_5} [ArTeX] \xrightarrow{KCN} \underset{\mathbf{7}}{ArTeCN} \quad (7)$$

(X= Cl, Br) Ar= C_6H_5, Ar= 4-$MeOC_6H_4$

The reaction conditions for the reaction in equation 7 are critical and need a long attention [21].

Treatment of sodium benzenetellurolate ($C_6H_5Te^-Na^+$) generated *in situ* by sodium borohydride reduction of diphenyl ditelluride, with cyanogen bromide gave phenyl tellurocyanate (**8**) in 58% yield [29], equation 8.

$$(C_5H_5)_2Te_2 \xrightarrow[C_2H_5OH]{NaBH_4} [C_6H_5TeNa] \xrightarrow[C_2H_5OH]{BrCN} \underset{\mathbf{8}}{C_6H_5TeCN} + NaBr \quad (8)$$

Aryl tellurocyanates were also obtained in low yields, as by product through the reaction of aryldiazonium tetrafluoroborates with potassium tellurocyanate in DMF or DMSO solution [30,31], equation 9.

$$ArN_2^+BF_4^- + KTeCN \longrightarrow \underset{\mathbf{7}}{ArTeCN} + \underset{\mathbf{9}}{Ar_2Te} + KBF_4 + N_2 \qquad (9)$$

Ar= 4-MeC_6H_4, 4-$MeOC_6H_4$, 2-$NO_2C_6H_4$, 4-$NO_2C_6H_4$, 2,6-$Me_2C_6H_3$, 2,4,6-$Br_3C_6H_2$, 2-$HOOCC_6H_4$, 2-PhC_6H_4, 2-NCC_6H_4.

This reaction (equation 9) produces the diaryl tellurides (**9**) as the major products. A mechanism of the disproprtionation of aryl tellurocyanates to diaryl tellurides and tellurium dicyanide was proposed [30] and is inconsistent with the stability of aryl and alkyl tellurocyanates, as no mention for such disproprtionation in their synthesis [18,19, 21, 29].

Treatment of lithium organyltellurolate (**10**) with cyanogen bromide to prepare the tellurocyanate **11** is another approach for the synthesis of organic tellurocyanates, equation 10. Tellurocyanate **11** was not isolated in pure form, but directly hydrolyzed to its corresponding ditelluride [26].

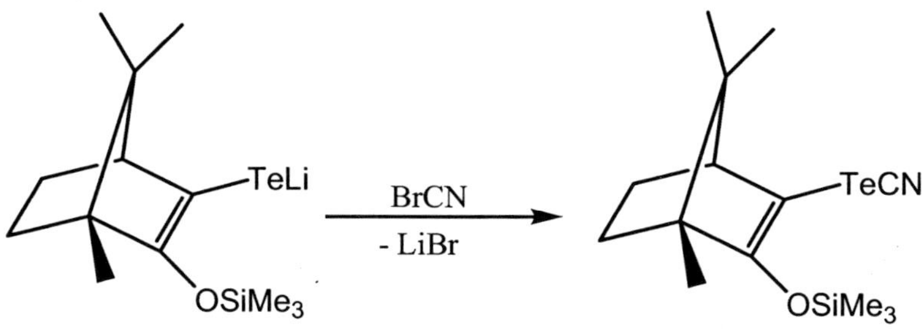

10 **11** (10)

Benzyl tellurocyanate (**12**) was the first alkyl tellurocyanate to be isolated in pure form as white needle with melting point of 126-127°C [32] by the reaction of benzyl chloride with potassium tellurocyanate in dry DMSO solution [32], equation 11.

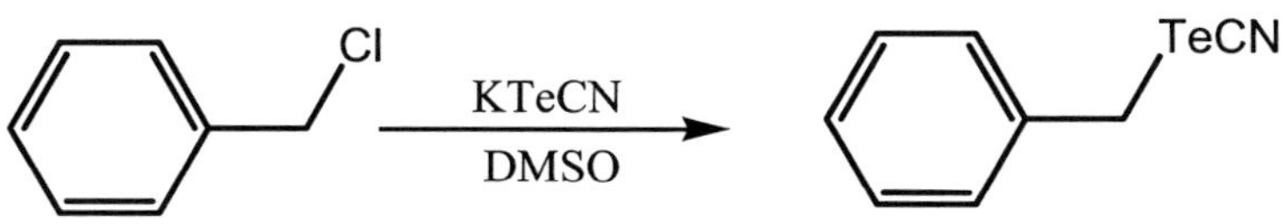

12 (11)

This method based on the nucleophilic substitution of chlorine atom in benzyl chloride by a CN anion, was successfully used to prepare several ring-substituted benzyl tellurocyanates [22, 33-38] in 48-85% yields. In contrast to the above reaction (equation 11), early efforts to prepare simple and pure benzyl tellurocyanate *via* reaction of benzyl bromide with tetraphenylarsonium tellurocyanate [39], tetraphenylammonium tellurocyanates [39] or with bis (triphenylphosphine)iminium tellurocyanate [40] gave complexes **13** and **14**, respectively in high yields [39,40] with exception of 3-nitrobenzyl tellurocyanates which was obtained in low yield [39], equations 12 and 13.

$$(C_6H_4)_4As^+TeCN^- + C_6H_5CH_2Br \longrightarrow [(C_6H_5)_4As]^+ [PhCH_2\overset{\overset{Br}{|}}{Te}\text{-}CN]^- \quad (12)$$

13

$$[(C_6H_5)_3P)_2N]^+TeCN^- + 4\text{-}O_2N\,C_6H_4CH_2Br \longrightarrow [(C_6H_5)_3P)_2N]^+[\,4\text{-}O_2NC_6H_4CH_2\overset{\overset{Br}{|}}{Te}CN]^- \quad (13)$$

14

The synthesis of 2-carbethoxybenzyl tellurocyanate (**17**) is a limited example of the halogenolysis of tellurophthalide followed by reaction with ethanol and then with potassium cyanide to afford 2-carbethoxybenzyl tellurocyanate (compound **17**) in 80% yield, m.p. 72-73°C [22], Scheme 2. Compound **17** was also prepared by treatment of 2-carbethoxybenzyl bromide with KTeCN in DMSO solution [22], Scheme 2.

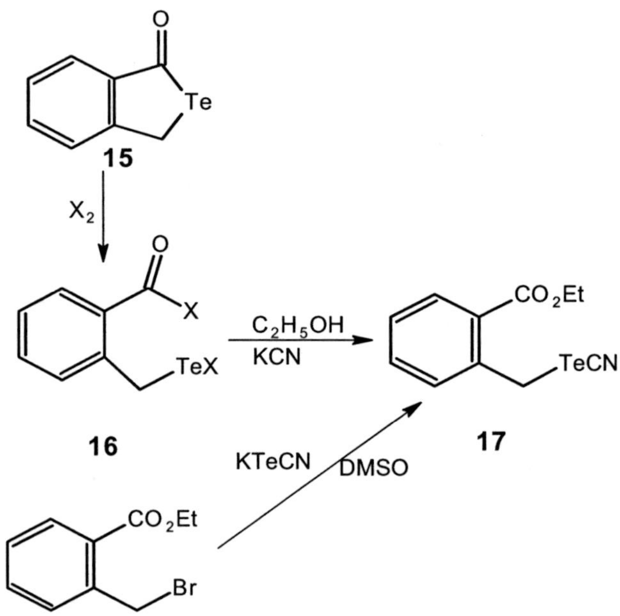

Scheme 2.

From above mentioned methods for the preparation of organic tellurocyanates, it seems likely that the most efficient reagent is potassium tellurocyanate dissolved in DMSO, which gave the corresponding organic tellurocyanate in fair to good yields.

This reagent (KTeCN) was also used to prepare 3-(*endo*-tellurocyanato)camphor (**18**), equation 14, in low yield[26].

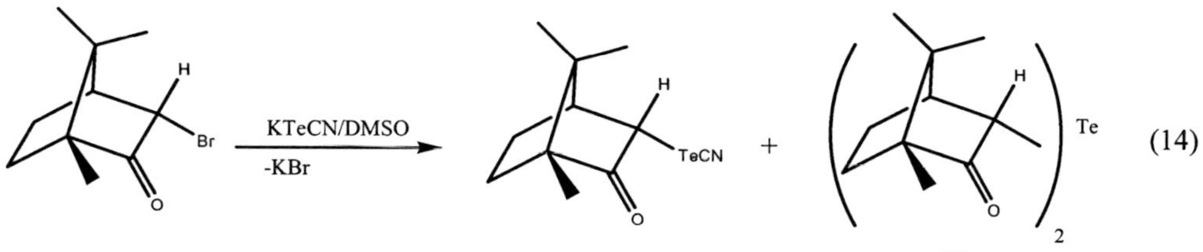

18 19

The main reaction product was di(*endo*-3-camphoryl)telluride (**19**), which was obtained in 51% yield while the tellurocyanate **18** was obtained in 15% yield[26].

On the other hand, reaction of potassium tellurocyanate dissolved in dry DMSO with α-bromobenzaldehyde arylhydrazones (**20**) gave the corresponding α-tellurocyanatobenzaldehyde arylhydrazones (**21**) in low yields as yellow solids of low melting points[41], equation 15.

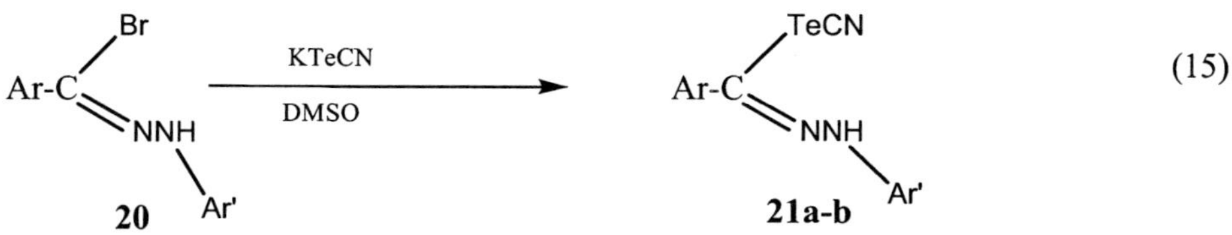

20 21a-b

a:Ar =Ar' = C_6H_5 ; **b**:Ar= 4-$NO_2C_6H_4$, Ar' = 2,4-$(O_2N)_2C_6H_3$

Recently, we reported the synthesis a series of α-tellurocyanatoketones (**22**) in fair yields (35-59%) by the reaction of potassium tellurocyanate dissolved in DMSO with α-bromoketones, equation 16. Compounds **22** were obtained as yellow solids of a moderate stability at room temperature after one recrystallization[42].

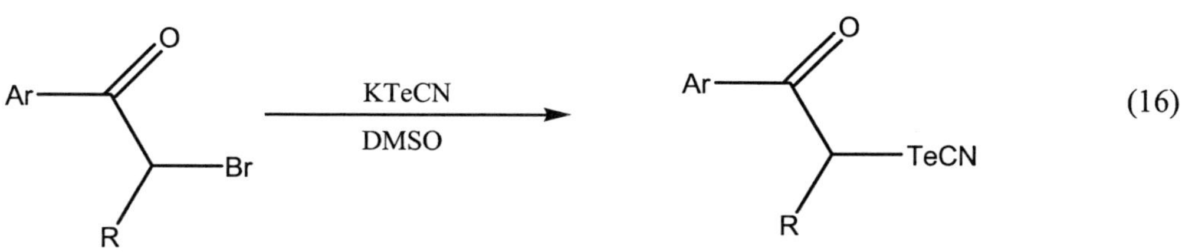

22a-d

a:Ar= C_6H_5, R=H; **b**:Ar= 4-BrC_6H_4, R=H;
c:Ar= 4-PhC_6H_4, R=H; **d**:Ar= C_6H_5, R=CH_3

We found that N-aryl acetimidoyl chlorides (**23**) react with potassium tellurocyanate in DMSO solution to form the corresponding N-aryl acetimidoyl tellurocyanates (**24**) in low yields(18-29%). These compounds are reported for the first time in this chapter. They are isolated and characterized in our laboratory [43], equation 17, Table 1.

(17)

23a-f **24a-f**

a: R = H; **b**: R= CH_3; **c**: R= NO_2; **d**: Cl; **e**: Br; **f**: biphenyl

The analytical and physical data for compounds **24a-24f** are listed in Table 1.

Table 1.

Compounds	M.p(°C)	Yield (%)	Color	^{1}H NMR, δ, ppm($CDCl_3$)	IR, ν(CN), cm^{-1}
24a	110-112	18	White	2.01(s,3H), 6.40-6.73(m, 3H); 7.05(t, 2H).	2130
24b	140-141	20	White	1.97(s, 3H); 3.22(s, 3H); 6.07(d,2H); 6.96(d, 2H).	2140
24c	126-127	29	Pale Yellow	2.02(s, 3H); 6.75(d, 2H); 7.98(d, 2H).	2120
24d	72	23	White	1.99(s, 3H); 6.45(d, 2H); 7.09(d, 2H).	2140
24e	76-77	34	Pale Yellow	2.00(s, 3H); 6.38(d, 2H); 7.21(d, 2H).	2130
24f	156-157	17	White	2.05(s, 3H); 7.30-7.72(m, 7H);7.87(s, 4H).	2130

Compounds **24a-24f** have been characterized by chemical analysis, IR and NMR, Table 1. They are white to pale-yellow solid of a reasonable stability at room temperature [43].

In spite of the fact that several organic tellurocyanates have been isolated and characterized by the above-mentioned procedures, there are still little known compared with well-studied selenocyanates[12]. This is probably due to the difficulties in synthetic methodologies and the instability of KTeCN compared with KSeCN and KSCN. Furthermore, these reagents (KSCN and KSeCN) are available commercially in contrast to KTeCN.

Reactions of Organic Tellurocyanates

Cyclization reactions of organic tellurocyanates can be considered as addition to the C≡N bond. There are only two examples dealing with such cyclization reactions, as we are aware. The reaction of hydrazoyl halides with potassium tellurocyanate in DMSO afforded the 2,4-disubstituted-5-imino-Δ^2-1,and 3, 4-telluradiazolines (**25**) together with α-tellurocyanatobenzaldehyde arylhydrazones (**21**) as major products [41], Scheme 3 and equation 15.

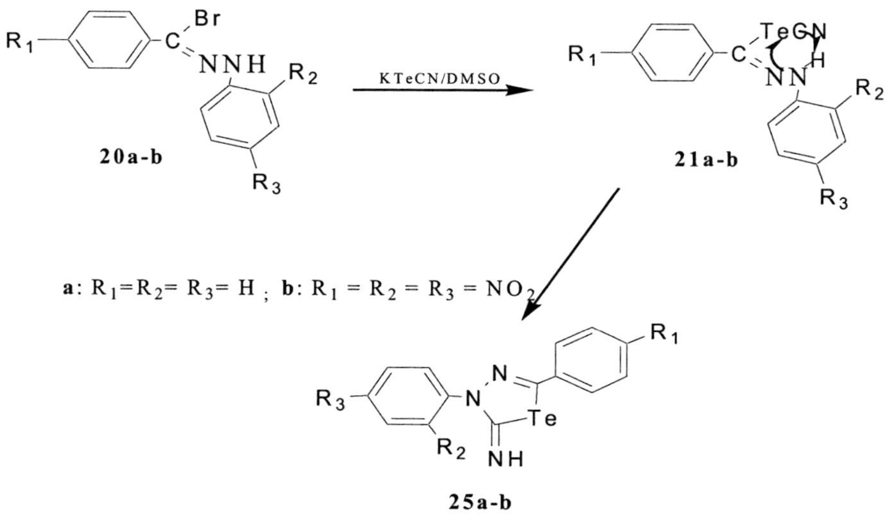

Scheme 3.

Unlike the similar reactions of KSeCN with hydrazoyl halides [44], which result in formation of only 1,3,4-selenadiazoles derivatives (**26**) (equation 18), the reactions of KTeCN with hydrazoyl halides (Scheme 3) produce α-tellurocyanatoketones (**21**) in low yields in addition to 1,3,4-telluradiazole derivatives (**25**) in fair yields[41].

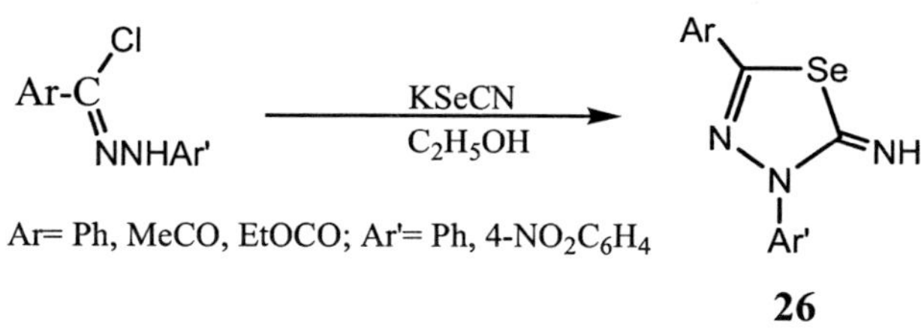

(18)

On the other hand, reaction of 2-aminobenzyl iodide (**27**) with potassium tellurocyanate in DMSO afforded the novel heterocycle 1,4-dihydro-2-imino-3-tellraquinoline (**29**) as the only isolated product. This yellow solid compound has a melting point of 75-77°C [41]. This indicates that the reaction may afford the organic tellurocyanate intermediate (**28**), which is not stable enough, and then cyclized spontaneously to give the new cyclic telluride (**29**). The yield of compound **29** was 56% yield [41], Scheme 4.

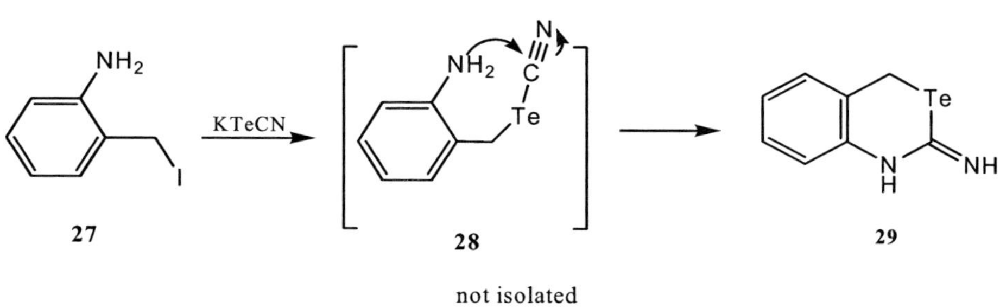

Scheme 4.

Acetyl derivatives of **26** and **29** were obtained by treatment of each of them with acetic anhydride [41] and as shown in Scheme 5.

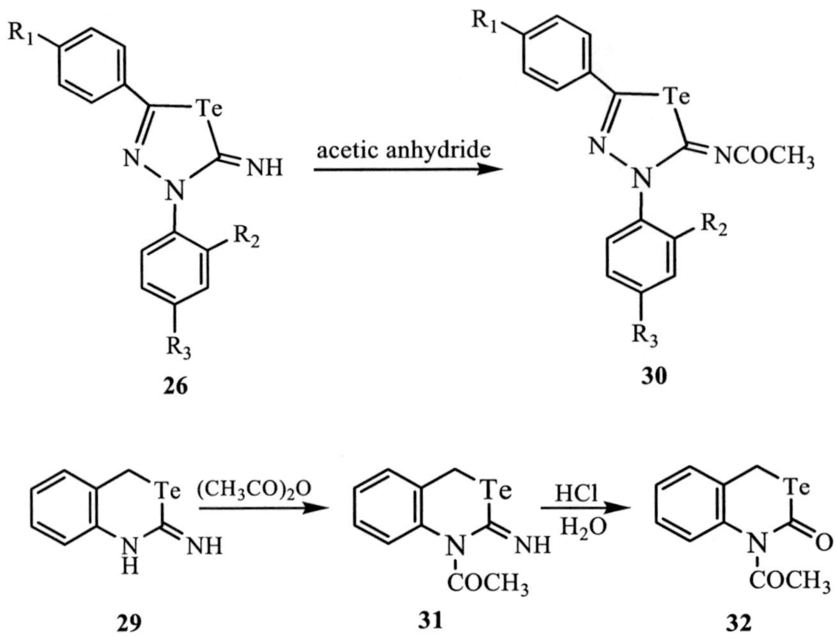

Scheme 5.

Substitution reactions of the C≡N bond can be illustrated by the reaction of organic tellurocyanates with halogens and thionyl chloride. A cyano group was replaced by the halogen to afford organyl tellurium trihalides [32, 42,45]. It should be mentioned that most reactions of organic tellurocyanates were studied mainly with benzyl tellurocyanate as an example [32]. The treatment of benzyl tellurocyanate with excess of bromine produces benzyl bromide *via* a benzyl tellurium tribromide intermediate [32], equation 19. It is worth noting that when dibenzyl ditelluride was treated with bromine, it exhibits the already seen sensitivity of tellurium-benzylic compound toward halogenolysis [46], equation 20.

$$C_6H_5CH_2TeCN \xrightarrow[CH_2Cl_2]{\text{excess } Br_2} [C_6H_5CH_2TeBr_3] \xrightarrow{\text{excess } Br_2} C_6H_5CH_2Br + TeBr_4 \quad (19)$$

$$(PhCH_2Te\text{-})_2 \xrightarrow[CCl_4]{Br_2} Te + (PhCH_2)_2TeBr_2 \xrightarrow{2Br_2} TeBr_4 + 2PhCH_2Br \quad (20)$$

α-Tellurocyanatoketones (**33**) were converted trichlorides, tribromides and triiodides in good yields by reacting them with thionyl chloride, bromine and iodine [42], respectively, equation 21.

$$\underset{\mathbf{33a\text{-}d}}{ArCOCH(R)TeCN} \xrightarrow[X= Br,\ I]{SOCl_2 \text{ or } X_2} \underset{\mathbf{34a\text{-}d}}{ArCOCH(R)TeX_3} \quad (21)$$

In contrast to the reaction of benzyl tellurocyanate with excess of bromine, treatment of **33** with excess of bromine did not produce the corresponding α-bromoketones and gave only the tribromide derivatives [42].

Treatment of benzyl tellurocyanate with methanolic sodium hydroxide (10%) afforded dibenzyl ditelluride (**35**) in 29% yield (equation 22). On the other hand, treatment of benzyl tellurocyanate with hypophosphoric acid improved the yield and gave the ditelluride **35** in almost quantitative yields [32], equation 22.

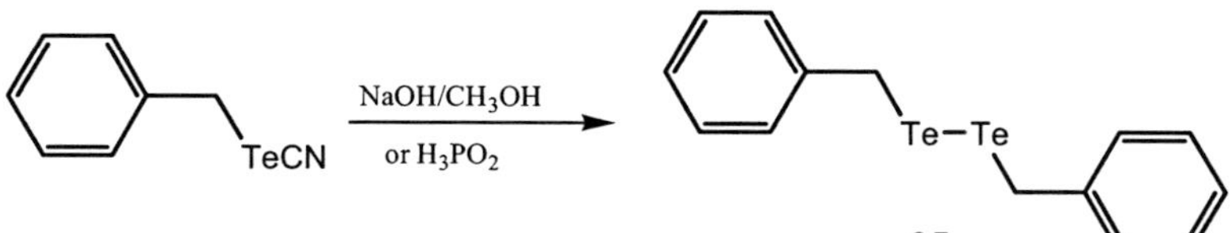

35 (22)

The reaction of 4-chloro-, 4-methoxy, 4-nitro- and 1, 2, 3, 4, 5-pentamethylbenzyl tellurocyanates with hypophosphoric acid gave the corresponding ditellurides in high yields [33,34]. The proposed mechanism for the formation of dibenzyl ditellurides from the reaction of benzyl tellurocyanates with H_3PO_2 depends on formation of benzyl tellurol ($C_6H_5CH_2TeH$), which in turn react, with another mole of benzyl tellurocyanate to give the corresponding dibenzyl ditelluride [32,33]. This proposed mechanism is consistent with the formation of dibenzyl thiotelluride when dibenzyl ditelluride (**35**) reacted with benzyl thiol, equation 23.

$$C_6H_5CH_2TeCN + C_6H_5CH_2SH \longrightarrow C_6H_5CH_2STeCH_2C_6H_5 \qquad (23)$$

Diphenacyl ditelluride(**36a**) and related ditellurides (**36b-36d**) were obtained in 39-76% yields by the reaction of α-tellurocyanatoketones with 10% of methanolic sodium hydroxide[42], equation 24.

$$ArCOCH(R)TeCN \xrightarrow{10\%\ NaOH} [\,ArCOCH(R)Te^{-}]_2 \qquad (24)$$

36

a: Ar=C_6H_5, R= H; **b**: Ar= 4-BrC_6H_4, R= H;
c: Ar= 4-PhC_6H_4, R= H; **d**; Ar=C_6H_5, R=CH_3

Compounds **36** are regarded as new compounds which can only be prepared by this new method [42].

Benzyl tellurocyanate is unstable in presence of air and underwent decomposition to benzaldehyde and benzyl alcohol in 60% and 40%, respectively with liberation of tellurium [32], equation 25.

$$PhCH_2TeCN \xrightarrow{O_2/\,h\nu} PhCHO + PhCH_2OH + Te \qquad (25)$$

The formation of benzaldehyde and benzyl alcohol may occur through the formation of benzylperoxy radical ($PhCH_2—O—O—TeCN$) which gave the benzyloxy radical($PhCH_2O^{\cdot}$) by O—O bond fission, equation 26.

$$PhCH_2TeCN \xrightarrow[h\nu]{O_2} [PhCH_2\text{-}O\text{-}O\text{-}TeCN] \longrightarrow [PCH_2O^{\cdot}] \longrightarrow PhCHO + PhCH_2OH \qquad (26)$$

The cleavage of Te-CN bond by carbanions can be illustrated by the reaction of 2-nitrophenyllithium with benzyl tellurocyanate [34], equation 27. This reaction afforded the unsymmetrical telluride **37** (*i.e.* 2-nitrophenyl benzyl telluride) in 72% yield, as an orange-red solid at -100°C.

$$PhCH_2TeCN + 2\text{-}O_2NC_6H_4Li \xrightarrow[-LiBr]{-100^{\circ}C} \underset{\mathbf{37}}{2\text{-}O_2NC_6H_4TeCH_2C_6H_5} \qquad (27)$$

Potassium tellurocyanate was also used as a cyclization agent to prepare heterocyclic telluride [47, 48]. Treatment of 2,2'-bis(bromomethyl) biphenyl and 2,2'-bis(bromomethyl)-1,1'-binaphthalene with KTeCN in DMSO solution gave 2,7-dihydro-1*H*-dinaphtho[c,e]tellurepin (**38**) [47] and 2,7-dihydro-1*H*-dinaphtho[c,e]tellurepin (**39**) [48] in 60 and 40% yields, respectively, equations 28 and 29.

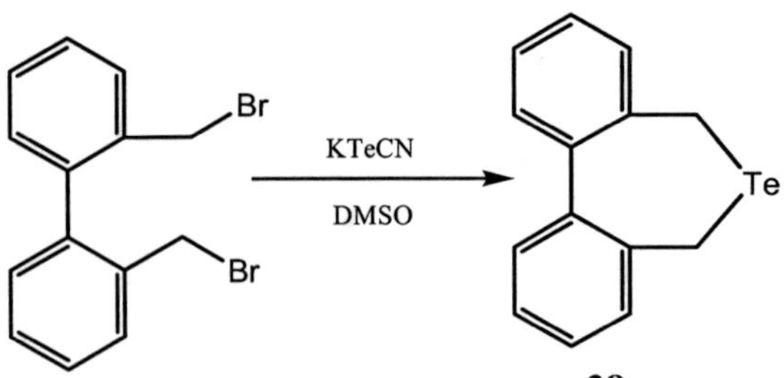

38 (28)

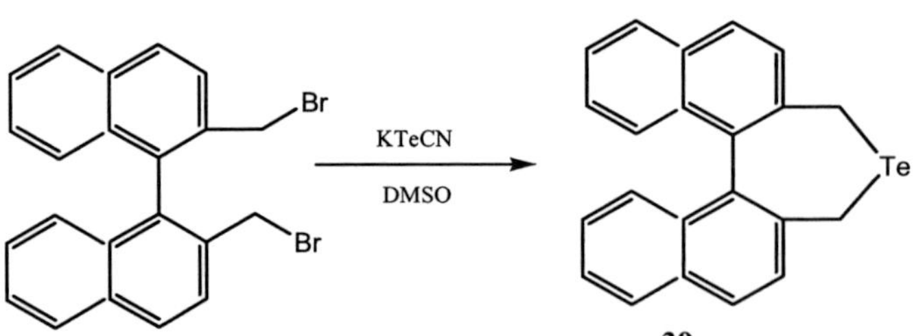

39 (29)

The preparation of compounds **38** and **39** (equations 28 and 29) by this new method directed our interest toward the reaction of potassium tellurocyanate with organic moieties containing bis (halomethyl) groups such as bis(2-bromoethyl) ether, bis(2-chloroethyl) amine and 1,2-bis(bromomethyl) benzene. In every case the corresponding heterocyclic telluride was obtained in fair yield [49]. Reactions and results are summarized in Schemes 6-8.

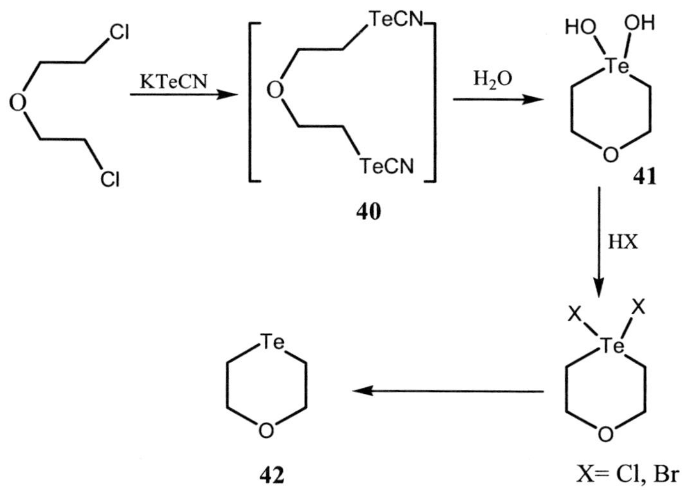

Scheme 6.

Compound **41** was separated during the work up. It seems to be that compound **42** was formed firstly and converted to **41** in aqueous medium, as it is well known that compound **42** is a moisture and air sensitive [50,51] and can be easily converted to compound **41**. In order to prove that compound **41** was formed, it was reacted with HX (X= Cl, Br) and gave the corresponding dihalides, Scheme 6. All these compounds are well known derivatives of compound **42** and were characterized by chemical analysis, IR, ^{1}H and ^{13}C NMR and mass spectroscopic data [49,51].

Reaction of bis(2-chloroethyl)amine with KTeCN gave the novel new heterocyclic telluride (**44**) which in turn was reacted with bromine to give its dibromo derivative, Scheme 7, Reaction of compound **44** with Co(II), Ni(II), Pt(II) and Pd(II) salts produced complexes of novel properties[49].

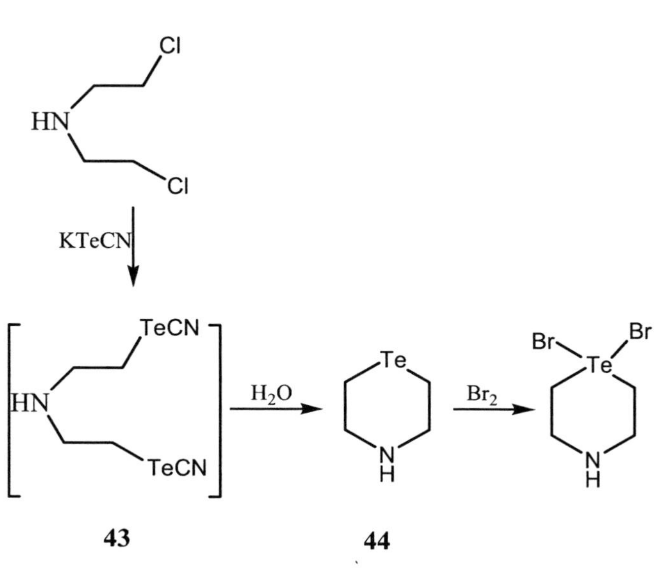

Scheme 7

The reaction of 1, 2-bis(bromomethyl)benzene with KTeCN merits special mention. The reaction gave the cyclic telluride (**46**) as a major isolated product in 58% along with a polymeric material which was formulated as [(-Te-C_8H_8-Te-)$_n$] [49].

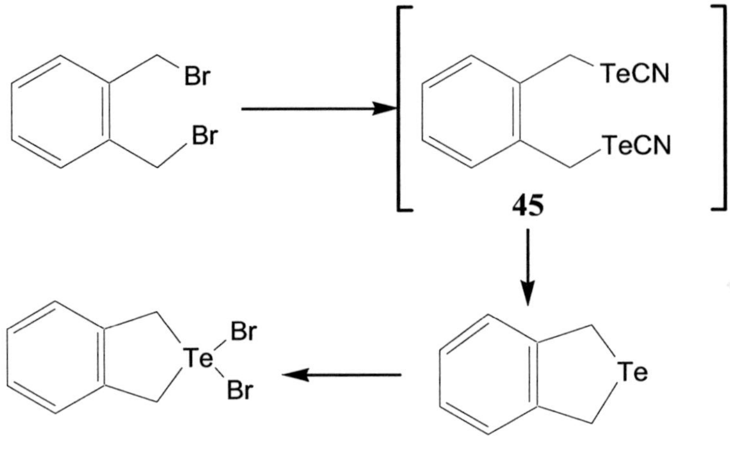

Scheme 8.

All cyclic tellurides (*i.e.* **42**, **44** and **46**) and their dihalo derivatives were characterized by CHN analysis and spectroscopic techniques [49]. Cyclization reactions of the bis(halomethyl) organic precursor with potassium tellurocyanate proceed under mild conditions, making it as a useful synthetic route to heterocyclic tellurides. In contrast to the reaction in Scheme 7, reaction of KSeCN with 1,2-bis(bromomethyl)benzene gave 1,2-bis(selenocyanato) benzene(**47**), which converted to selenocyclophane(**48**) by alkaline hydrolysis [52,53], equation 30. Compound **48** was isolated and purified unlike its analogue tellurocyanate (**45**) which converted spontaneously to the cyclic telluride **46**[49].

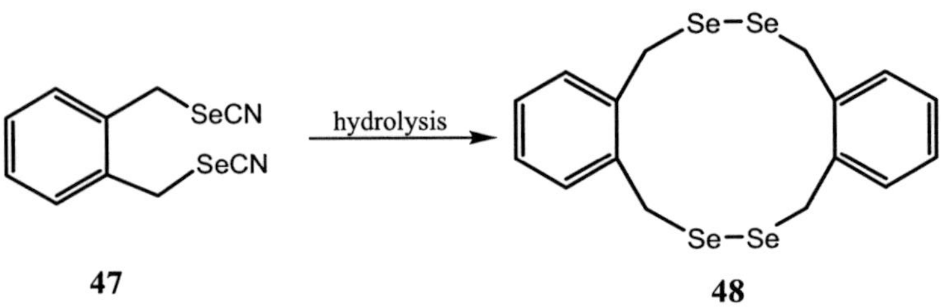

47 **48** (30)

The formation of cyclic tellurides **42**, **44** and **45** could be explained by the unstability of the intermediates **40**, **43** and **45**(Schemes 6,7 and 8) and their disproportionation to the cyclic tellurides and tellurium dicyanide and the latter decomposed to tellurium and dicyanogen, as shown in Scheme 9.

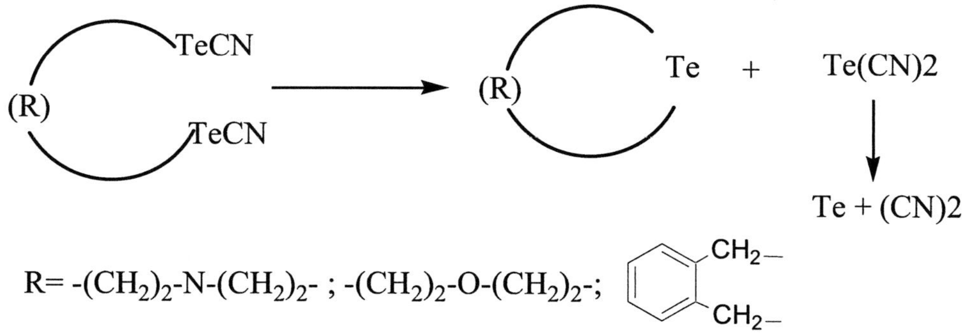

Scheme 9.

The aryl tellurocyanates could be converted to alkyl aryl tellurides(**49**) when reacted with primary and secondary alcohols in dichloromethane or tetrahydrofuran at room temperature in the presence of tributylphosphine [29,30], equation 31. Such a transformation did not occur by using triphenylphosphine or triethylphosphite [29, 30].

$$\text{ArTeCN} + \text{ROH} \xrightarrow{Bu_3P} \underset{\mathbf{49}}{\text{Ar-Te-R}} \qquad (31)$$

We applied the reaction shown in equation 31 to the reaction of phenacyl tellurocyanates with 1-(hydroxymethyl)pyrazole or 1-(hydroxymethyl)-3,5-dimethylpyrazole in the presence of tri-n-butyl phosphine and gave the unsymmetrical tellurides **50**, Scheme 10. These new unsymmetrical tellurides containing pyrazole moieties were obtained in low yield (25-39%), Scheme 10.

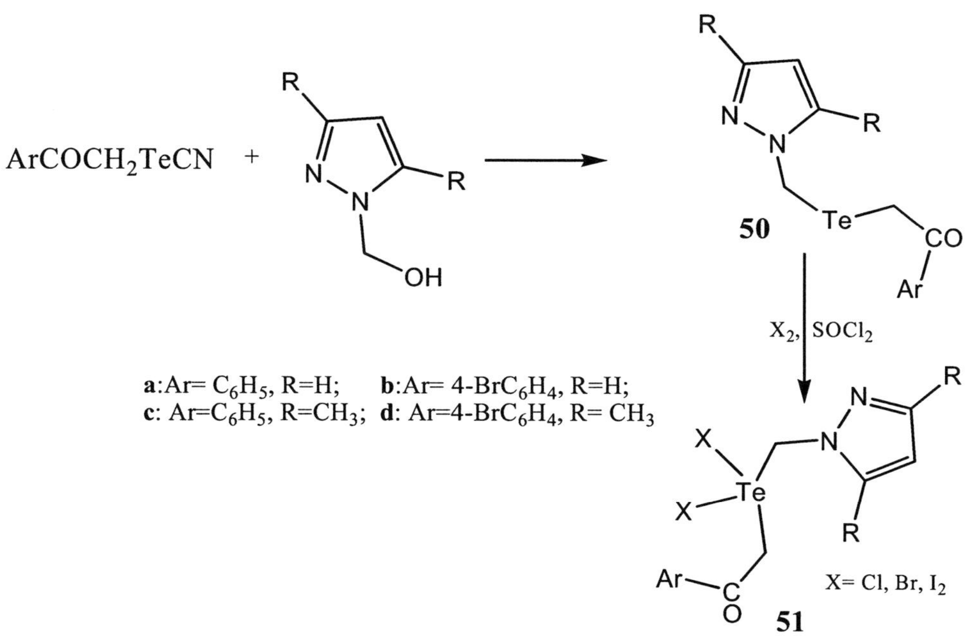

Scheme 10.

The tellurides **50a-d** were characterized by analytical and spectroscopic data [54]. They are stable enough to handle at room temperature. They gave the dichloro, dibromo and diiodo derivatives on treatment with thiony chloride, bromine and iodine, respectively and as shown in Scheme 10. Some of physical and analytical data of these new unsymmetrical tellurides **50a-d** are summarized in Table 2.

Table 2. Some physical and analytical data for compounds 50a-d

Compounds	M.p (°C)	Yield (%)	Colour	^{1}H NMR, δ, ppm($CDCl_3$)
50a	45-47	18	White	3.65(s, 2H, N-*CH₂*-Te); 4.42(s, 2H, CO-*CH₂*-Te); 6.15(m, 1H,Pz); 7.30-7.54(m, 5H, Pz+ Ar-H); 7.92-8.02(m, 2H, Ar-H).
50b	64-66	20	White	3.72(s, 2H, N-*CH₂*-Te); 4.52(s, 2H, CO-*CH₂*-Te); 6.07(m, 1H, Pz); 7.35-7.60(m, 4H, Ar-H+Pz); 7.85(d, 2H, Ar-H).
50c	97-98	29	Yellow	2.20(d, 6H, *CH₃*-Pz); 3.70(s, 2H, N-*CH₂*-Te); 4.48(s, 2H, CO-*CH₂*-Te); 5.80(s, 1H, Pz); 7.30-8.18(m, 5H, Ar-H).
50d	72-74	23	White	2.23(d, 6H, *CH₃*-Pz); 3.65(s, 2H, N-*CH₂*-Te); 4.45(s, 2H, CO-*CH₂*-Te); 6.15(s, 1H, Pz); 7.45(d, 2H, Ar-H); 7.82(d, 2H, Ar-H).

Compounds **51** were also characterized by elemental analysis, NMR and mass spectroscopic data[54]. A further work is carring out in our laboratory in order to explore their reaction with transition metal salts and their uses in organic synthesis[54].

The mechanism of the above reactions was not studied. The reaction presumably may proceed through the substitution by a phosphine atom as a first step as shown in Scheme 11.

$$RTeCN + Bu_3P \longrightarrow RTe\overset{+}{P}Bu_3\ \overset{-}{CN}$$

$$RTe\overset{+}{P}Bu_3\ \overset{-}{CN} + RCH_2OH \longrightarrow RCH_2O\overset{+}{P}Bu_3 + HCN + R\overset{-}{Te}$$

$$RCH_2O\overset{+}{P}Bu_3 + R\overset{-}{Te} \longrightarrow RCH_2TeR + Bu_3PO$$

Scheme 11

Phenyl tellurocyanate was used as a source of introducing phenyltellurium species to other organic compounds. The treatment of cyclohexene with phenyl tellurocyanate in methanol in the presence of copper(II) chloride afforded (2-methoxycyclohexyl) phenyltellurium dichloride (**52**) in 62% yield[55,56], equation, 32.

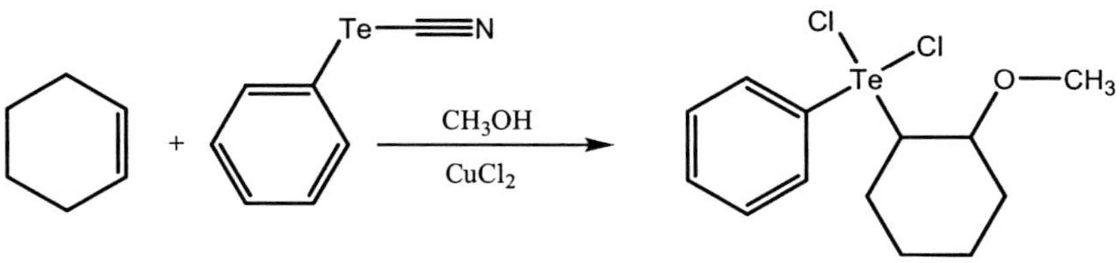

(32)

52

The same compound (*i.e.* phenyl tellurocyanate) was treated with γ or β-halogenoamino acids in the presence of sodium borohydride in ethanol to give phenyltelluroamino acids (**53**) in ~ 30% yield [57], equation 33 .

$$X(CH_2)nCH(NH_3Cl)CO_2H + PhTeCN \xrightarrow[C_2H_5OH]{NaBH_4} PhTe(CH_2)nCH(NH_3Cl)CO_2H \quad (33)$$

X= Cl, Br; n= 1, 2. **53**

By applying the reaction in equation 32 to the reaction between phenacyl tellurocyanate and cyclohexene in ethanol in presence of copper(II) chloride and under same reaction conditions[58], we found that the reaction afforded (2-ethoxycyclohexyl)phenacyltellurium dichloride (**54**) in 37% yield[58], equation 34.

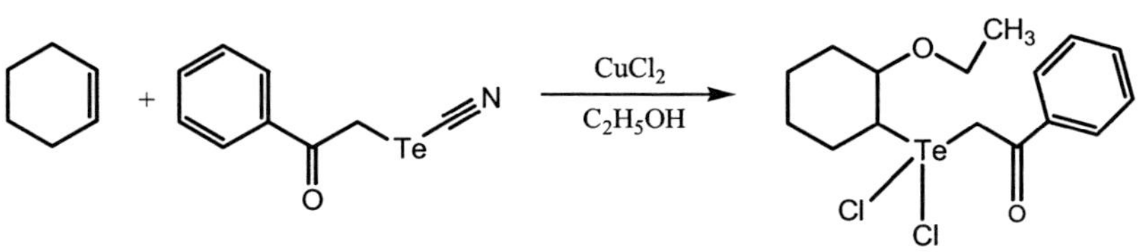

(34)

54

Compound **54** was obtained as pale- yellow solid which melted at 90-92°C. Chemical analysis and spectroscopic data for this compound are in agreement with its structure [58]. A further work is going on in our laboratory to investigate the above reaction (equation 34) using different olefinic compounds.

Physical and Structure Properties of Organic Tellurocyanates

In general, the infrared spectra of tellurocyanates show that ν(CN) are within range of 2110-2180 cm^{-1}[18-22, 32, 33, 42]. The nature of organic substituent appears to have effect on ν(CN) of organic tellurocyanate. This clearly reflects on the IR spectra of α-tellurocyanatobenzaldehyde arylhydrazones (**21a-b**, equation 14) which showed ν(CN) at ~2220 cm^{-1}[41]. IR study is needed to be explored for organic tellurocyanates and further investigations also needed to study the effect of substituent of organic group in tellurocyanates on the υ(CN).

The X-ray crystal determination of 4-nitrobenzyl tellurocyanate (**55**) [36,59] and 2-nitrobenzyl tellurocyanate (**56**)[35] revealed that the Te—CN bond lengths are 2.060 and 2.095 Å, respectively. The main feature in these structures is the secondary intermolecular interactions between the heteroatoms(N or O) and the tellurium atoms [32,34,35]. In compound **55**, two strong secondary interactions with oxygen atoms of nitro groups of neighboring molecules are observed[36,59]. The Te—O distances (2.949 and 3.182 A°) are significantly shorter than the sum of van der Waals radii (3.58 A°)[60]. Compound **55** may be considered as a distorted square-planar tellurium(II) complex. The coordination geometry around tellurium in 2-nitrobenzyl tellurocyanate(compound **56**) is similar to that of **55**. The tellurium atom in compound **53** forms two Te—N bond distances with tellurocyanate nitrogen atoms of two neighboring molecules(2.889 and 3.382 A°)[35]. The Te—N intermolecular distances are considerably shorter than van der Waals distance of these two atoms[60]. The stability of compounds **55** and **56** in crystals and in solution in compare to benzyl tellurocyanate, may be due to the secondary interaction between oxygen(nitrogen) and tellurium atoms in both compounds (*i.e.* compounds **55** and **56**).

4-Nitrobenzyl selenocyanate (**57**) has been shown to have a similar structure of compound 55. In both compounds (**55** and **56**) the telluro- and selenocyanate groups are linear and form strong secondary interactions[59].

The ^{125}Te NMR spectra of a limited number of organic tellurocyanates were determined. The ^{125}Te chemical shifts for organic tellurocyanates were found to vary within range of 652-692 ppm[37]. No information is available on the effect of the medium or the organic substituent on the ^{125}Te chemical shifts of the organic tellurocyanates. On the other hand, the ^{125}Te chemical shifts for ionic salts such as KTeCN, $(C_6H_5)_4AsTeCN$ and $((C_6H_5)_3P)_2NTeCN$ were found to be –509.5, -511.1 and –568.8 ppm, respectively. These values are attributed to the nature of cation [37].

The ^{13}C NMR for a number of tellurocyanate revealed the presence of Te—CN in range of 115.08-125.08 ppm.[42].

In conclusion, there are few isolated organic tellurocyanate in pure form in compare with thio- and selenocyanate. The general method of preparing thiocyanate and selenocyanate derivatives from diazonium salts is not suitable for the preparation of aryl tellurocyanates. The chemistry of organic tellurocyanates still largely remains unexplored and new synthetic methods are needed. It seems likely that some reactions of organic tellurocyanates proceed in a similar way with that of the corresponding reactions of their selenium analogs.

Organic Isotellurocyanates

There were many attempts to prepare organic isotellurocyanates from the reaction of tellurium with isonitriles or with organocyanosilanes [61-63], but none of them has been succeeded. Photoinduced isomerization of organic tellurocyanates was also applied to prepare organic isotellurocyanates . The photolysis of the tellurocyanates gave only tellurium[34]. This could be explained as a result of formation of alcohol and aldehyde in presences of oxygen (see equations 25 and 26). The instability of organic isotellurocyanates makes it difficult to prepare and to introduce the NCTe group into organic molecules. Further attempts are needed to prepare these unknown organic isotellurocyanate compounds, R—N=C=Te.

Formation of Metal Complexes

The ligand chemistry of organic tellurocyanate compounds have not been studied , as we are aware. No work with this class of ligands has been reported, primarily owing to their inherent poor stability. Thus, no information is available on the reaction of organic tellurocyanates with metal salts. In contrast, organic selenocyanate complexes, such as $M(CO)_5(RSeCN)$ (M= Cr, Mo), are known [64]. The palladium and platinum complexes of formula $PdCl_2.RSeCN$ or $PdCl_2.2RSeCN$ [65] and of formula $[Pt(SeCN)_4[n\text{-}Bu_4N]_2$ and $[Pd(SeCN)_4[n\text{-}Bu_4N]_2$[66] are known in literature[65,66]. Synthesis and characterization of triamine complexes of nickel(II) selenocyanate , *i.e.* $Ni(dpt)(SeCN)(H_2O)]$ and $[Ni(aepn)(SeCN)_2(H_2O)$; dpt= bis(3-aminopropyl)amine and aepn= (2-aminoethyl)(3-aminopropyl)amine, are reported [67] and their structures were determined by X-ray analysis [67]. The synthesis of copper(II) and gold(I) complexes containing selenocyanate ligands have been reported recently [68,69].

As we mentioned before in this chapter that attempts to separate stable solution of sodium and potassium tellurocyanates by addition of anhydrous acetone or ether were unsuccessful due to immediate liberation of metallic tellurium. Although simple salts of tellurocyanate have now been isolated (e.g. A[TeCN], A= Et_4N^+[39], Ph_4As^+[39], $Ph_3P{=}N{=}PPh_3^+$[40]), only few examples of coordinated tellurocyanate has been reported. Thus, we will give brief details about the reaction of tellurocyanate anion with some transition metal complexes. The reaction of NaTeCN with $Fe_3(CO)_{10}(\mu_3\text{-PPr-i})$ followed by alkylation of the intermediate anionic adduct (**58**) with triethyloxonium fluoroborate resulted in formation of complex **59** in high yield [70],Scheme12.

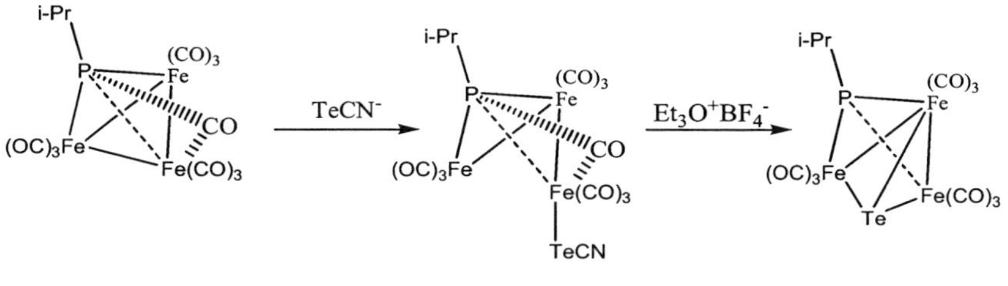

58 59

Scheme 12.

The structure of complex **59** was confirmed by X-ray diffraction analysis[70].

The formation of telluroketones was observed as the metal complex with tungsten carbonyl. Treatment of $[(PhRC{=})W(CO)_5]$ (where R= H or C_6H_5) with tellurocyanates anion ($TeCN^-$) derived from bis(triphenylphosphoranylidene)ammonium gave compounds **60** [71,72], where tellurium was inserted to the tungsten-carbene bond to produce complexes **60** , as shown in equation 35.

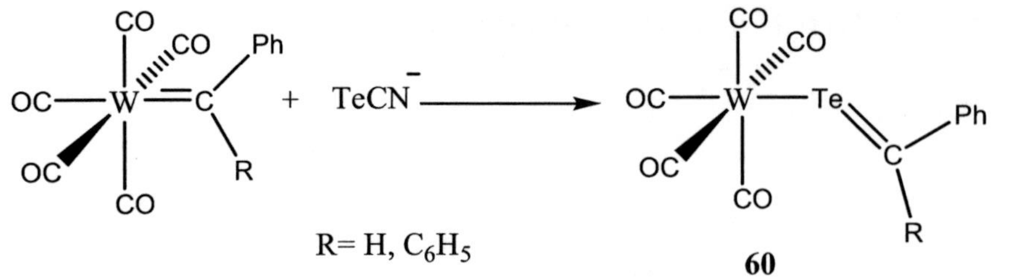

60

(35)

It worth noting that the cationic selenoformaldehyde complex (i.e. { $[(\eta^5$-$C_5H_5)(NO)(Ph_3P)Ru\ (\eta^2$-$Se{=}CH_2)]^+PF_6^-$}) was prepared by the reaction of rhenium complex { $[(\eta^5$-$C_5H_5)(NO)(Ph_3P)Ru{=}CH_2]^+PF_6^-$} with potassium selenocyanate [73] and elimination of the cyano group occurred in such reactions[73]. The reaction of $FeSO_4.7H_2O$ with trans-1,2-bis(4-pyridyl)ethylene(tvp) and NaCCSe in water/ethanol gave complex $[Fe(tvp)_2(NCSe)_2(H_2O)_2]$. The ^{57}Fe Mössbauer spectrum revealed that the complex is in Fe^{II} high-spin state [73].

The ligand chemistry of tellurocyanates is much less developed than that of selenium counterpart owing to their poor stability, and future work in this area can be expected to be aimed at the synthesis and isolation of tellurocyanate compounds, for which well-established coordination chemistry has been developed for sulfur and, to lesser extent, selenium.

REFERENCES

[1] Wöhler; F. *Liebig's Ann. Chem.* 1840, 35 ,111-117.

[2] *The Chemistry of Organic Selenium and Tellurium Compounds*; Patai, S.; Rappoport, Z. Eds.; Wiley; Chichester, 1986(vol. 1), 1987(vol. 2).

[3] *Organicselenium Chemistry- A Practical Approach*; Back, T.G., Ed.; Oxford University, Press; Oxford, 1999.

[4] Hope, E.G.; Levason, W., *Recent Developments in the Coordination Chemistry of Selenoether and Telluro ether Ligands* 1993, 122, 109-170.

[5] Arnold, J. in The Chemistry of Metal Complexes with Selenolate and Tellurolate Ligands; Karlin, K.D.; Ed.; John Wiley and Sons, NY, 1995,vol 43, pp 353-417.

[6] Singh, A.K.; Sharma, S. *Coord. Chem. Rev.* 2000, 209, 49-98.

[7] Sudha, N.; Singh, H.B. *Coord. Chem. Rev.* 1994, 135/136, 469-515.

[8] O'Regan, M.B. In *Tellurium-Containing Heterocycles*; Taylor, G. Ed.; John Wiley and Sons, 1994; pp. 425-490.

[9] Irgolic, K.J., *The Organic Chemistry of Tellurium*, N.Y.; Gordon and Breach, 1974.

[10] a)Irgolic, K.J., In Houben-Weyl Methoden der Organischen , Müller,E. Ed., vol. IX, G. Tieme, Stuttagart, 1990.*Topics in Current Chemistry*, ed. T. Wirth, vol. 208, Springer-Verlag, Berlin, 2000.

[11] Downs, A.W. *J. Chem. Soc., Chem. Commun.*, 1968, pp. 1290-1291.

[12] Toshimitsu, A.; Uemura, S. *Organic Selenocyanates and Tellurocyanates and Related Compounds*; In *The Chemistry of Organic Selenium and Tellurium Compounds*, Patai, P. Ed.; John Wiley and Sons, N.Y., 1987, vol. 2, pp. 541-590.

[13] Skopenko. V.V.; Ivanovs. E.I. *Ukr. Khim. Zh.*, 1990, 56, 339-343.
[14] Sadekov, I.D.; Skopenko, V.V.; Buriov, A.S.; Ivanova, E.I.; Garnovskii; A.D. *Russ. J. Gen. Chem.*, 1997, 67, 1058-1067.
[15] Barglin, J.N.; *U.S. pat.*, 22091, 84; *Chem. Abst.* 1941, 35, 272.; *U.S. pat.*, 2217611, 2217612, 2217613, 2217614, 2217615; *Chem. Abst.*, 1941, 35, 1156.
[16] Von Wasserman, A. *Br. Pat.*, 978; *Chem. Abst.*, 1913, 7, 2286.
[17] Nefedov, V.A. *Zh. Obshch. Khim.*, 1968, 38, 2191-2193.
[18] Piette, J.L.; Thibui, P.; Renson, M. *Tetrahedron*, 1978, 34, 655-659.
[19] Piette, J. L.; Lysy, R.; Renson, M. *Bull. Soc. Chim. Fr.*; 1972, 3559-3561.
[20] Walter, F.; Sladky, F. *Chem. Ber.*, 1991, 124, 1131-1133.
[21] Falcon, J.; Cava, M.P. *J. Org. Chem.*, 1980, 45, 1044-1045.
[22] Engman, L., Cava, M.P. *J. Org. Chem.*, 1981, 46, 4194-4197.
[23] Sadekov, I.D.; Minikin, V.I.; *Usp. Khim.*, 1995, 64, 527-560
[24] McWhinnie, W.R.; Sadekov, I.D.; Minikin, V.I. *Sulfur Rep.*, 1996, 18, 295-335.
[25] Hugesh, G.; Singh, H.B. *Accounts Chem. Res.*, 2003, 10, 226-236.
[26] Back,T.C.; Dyck,B.P.; Parvez, M. *J.Org.Chem.*, 1995,60, 4657-4659.
[27] Al-Rubaie, A. Z.; Unpublished results. M.p. 112-113°C.
[28] Talbot, M.; Piette, J.L.; Renson, M. *Bull. Soc. Chim.Fr.*, 1976, 294-299.
[29] Ogura, F.; Yamaguchi, H.; Otsubo, T.; Chikamatsu, K. *Synth. Commun.*, 1982, 12, 131-136.
[30] Engman, L. *J. Org. Chem.*, 1983, 48, 2920-2922.
[31] Dunne, S.J.; Summers, Z.A.; von Nagy-Felsofuki, F. *J. Heterocycl. Chem.*, 1993, 30, 409-412.
[32] Spencer, H.K.; Lakshmikantham, M.W.; Cava, M.P. *J. Am. Chem. Soc.*, 1977, 99, 1470-1473.
[33] Maartmann-Moe, K.; Sanderud, K.A.; Songstad, J. *Acta Chem. Scand. Ser. A*, 1981, 35, 157-153.
[34] Suzuki, H.; Miyoshi, K.; Osuka, A. *Nippon Kagaku Kaishi*, 1981, 472-478.
[35] Grung, K.E.; Romming, C.; Songsted, J. *Acta Chem. Scand. Sect. A*, 1989, 43, 1518-1526.
[36] Maartmann-Moe, K.; Sanderud, K.A.; Songstad, J. *Acta Chem. Scand. Ser. A*, 1984, 38, 187-200.
[37] Jones, C.H.W.; Sharma, R.D. *Organometallics,* 1978, 6, 1419-1423.
[38] Maretinsen, A.; Songstad, J. *Acta Chim. Scand. Ser. A*, 1977, 31, 645-650.
[39] Austad, T.; Esperas, S.; Songstad, J. *Acta Chim. Scand. Ser. A*, 1973, 27, 3594-3596.
[40] Martinsen, A.; Songstad, J. *Acta Chem. Scand., Sect. A*; 1977, 31, 1470-1473.
[41] Al-Rubaie, A. Z.; Al-Jadaan, S. A. N. *Polyhedron*, 1997,16, 1241-1247.
[42] Al-Rubaie, A. Z.; Yousif, L.Z.; Al-Baa'aj, A.K. *J. Organomet. Chem.*, 2003, 673, 40-46.
[43] Al-Rubaie, A. Z.; Al-Jadaan, S.A. N.; Al-Masoudi, W. A.; Unpublished data.
[44] Hassaneen, H.M.; Shetta, A.; Shawali, A.S. *J. Heterocycl. Chem.*, 1980, 17, 1185-1189.
[45] Wiriyachitra, P.; Falcon J. S.; Cava, M.P. *J. Org. Chem.*, 1979, 44, 3957-3963.
[46] Comasseto, J.V. *Phosphorous, Sulfur and Silicon*, 1992, 67, 183-201.
[47] Al-Rubaie, A.Z.; Al-Marzook, A. Y.; Al-Jadaan, S. A. N. *Rec. Trav. Chim. Pays-Bas,* 1996, 115, 427-432.
[48] Al-Rubaie, A.Z.; Fahad, T.A.; Al-Jadaan, S.A.N.; Aboud, N.A. *J. Organomet. Chem.*, 2004, 687, 2377-2381.

[49] Al-Rubaie, A.Z.; Al-Jadaan, S. A. N. Unpublished data.
[50] Farrar, W. V.; Gulland, M.J. *J.Chem.Soc.*, 1945, 11-15.
[51] Al-Rubaie, A.Z.; Al-Shiryda,H.A.; Auoob, A.I. *Inorg. Chim. Acta*, 1987, 134, 139-144.
[52] Esaki, N.; Tanaka, H.; Uemura, S.; Suzuki, T.; Soda, K. *Biochemistry*, 1979, 18, 407-414.
[53] Mitchell, R.H. *Can. J. Chem.*, 1976, 54, 238-
[54] Al-Rubaie, A. Z.; Al-Jadaan, S.A.N.; Ajadaad, A. R.; Unpublished data.
[55] Uemura, S.; Fukuzawa, S.; Tomshimitsu, A. *J. Organomet. Chem.*,; 1985, 250, 203-211.
.
[56] Uemura, S.; Fukuzawa, S.; Tomshimitsu, A.; Okano, M. *Tetrahedron Lett.*, 1982, 23, 1177-1178.
[57] Ajinomoto Co.; *Jpn. Kokai Tokkyo Koho*, JP 59 216868 [84216868]; *Chem. Abstr.*; 1985, 102, P185497h.
[58] Al-Rubaie, A. Z. Unpublished data. Compound 54: M.p. 90-92°C. ^{1}H NMR ($CDCl_3$) ppm: 1.22(t, 3H, CH_3), 1.31-1.42 (complex signal, 6H, Cyclohexane); 2.11(complex, 1H, Cycl.); 3.12 (s, 2H, CO-CH_2-Te); 3.78(q, 2H, CH_2-O); 7.30-7.45(complex, 3H, Ar-H); 7.80-7.97(complex, 2H, Ar-H).
[59] Maartmann-Moe,K.; Sanderud, K.A.; Songstad, J. *Acta Chem. Scand. Sect. A*, 1981, 33, 151-153.
[60] Bondi, A. *J. Phys. Chem.*, 1964, 68, 441-451.
[61] Lipp, M.; Dallacker, F.; Meeier zu Kocker, I. *Monatsch. Chem.*, 1959, 90, 41- .
[62] Franklin, W. J.; Werner, R. L. *Tetrahedron Lett.*, 1965, 3003- .
[63] Thayer, J. S.; *Inorg. Chem.*, 1968, 7, 2599-2604.
[64] Quick, M. H.; Angelici, R. J. *Inorg. Chem.*, 1976, 15, 160-165.
[65] Uemura, S.; Toshimitsu, A.; Kozawa, Y.; Okano, M. *Bull. Inst. Chem. Res. Kyoto Univ.*, 1981, 59, 350-357.
[66] Gery, J. K.; Butler, I. S.; Reber, C. *Inorg. Chem.*, 2003, 42, 6503-6518.
[67] Maji, T. K.; Mostafa, G.; Mukherjee, P. S.; Mondal, A.; Welch, A. J.; Okamoto, K.I.; Ray Chaudhuri, N. *Polyhedron,* 2000, 19, 1903-1915.
[68] Ribas, J.; Diaz, C.; Solans, X.; Fort-Bardia, N. *J. Chem. Soc., Dalton Trans*, 1997, 35-38.
[69] Baker, M. V.; Barnard, P. J.; Brayshaw, S. K.; Hickey, J. L.; Skelton, B. W.; White, A. H. J. *Chem. Soc., Dalton Trans.*, 2005, 37-43.
[70] Buchholz, D.; Muttner, G,; Zholhai, L.; Imhof, W. *J. Organomet. Chem.*, 1989, 377, 25-41.
[71] Fisher, H.; Kalbas, C.; Hoffmam, I. *J. Chem. Soc. Chem. Commun.*, 1992, 1050-1051.
[72] Fisher, H.; Fruth, A.; Troll, C. *J. Organomet. Chem.*, 1991, 415, 211-221.
[73] Nakashima,S; Yamamoto, A; Asada, Y, Nobuyoshi Koga; Okuda, T. *Inorg. Chim. Acta*, 2005, 358, 257-264

AKNOWLEDGEMENTS

The author is grateful to Professor Sakea Uemura, Kyoto University, Kyoto, Japan for his helpful discussion and valuable comments. Thanks also due to Dr. Shaker A.N. Al-Jadaan,

Chemistry Department, College of Science, University of Basrah, Basrah, Iraq for his assistance and discussion. The author acknowledges the assistance of colleagues whose names appear on many of the cited articles from College of Science, University of Basrah, Basrah, Iraq.

In: Focus on Electrochemistry Research
Editor: Magdalena Nunez, pp. 93-146

ISBN: 1-59454-545-6

Chapter 4

COPPER ELECTROCHEMICAL POLISHING FOR ULSI FABRICATIONS

J. Huo[1], R. Solanki[2] and J. McAndrew[1]
[1] Electronics R&D Group, Air Liquide, Countryside, Illinois, USA
(Contact e-mail: Jinshan_Huo@Yahoo.com)
[2] OGI School of Science and Engineering, OHSU, Beaverton, Oregon, USA

ABSTRACT

As device dimensions shrink to sub-micron regime, signal propagation delay in the interconnects becomes a bottleneck against increasing device speed. In order to reduce this delay, besides using copper as interconnects, low-k and ultra low-k materials have to be used as inter-layer dielectrics. Unfortunately, these dielectric materials tend to be soft, hence difficult to integrate with chemical mechanical polishing (CMP). As an alternative to CMP, electrochemical polishing (ECP) is under serious consideration. However, this process faces several challenges, especially for planarizing patterned Cu films on Si wafers (Cu/Si) using conventional ECP solutions for Cu bulk materials. This work presents results of an investigation to determine optimal ECP conditions for Cu/Si wafers, including two proved-successful approaches for ECP of Cu/Si wafers. These are based on investigation of the basic ECP processes and mechanisms.

A measure of the progress in microelectronics is often gauged by the size and the number of transistors incorporated on a given chip. Currently, the most advanced microprocessors consist of over 100 million transistors and this number is projected to reach one billion by 2010 [1]. Each of these transistors has to be connected to provide power and to be functional. Although the speed of the individual transistors is very fast, the overall speed of these devices is limited by the signal propagation delay or the RC (where R is the resistance and C the capacitance) constant of these interconnects. In order to reduce the RC value of these interconnects, one has to reduce their resistance and the inter-and intra-layer capacitances. Although Al has served well as the interconnect material, it can no longer meet the demands of current microelectronic devices and is gradually being replaced by Cu. Copper interconnects have been shown to provide significant advantages over Al, including higher speed (due to lower resistance) and higher resistance to electromigration.

Unlike Al, there is no practical process for dry-etching Cu, hence Cu interconnects are fabricated using a process called damascene. This process consists of several steps: (a) etching a trench in the interlayer dielectric, (b) depositing a thin conformal layer of a barrier material over the surface, (c) depositing a thick copper layer, and (d) removing the excess Cu layer by employing chemical mechanical polishing (CMP) so that Cu remains only in the trenches that define the interconnects.

As mentioned above, the device speed is limited by the RC bottleneck. Therefore, besides reducing the resistance (by switching from Al to Cu), the RC value can further be reduced by employing a low dielectric (low-k) constant material between the interconnects. However, the low-k dielectric tends to be soft, hence difficult to integrate with CMP. A possible solution is to employ electrochemical polishing in place of CMP to remove the excess Cu. This would allow integration of mechanically less stable interlayer dielectrics. Before such a process can be employed, several issues related to electropolishing need to be resolved. In the following sections, the basic mechanisms, parameters, chemicals, and geometries pertaining to electropolishing of copper are presented and discussed.

1. Fundamental Aspects of Electrochemical Polishing

The process of electrochemical polishing (ECP) involves interaction between metal surface and electrolyte, electrochemical reaction and electron transfer, ion / molecule motion, and convection. To better understand ECP mechanisms, it is necessary to know the state of metal surface, the chemistry of solid-liquid interface, and the electrochemical processes.

1.1. Solid Surface

Bulk metal materials are normally polycrystalline. Their crystal periodicity is disrupted at dislocations, grain boundaries and metal surfaces. Atoms at these locations exist in higher energy states than inner atoms. In addition, the packing density variety between different crystal planes apparently affect the physical and chemical properties as well. For example, on copper, (100) planes have the lowest and (111) planes have the highest dissolution rate in the active potential region of an electrochemical process [2].

Copper is the second best electrical conductor among single-element materials. It is chemically stable in air and pure water [3]. In contact with electrolytes, a series of physical, chemical and electrochemical reactions may occur between copper and electrolyte solutions. It can be oxidized in oxygen-containing acid solutions. Complex copper compounds may form in organic solutions under electrochemical conditions.

On the macro scale, a surface profile can be described by mean roughness or Fourier sine series. Pure metals, such as copper and silver, generally exhibit low strength, high plasticity and toughness. A mechanically polished surface of copper bulk material displays characteristics similar to a tough fracture surface (Fig. 1a), which can be described by short wavelength components of the Fourier series. In contrast, the surface profile of a copper film deposited on a substrate has gentler geometrical variation (Fig.1b). If the film is electroplated on a patterned wafer, bumps and pads (0.2 ~ 0.5 μm high) form on the surface (Fig.1c) due to the bottom-up filling mechanism. Such surfaces are described by long wavelength components.

1.2. Aqueous Electrolyte Solution

An electrolyte is a substance in a solution or melt that is at least partly in the form of charged species (ions) and thus conducts electricity. There are solid and liquid electrolyte solutions. The latter may be further classified as aqueous and non-aqueous. In an aqueous electrolyte solution, water (H_2O) is used as the solvent.

a

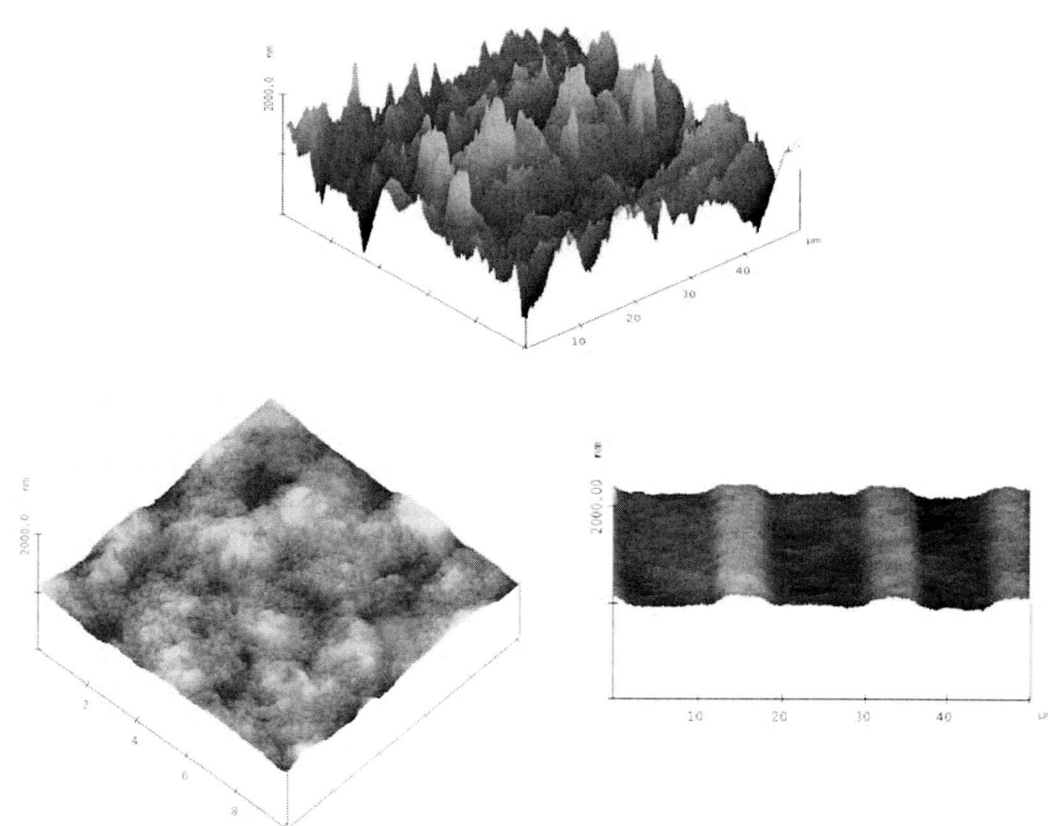

c

Figure .1 Surface topography on micro scale: (a) copper bulk material, (b) copper thin film electroplated on blanket wafer, and (c) on patterned wafer.

Water is the most common solvent. The two O-H bonds in a H_2O molecule are oriented at an angle of 105°. The center of negative charge (-e) of O^{2-} and the center of positive charge (+e) of $2H^+$ are about 0.97 Å apart and thus form dipole [4]. Such a molecular structure may break the chemical bonds of certain substances when it interacts with them. For example, sodium chloride (NaCl) can be easily dissolved in water due to the interaction between H_2O dipole and Na^+-Cl^- bond. The dissolution of the NaCl crystal in water is achieved when the chemical bonding between Na^+ and Cl^- ions is overcome by the electrostatic attraction and thermal vibration energy of Na^+ and Cl^- ions in crystal.

In an electrolytic solution, water molecules exhibit long-range disorder, i.e., although there is some short-range structure, especially near ions, on average the water molecules are randomly orientated. Ions and water molecules move around randomly in the solution. Overall, the solution is electrically neutral (i.e., the net charge is zero) because the total charge of the positive ions is equal to the total charge of the negative ions. If there is a difference in the concentration of ions in different regions of the solution, the resulting concentration gradient produces a flow of ions known as *diffusion*. If there are differences in electrostatic potential, at various points in the electrolytic solution, then the resulting electrical field produces a flow of charge in the direction of the field. This is termed *migration* or *conduction*. The conductivity of an electrolytic solution depends on the electrolytes and their concentration in the solution [5]. Finally, if a difference of pressure, density, or temperature exists in the solution, the liquid will move as a whole or parts of it move relative to other parts. This is *hydrodynamic* flow is termed as convection. These motions usually occur simultaneously in a real electrochemical system.

1.3. Solid-Liquid Interface

In an electrochemical system, electrodes are indispensable components and have to contact the electrolyte so that electrochemical processes can occur. The most important electrochemical processes, charge transfer and chemical reactions, occur at the electrolyte-electrode interfaces.

As stated above, an aqueous electrolytic solution exhibits electroneutrality and random orientation of water dipoles. Thus, on a time average, the ions and water molecules in an aqueous solution experience forces that are independent of direction and independent of position in the electrolyte [6]. On the other hand, in the metal (electrode), the valence electrons (free electrons) are highly mobile. The free electrons and atoms at the solid surface exist in higher energy states. Thus some of the atoms on the metal surface may lose electrons to form ions (M^{n+}). The ions may also recombine with electrons and become atoms. The ionization process may be promoted when the metal is in touch with aqueous solution due to the facts that electrons can freely move in metal and water dipoles and negative ions can solvate metal ions in solution[7, 8]. As more surface metal atoms ionize, more extra electrons remain inside the metal solid. The electrostatic attraction from these electrons holds the metal ions at the metal surface. If a positive potential is applied to the metal, the ionization of surface metal atoms will be enhanced and thus more metal ions are produced at the surface and more electrons inside the metal.

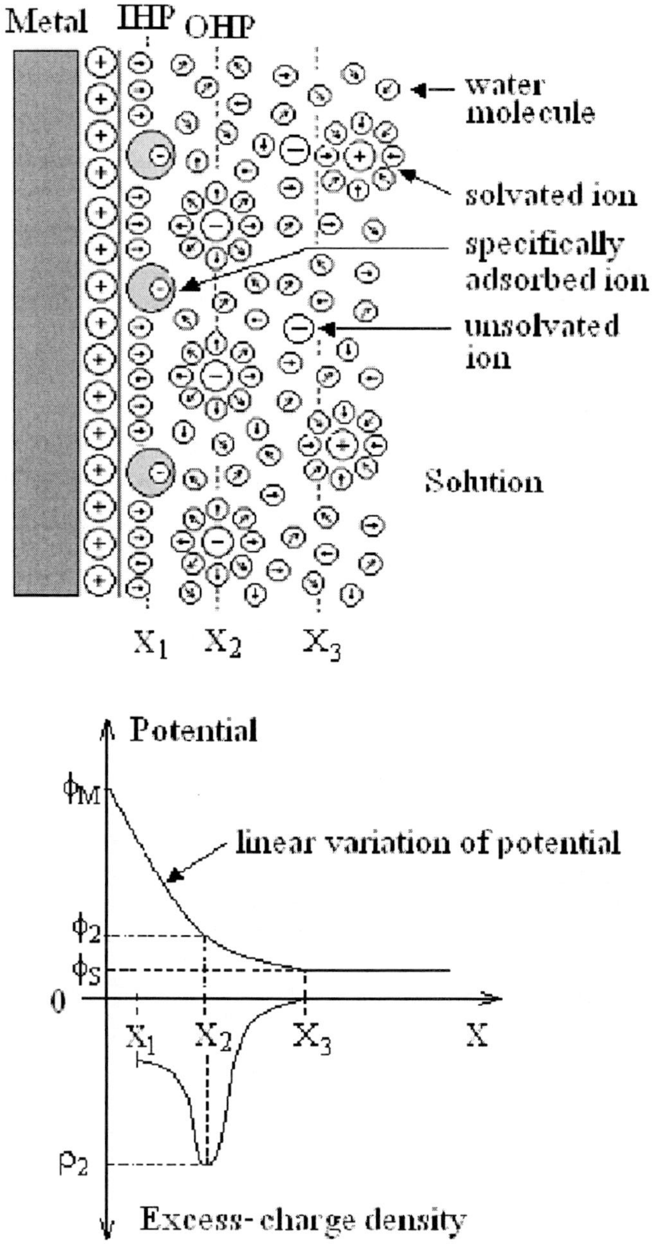

Figure 2. Double layer formed at the metal-solution interface (a) and the potential variation across the double layer (b).

Positive charges at the metal surface attract nearby water dipoles in a preferential direction, which move towards the metal surface. This forms the first layer (*hydration sheath*) – *inner Helmholtz layer*. The center of the layer is called *inner Helmholtz plane* (IHP), as shown in Fig.2a. The solvated negative ions are also attracted by the surface M^n ions to a range of $X_2 \sim X_3$, which is called *diffuse layer* ($\delta_d = X_3 - X_2$). Ions are free to move around, in and out of this layer. The first plane of the layer (X_2) is called *out Helmholtz plane* (OHP) [6, 9].

Some of the ions in the diffuse layer are positive, but the net charge of the layer is negative. The excess-charge density in the diffuse layer decreases with the distance from OHP (ρ_2 at OHP) to the bulk of solution (zero at X_3) where positive and negative ions are equally likely in any region – thermal motion reigns supreme. The total excess-charge in the diffuse layer is equal to that of the adsorbed fixed layer [6, 8]. Likewise, the potential falls off into the solution (Fig.2b).

The adsorbed fixed layer and the diffuse layer together are the so-called *electrical double layer*, which acts as a capacitor with a 'fixed plate' (the adsorbed fixed layer) and a 'moveable plate' (the diffuse layer). Its structure may significantly affect the electrode processes (refer to sections 4 and 5).

1.4. Electrochemical Cell and Electropolishing

1.4.1. Electrochemical Cells

As stated above, the surface metal atoms can be ionized (oxidized) when a metal piece is dipped into water or electrolytic solution. Different metals have different tendencies of oxidation. However, the process can be promoted by applying an external potential to the metal.

Fig.3 shows the connections of an electrochemical cell. An external power supply is connected to two Cu bars immersed in electrolyte solution. When the applied voltage is high enough ($>E^{0'}$), electrons in the anode may be pumped out and the metal atoms on the *anode* (connected to the positive pole of the power supply) surface will be oxidized (e.g., $Cu - 2e \rightarrow Cu^{++}$) and dissolved into the electrolyte solution. Under electrical field, the positive ions (*cations*) move towards cathode and negative ions (*anions*) move towards anode. The cations may get electrons and be reduced to atoms (e.g., $Cu^{++} + 2e \rightarrow Cu$) again at the cathode surface. When the working electrode (WE) is set to be anode, dissolution can be processed. Likewise, when the working electrode is set to be cathode, deposition can be processed. For electropolishing of copper, the copper part to be polished is set to be anode while the cathode can be any conductive material.

The critical potential ($E^{0'}$), at which the oxidation / reduction starts to occur is related to the *standard redox potential* (E^0) of the electrode materials and the mean activity of the oxidation / reduction products. E^0 is a measure (in volts) of the affinity of a substance for electrons (electronegativity) compared with hydrogen (which is set at 0). For pure copper, (Cu / Cu^{2+}: $E^0 = 0.34$ V).

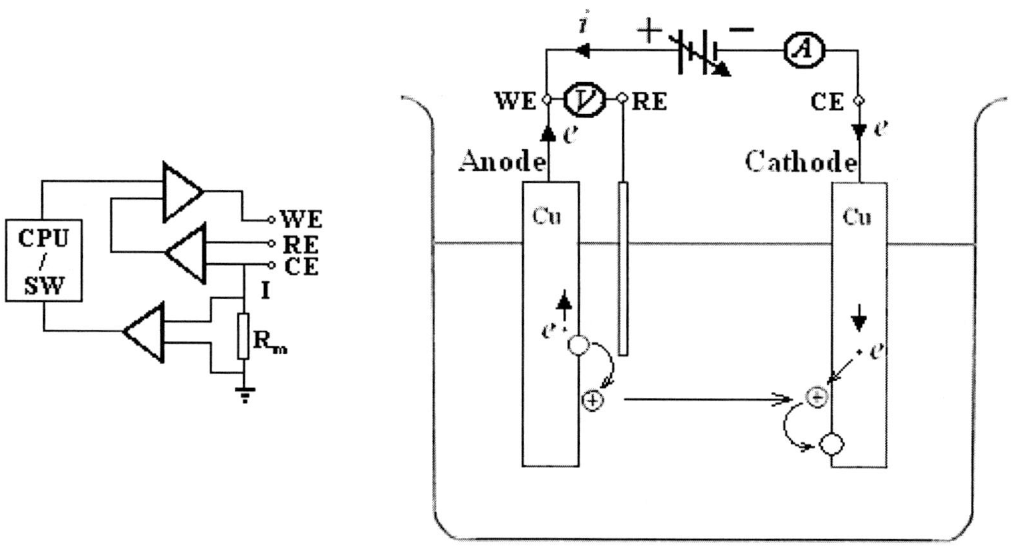

Figure 3. A copper electrochemical deposition / polishing cell.

1.4.2. Kinetics of Electrode Reactions

For an electrochemical process involving oxidized species “ox” (e.g., Cu^{2+}) and reduced species “re” (e.g., Cu^0)

$$\text{ox} + n\,e \rightleftharpoons \text{re} \tag{1}$$

The dynamics of the process in equilibrium may be characterized by the *Nernst equation* [4, 10]

$$E = E^0 + \frac{RT}{nF}\ln\frac{a_{ox}}{a_{re}} \tag{2}$$

where E is the electrode potential, R = 8.317 J mol^{-1} K^{-1} is the gas constant, T is the temperature of the system in kelvins (K), n is the number of electrons transferred in the electrochemical reaction, F = 9.64853 × 10^4 C is Faraday constant, a_{ox} and a_{re} are the activities of ox and re.

Driven by the applied electrical field, the ionic products, together with other existing ions in the solution, move towards electrodes according to the signs of their charges. The ion drift forms a current flow in the solution. Electrons are transferred at the electrode surfaces through electrode reaction, and travel along the conductive wires in the closed circuit of the cell. As a result, an electric current flow is realized under the externally applied voltage.

Under dynamic conditions, i.e., at low current and efficient stirring, in which mass transport is not a factor determining the current, the current-potential relation is given by the Tafel equation [10]

$$\eta = E - E^{0\,\prime} = a + b\log i \tag{3}$$

where η is the over potential, $E^{0'} = E^0 + \frac{RT}{nF}\ln\frac{\gamma_{ox}}{\gamma_{re}}$ is the formal potential at which the interface is at equilibrium with a solution (i.e., oxidation rate equal to reduction rate), *i* is current, *a* and *b* are constants.

Introducing the correlation between concentrations C_{ox} / C_{re} and current (*i*) into the above equation leads to the well known Butler-Volmer formulation of electrode kinetics [9]

$$i = i_0[\frac{C_{ox}(0,t)}{C_{ox}}e^{-\alpha f\eta} - \frac{C_{re}(0,t)}{C_{re}}e^{(1-\alpha)f\eta}] \quad (4)$$

with $i_0 = F\,A\,k^0\,C_{ox}^{(1-\alpha)}\,C_{re}^{\alpha}$ and $f = \frac{F}{RT}$, where C_{ox} and C_{re} are bulk concentrations, $C_{ox}(0, t)$ and $C_{re}(0, t)$ are the concentrations of ox and re at the electrode / solution interfaces at time t, α (=0 ~ 1) is transfer coefficient, A is the area of the electrode, and k^0 is standard rate constant (i.e., the reaction rate constant at E^0 at which the rate of the forward process (v_f, M/s) and backward process (v_b, M/s) are equal: $v_f = k_f C_{re} = v_b = k_b C_{ox}$ and thus $C_{re} = C_{ox}$ and $k_f = k_b = k^0$).

Eq. (4) gives a quantitative relation between current and overpotential, which reflects the dynamics of electrode reaction at low current. A graphic illustration of the equation is shown in Fig.4 with constant vales: $i_0 = 10^{-7}$ A/cm^2; α = 0.3, 0.5; T = 298 K; F = 96500 C/mol, and R = 8.3144 J/(mol K). The condition "low current" means that mass transport is not the limiting process and interfacial electrochemical reaction process is dominant.

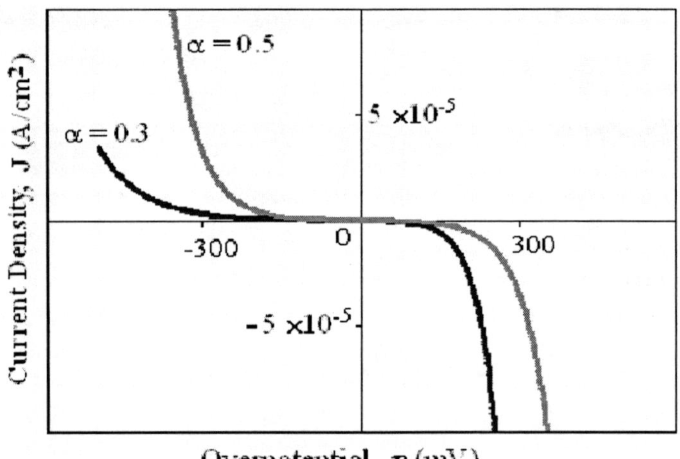

Figure 4. Current-overpotential curves for redox reaction (3) with $i_0 = 10^{-7}$ A/cm^2 at 25 ^{0}C.

1.4.3. Mass Transport in Electrolyte Solution

Mass transport processes are crucial to ECP. With stirring or other type of forced circulation, mass transport occurs by three mechanisms in an electrochemical cell:

(1) Migration, motion of ions driven by electrical field strength $\frac{\partial \phi}{\partial x}$;

(2) Diffusion, motion of species (ions, atoms, molecules) driven by chemical potential due to concentration gradient $\frac{\partial C}{\partial x}$; and

(3) Convection, driven by pressure (P) and/or temperature gradient ($\frac{\partial T}{\partial x}$).

Thus, in general, a flux J_j (mol s^{-1} cm^{-2}) of species j will occur due to the above mass transport mechanisms. The flux may be described by Nernst-Planck equation [9]

$$J_j = -D_j \frac{\partial C_j(x)}{\partial x} - \frac{z_j F}{RT} D_j C_j \partial \frac{\phi(x)}{\partial x} + C_j \, \mathrm{v}(x) \tag{5}$$

where C_j, z_j, and D_j are respectively the concentration, number of valence electrons, and diffusion coefficient of species j; and $\mathrm{v}(x)$ is the velocity of solution convection in x direction.

If j is an ionic species, then the flux J_j is equivalent to a current density. Thus, the current (i, C s^{-1}) for a solution flow through a cross-section area, A, normal to the axis of mass flow, is

$$i_j = i_{d,j} + i_{m,j} + i_{c,j} = J_j \, z_j \, F \, A \tag{6}$$

where $i_{d,j}$ is diffusion current, $i_{m,j}$ is migration current, and $i_{c,j}$ is convection current, which represent the contribution of diffusion, migration and convection respectively.

In bulk solution (away from electrode), concentration gradients, $\frac{\partial C_j(x)}{\partial x}$, are generally small, and the current is carried mainly by migration, i.e.,

$$i_j = i_{m,j} = \frac{z_j^2 F^2 A D_j C_j}{RT} \cdot \frac{\partial \phi}{\partial x} \tag{7}$$

and the total current is

$$i = \sum_j i_j \tag{8}$$

The transference number of species j, defined as the fraction of total current that ionic species j carries is given by [9]

$$t_j = \frac{i_j}{i} = \frac{\kappa_j}{\kappa} = \frac{|z_j| C_j \lambda_j}{\sum_k |z_k| C_k \lambda_k} \quad (9)$$

where κ is the conductivity of the solution, κ_j and λ_j are the ionic conductivity and mobility for ion j.

For an electrochemical cell shown in Fig.5, assume: the electrolyte solution contains 0.3M $CuSO_4$ and 0.001M H_2SO_4; the mobilities of Cu^{2+}, SO_4^{2-}, and H^+ are $\lambda_{Cu}^{++} = 50$, $\lambda_{SO4}^{--} = 80$, and $\lambda_H^+ = 350$ respectively; an applied potential difference (voltage) ΔE produces a total current equivalent to 200 electrons per unit time passing through the cell, i.e., $i = 200$ eq/s; and no convection (i.e., $i_{c,j} = 0$). Then from Eq. (9), $t_{Cu}^{2+} = 0.380$, $t_{SO4}^{--} = 0.611$, and $t_H^+ = 0.009$; and $i_{m,\,Cu}^{2+} = i_{Cu}^{2+} = 76$ eq/s, $i_{m,\,SO4--} = i_{SO4--} = -122$ eq/s, and $i_{m,\,H}^+ = i_H^+ = 2$ eq/s, as shown in Fig. 5. Accordingly, the diffusion current of reactant species, Cu^{++}, may be calculated from Eq. (6): $i_{d,\,Cu}^{2+} = i - i_{m,\,Cu}^{2+} = 124$ eq/s. However, for non reactant species SO_4^{2-} and H^+, to maintain the electroneutrality and the balance of concentration, same amount of ions have to diffuse to the opposite direction, i.e., $i_{d,\,SO4--} = -\, i_{m,\,SO4--} = 122$ eq/s, and $i_{d,\,H}^+ = -\, i_{m,\,H}^+ = i_H^+ = -\,2$ eq/s. See Fig. 5.

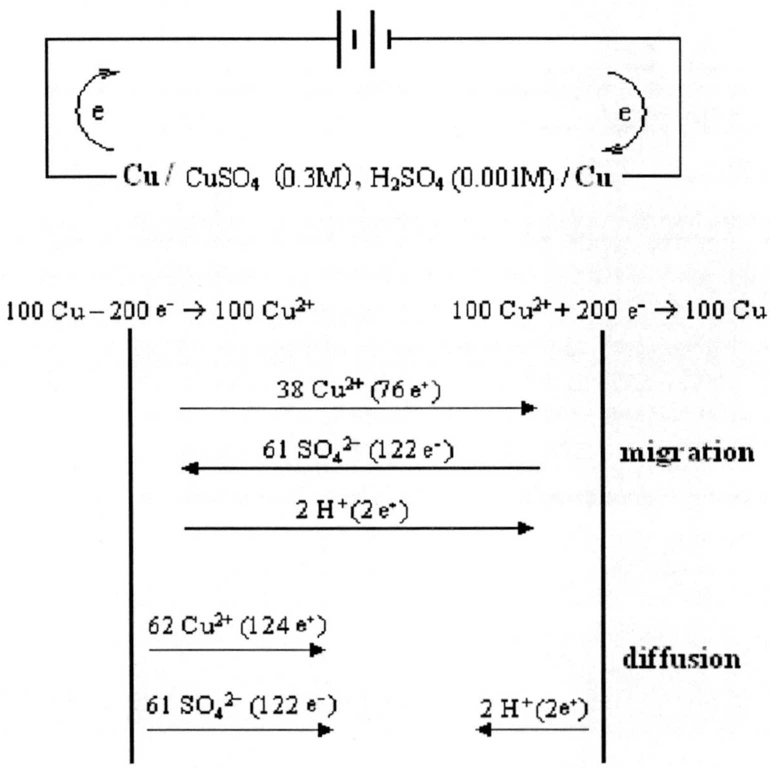

Figure 5. Schematic of ion motions an electrochemical cell.

In the above example, convection is neglected. When a forced convection is introduced by methods such as rotating electrode, stirring, and circulation, mass transfer is usually dominated by convection. In this case, the velocity of solution flow is generally a constant

(v_∞) in bulk solution and decreases to zero while approaching the solid surfaces due to the friction between the fluid and the solid surface. The *hydrodynamic* (or *Prandtl*) *boundary layer* has a thickness [11]

$$\delta = \frac{5x}{\sqrt{\mathrm{Re}_x}} \tag{10}$$

where $\mathrm{Re}_x \equiv \frac{\rho v_\infty x}{v}$ is the Reynolds number (in range of $10^5 \sim 3 \times 10^6$ for flow over flat plate), ρ is the density (kg/m^3) of the solution, x is the characteristic length in the flow direction, and v is the viscosity (kg/s-m) at x.

Inside the boundary layer, convection is limited. The main mass transport mechanism is diffusion. Outside the boundary layer, material flux under forced convection can be described by [4]

$$\mathbf{J}_d + \mathbf{J}_c = - D \text{ grad } C + C\, \mathbf{v} \tag{11}$$

where C is the material concentration, **v** is the velocity of solution flow, $\mathbf{J}_d$ is the diffusion and $\mathbf{J}_c$ the convection material flux. In case of rotating disk electrode (RDE),

$$J = -D\left(\frac{dC}{dy}\right)_{y=0} = 0.62\, D^{2/3}\, v^{-1/6}\, \omega^{1/2}\, (C^o - C^*) \tag{12}$$

Thus, the current (for ionic species),

$$i = n F A J_L = 0.62\, n F A D^{2/3}\, v^{-1/6}\, \omega^{1/2}\, (C^o - C^*) \tag{13}$$

where *i* is current, ω is the angular rotation rate of the electrode (radians/sec) and *v* is the kinematic viscosity of the solution (cm^2/sec), C^o and C^* are the concentration of the mass transport limiting species in bulk solution and at the anode surface respectively.

Under *mass transport controlled* condition, i.e., when the applied potential is so high that the oxidized ions produced every second are more than the diffusion can remove from the anode surface into the bulk solution, the concentration at the anode surface is higher than that in the bulk solution, as illustrated in Fig.6. The concentration variation occurs only in a thin layer (*Nernst layer*, δ_c) within the Prandtl layer (δ). δ_c is given by [9]

$$\delta_c = 1.61 D^{\frac{1}{3}} v^{\frac{1}{6}} \omega^{-\frac{1}{2}} \tag{14}$$

This layer has significant importance to electrochemical polishing and will be further discussed in section 4. The current in the condition of mass transport controlled is termed as *limiting current.*

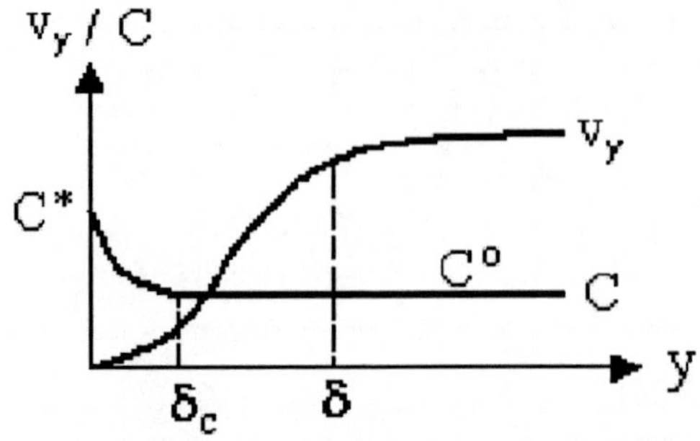

Figure 6. Concentration and velocity profiles from anode surface into bulk solution.

1.4.4. Condition for Electrochemical Polishing

For a copper electrochemical polishing system, if the current flow through the cell is known, one can calculate the rate of copper anodic dissolution (Assuming that Cu – 2e → Cu^{++} occurs at Cu anode surface)

$$R_d = \frac{M}{2\,FdA} i \tag{15}$$

where R_d is the dissolution rate in thickness (cm/s), M and d are respectively the molecular weight and density of the anode (Cu), A is the area of the electrode surface.

From Eq. (3) we can see that current increases with increasing overpotential ($\eta = E - E^{0\prime}$). Therefore, for a certain applied potential E, higher standard redox potential E^0 and so higher $E^{0\prime}$ will result in lower current. According to Eq. (15), this will lead to lower dissolution rate.

As stated in section 1.1, on a polycrystalline metal surface there are sites, such as defects and grain boundaries, where atoms are at higher energy states. In addition, due to arbitrary crystal orientation, there are different crystalline planes with different energy states of atoms on the electrode surface. Thus, atoms at all these different sites and planes have different standard redox potential E^0, and therefore have different dissolution rate according to Eq. (15). Such an anodic dissolution will not lead to polishing. Instead, a *crystallographic etching* is produced [2, 12-14]. This is true at lower current (lower applied potential).

However, at higher potential, the current is a constant (Fig. 7) and the anodic dissolution process is controlled by the mass transport in the anodic layer(s). Thus the crystallographic etching is suppressed and an ECP effect can be achieved. Further discussion will be presented in the following sections.

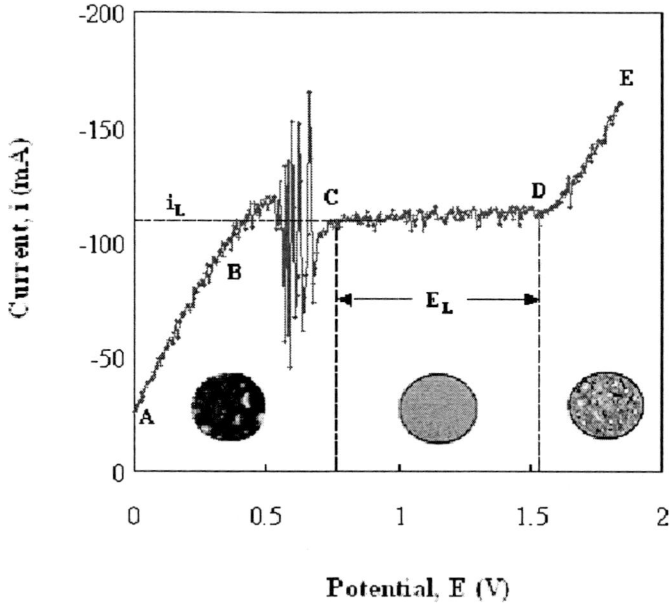

Figure 7. Anodic polarization curve and surface topography after anodic dissolution at different potential range.

With continuing increase of applied potential, other reactions than Cu oxidation / reduction may occur and contribute to the increasing current. These reactions produce H_2 and O_2 bubbles, which occur at or reach the anode and produce a rough surface. A whole range of *i* ~ E curve (*polarization curve*) and surface characteristics after anodic dissolution are shown in Fig. 7, and will be further discussed in next section.

2. Copper Anodic Behavior and Polarization Curves

As stated above, electropolishing can be achieved only in the range of the limiting current plateau on a polarization curve. Therefore, a practical search for electrolytes and favorable conditions of electropolishing is best started with the determination of current-potential curves under controlled mass transport conditions, e.g., by using a rotating disk electrode (RDE) [2]. This section will discuss and present the anodic behavior and polarization curves of copper rotating disks in various aqueous electrolyte solutions.

2.1. Copper Anodic Behaviors and Polarization Curve in General

When a voltage is applied to an electrochemical system (anode-electrolyte solution-cathode-power supply / control), a current ~ anode potential curve (*anodic polarization curve*) may be recorded. A typical anodic polarization curve (Fig. 7) consists of linear region AB, limiting current plateau CD, transition region BC, and another linear region DE.

At low potential (in region AB), the electrochemical process is kinetics dominated. The current is linear in the electrode potential, obeys the Butler-Volmer equation Eq. (4), and may be approximated by [4]

$$i = \frac{\eta}{R_p} \tag{16}$$

with $$R_p = \frac{RT}{nFi_0} \tag{17}$$

where i_o is the equilibrium current (i.e., when the cathodic current and anodic current have equal amplitude and therefore the net current is zero), R_p is the *polarization resistance*.

In this region, the reactions

$$2H^+ (aq) + 2e^- \rightarrow H_2 (g); E^o = 0.00 \text{ V} \tag{18}$$

may occur at the cathode and

$$Cu + 2OH^- \rightarrow CuO (c) + H_2O + 2e^-; E^o = 0.29 \text{ V} \tag{19}$$

at the anode [16]. Due to the low redox potential (0V), reaction (18) occurs and so hydrogen bubbles can be observed on the copper cathode all the time. Black CuO films can also be observed in this region for certain solutions [17].

In BC region, current – potential deviates from linear relationship. As the potential continues to increase, current increase slows down until it reaches a maximum and then decreases to a minimum point, and finally increases to the limiting current plateau. This is a transition from kinetics (of charge transfer / chemical reaction) domain to mass transport domain. When a black solid film formed in AB region, the film was observed to break down and strip off in this region. For most phosphoric acid containing solutions, a current – potential oscillation was observed (See Fig. 7). The oscillations are very likely related to the formation and breakdown of the black film. Experiments indicated that the oscillations depend on potential and the amount of phosphoric acid and water [17]. More details of the oscillation phenomena in copper - phosphoric acid system have been discussed in the literature [18-19].

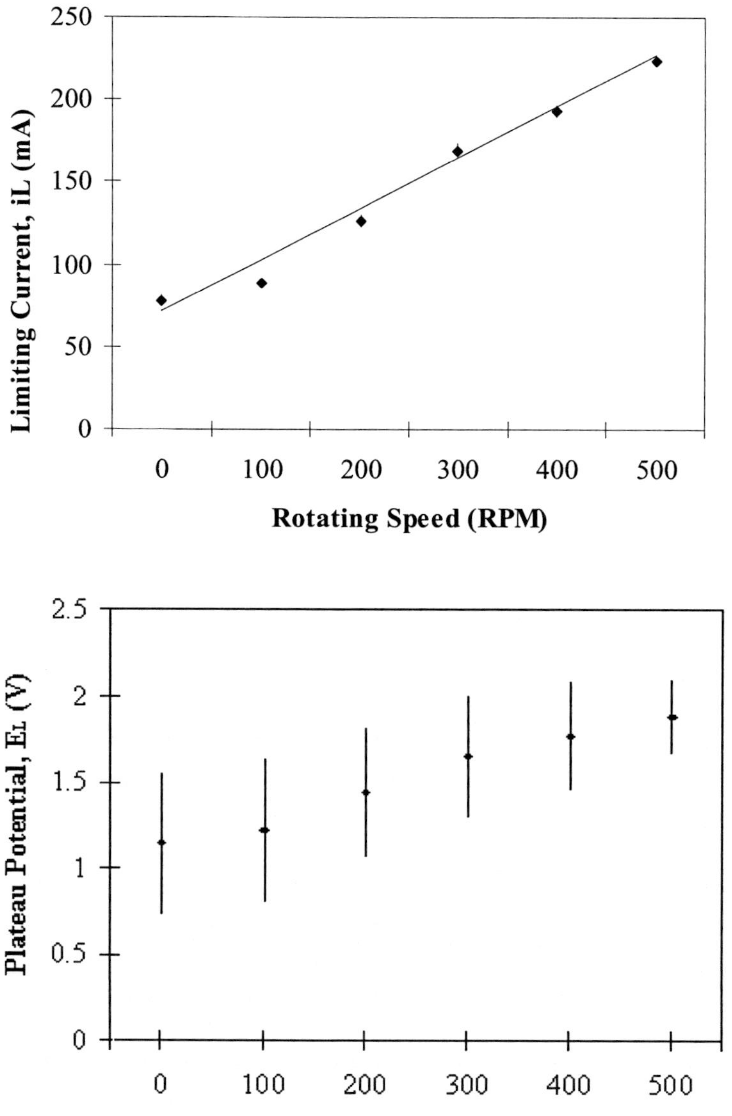

Figure 8. Influence of rotating speed on limiting current i_L (a) and the potential range E_L of limiting current plateau (b). Solution: 54% H_3PO_4 + 1.2M CuO + 46% H_2O.

CD region is the *limiting current plateau*. In this region, the process is mass transport controlled. The limiting current (i_L) depends on the diffusion/migration of Cu^{++} ions from anode surface to bulk solution across anodic layers and obeys *Levich equation* [4, 21]:

$$i_L = 0.62 \text{ n F A } D^{2/3} \nu^{-1/6} \omega^{1/2} C^* \quad (20)$$

$$= 0.62 \text{ n F A D} \frac{C^*}{\delta_c}$$

which describes the influences of concentration gradient and diffusion coefficient of Cu^{++}, viscosity of solution, and disk rotating speed on limiting current.

Fig. 8 shows the influence of disk rotating speed (RPM) on i_L and potential range of limiting current plateau (E_L). Increasing rotating speed results in higher limiting current and narrower E_L due to the decreasing thickness of the boundary layer. When the speed of the rotating disk is so high that the anodic layers vanish, the limiting current plateau will no longer exist. As shown in Fig. 9, well-defined limiting current plateaus were obtained from phosphoric acid solution at disk rotating speed lower than 200 rpm. As the speed of rotating disk increases, the limiting current plateaus shift to higher potentials and tilt away from the horizontal orientation. This indicates that the diffusion layers are critical for holding the process in mass transport controlled regime.

Study indicated that inter-electrode distance (l) did not have an obvious influence to polarization curves over the range of 20 cm (Fig. 10).

As stated in earlier, electrochemical polishing can be achieved only under mass transport limited condition in the potential range of limiting current plateau. However, the presence of a current plateau with an anodic film is not a sufficient criterion for microsmoothing [2]. For example, transpassive nickel dissolution in sulfuric acid yields well-defined current plateaus but only in a certain concentration and temperature range are mass transport controlled [2, 22].

Levich plot can be used to validate the mass transport condition. According to Eq. (20), limiting current (i_L) is directly proportional to the square root of RDE angular rotating speed ($\omega^{1/2}$). So, Levich plot ($i_L \sim \omega^{1/2}$) or Levich-Koutecky plot ($1/i_L$ vs $1/\omega^{1/2}$) is a straight line if the process is mass transport controlled. Fig. 11 shows Levich-Koutecky plots obtained from Cu – phosphoric acid and Cu - Hydroxyethylidenediphosphonic acid (HEDP) systems. Both of the plots fall on a straight line, which indicate that they are mass transport controlled in the potential range of limiting current plateaus. In addition, the plot obtained from phosphoric acid solution passes through the origin. The plot obtained from HEDP solution has a nonzero intercept with $1/i_L$ axis, which indicates the existence of additional series resistance to mass transport [23]. This was further proved by electrochemical impedance spectroscopy (refer to section 3).

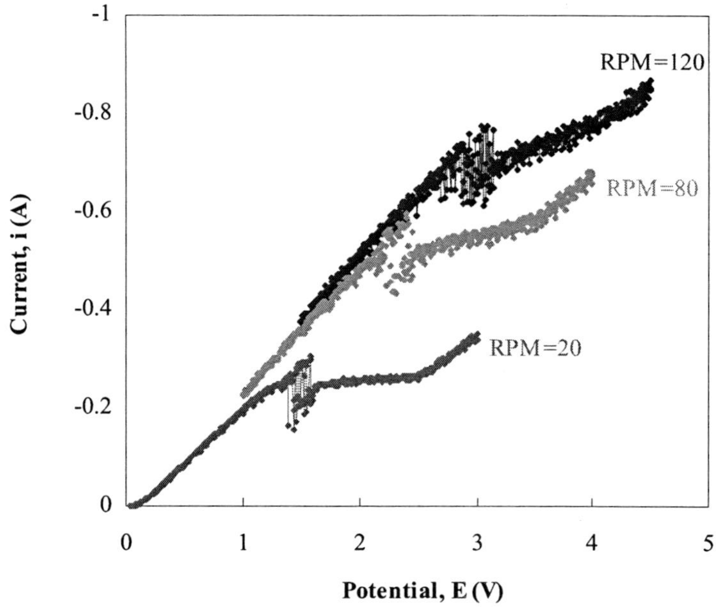

Figure 9. Copper anodic polarization curves obtained from solution: 37% H_3PO_4 + 63% H_2O, with rotating speed 200, 800, and 1200 rpm.

In DE range of the polarization curve, current increases again almost linearly with increasing potential. This is most likely due to the occurrence of reactions such as [24]

$$4OH^- - 4e^- \rightarrow O_2 + 2H_2O \qquad (21)$$

in addition to faster copper dissolution. All the reactions contribute to charge transfer and therefore the increasing current.

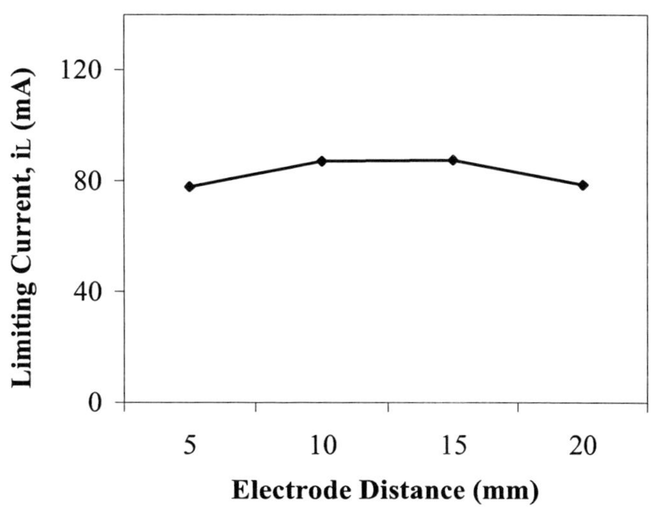

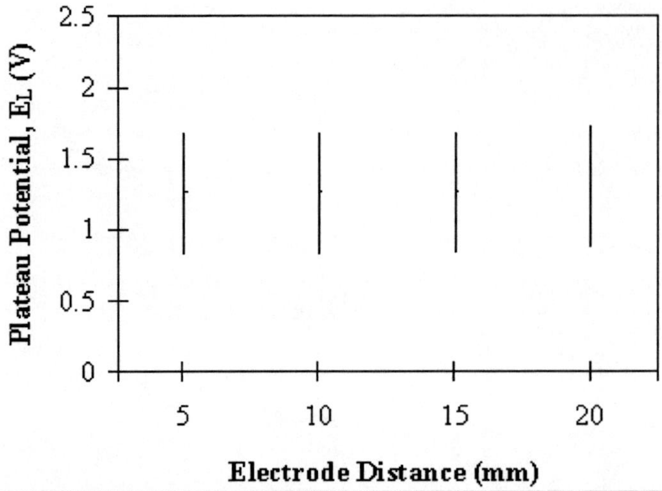

Figure 10. Influence of inter electrode distance on limiting current i_L (a) and the potential range E_L of limiting current plateau (b). Solution: 54% H_3PO_4 + 1.2M CuO + 46% H_2O.

These gas (oxygen or hydrogen) bubbles on anode surface may block Cu^{++} ion transport and therefore terminate the electrochemical dissolution process in the area under the bubbles, as illustrated in Fig. 12. However, the residual solution on the surface area inside the bubbles may react with Cu atom and result in chemical etching. Depending on the chemical property of the electrolyte solution and the value of current density at which the electrochemical dissolution occurs, the etching speed can be higher than the rate of electrochemical dissolution. In this case, pits will be produced on the anode surface, as shown in Fig. 13. If etching does not occur inside the bubbles, or if its speed is slower than that of electrochemical dissolution process, the area inside the bubbles will remain and appears as protruding particles after the electrochemical dissolution process. In either case, a rough surface is produced.

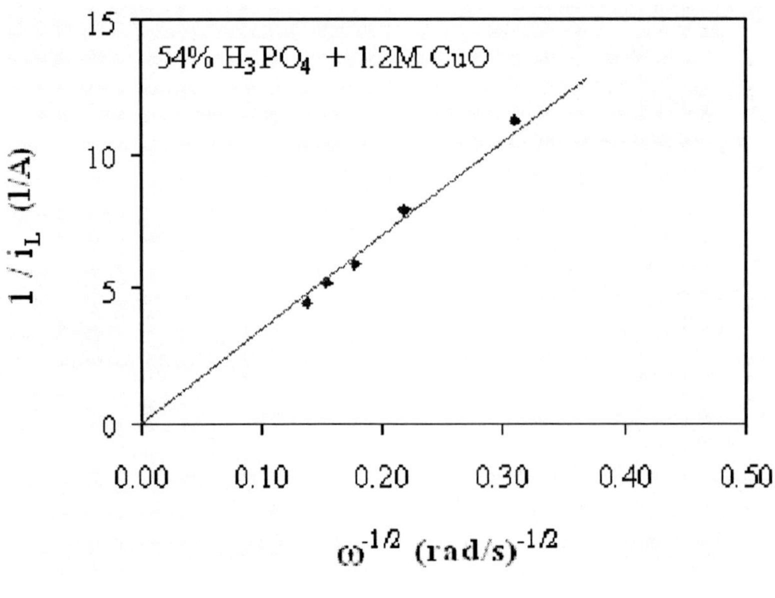

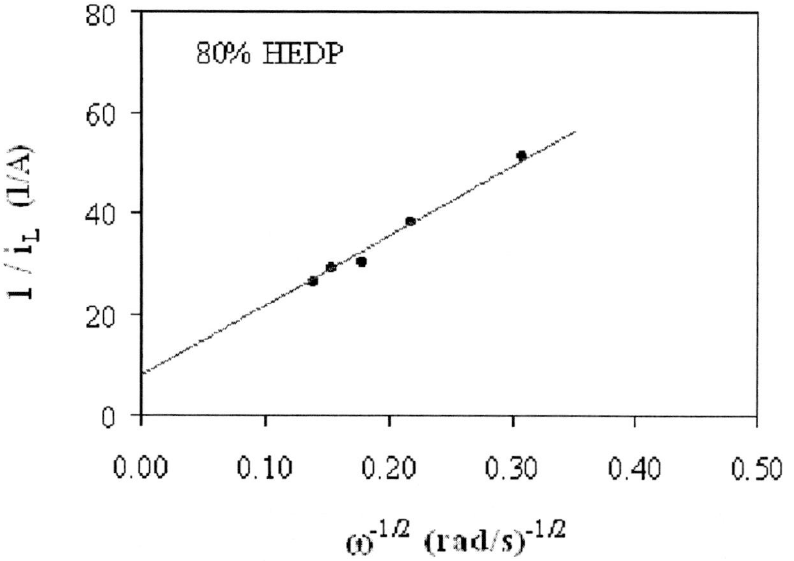

Figure 11. Levich-Koutecky plot of solutions: (a) 54% H_3PO_4 + 1.2M CuO, (b) 80% HEDP (60% $C_2H_8O_7P_2$)

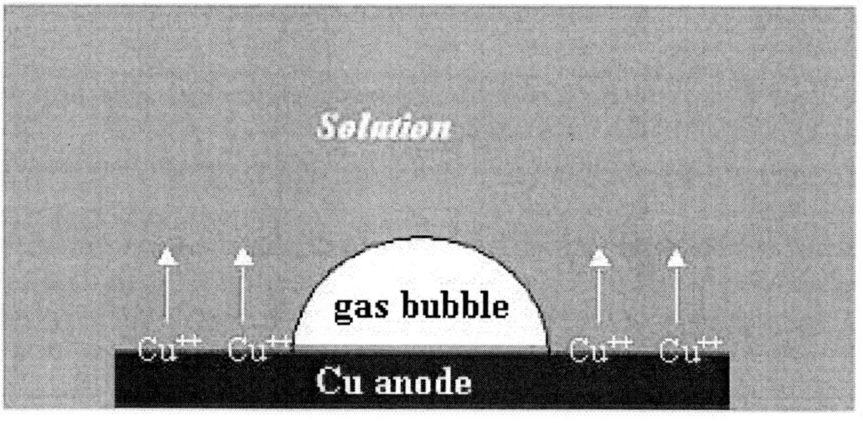

Figure 12. Gas bubble formed on anode surface terminates electrochemical dissolution process and produces rough surface.

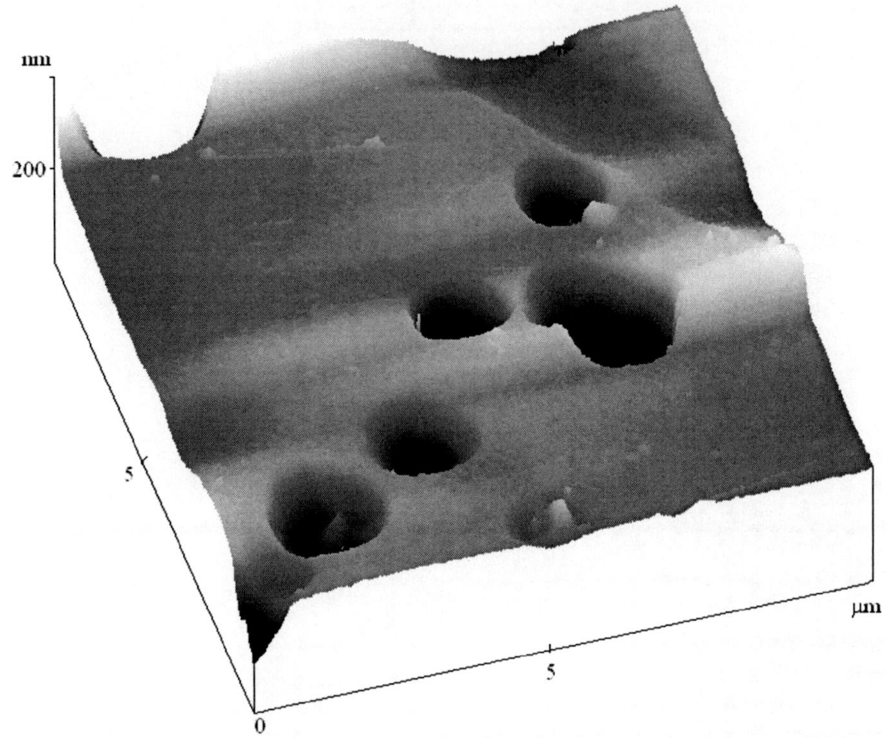

Figure 13. AFM topography image of the copper rotating disk after 20 seconds of electropolishing at 2V in 70 % phosphoric acid + 2.0 M CuO.

2.2. Polarization Curve Measurement and Some Results

The apparatus used for anodic polarization curve measurements is usually comprised of a rotating disk electrode as anode (working electrode, W), a counter electrode (C), a reference electrode (R), and a potentiaostat, as shown in Fig. 14. The method used for the measurements is Linear Sweep Voltammetry (LSV). The authors measured copper anodic polarization curves in various electrolytes (listed in Table 1) with the following conditions: scan range of potential, 0 V ~ 4 V; scan rate, 5 mV/s; disk rotating speed, 100 rpm, inter-electrode distance, 15 mm; and room temperature.

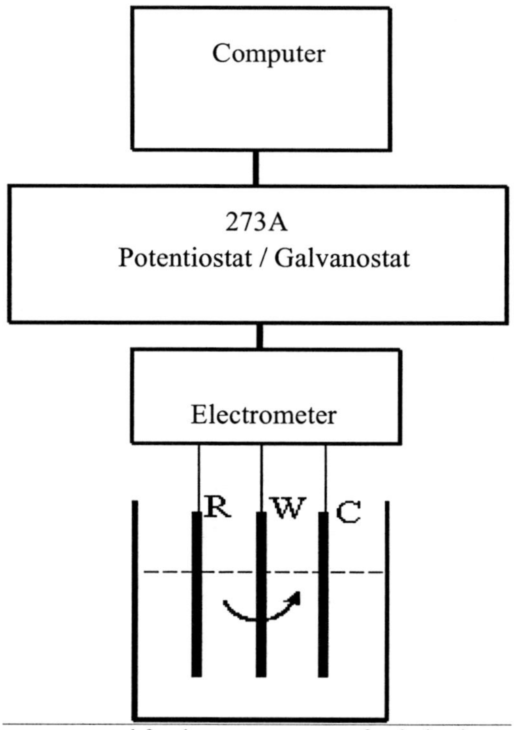

Figure 14. A schematic of the system used for the measurement of polarization curves. R: reference electrode. W: working electrode. C: counter electrode.

Details of the measurements and data are available in [25]. A summary of the i_L and E_L data obtained from studied solutions and the influence of the electrolytes on i_L are listed in Table 1.

Overall, the values of i_L and E_L of the limiting current plateau and the shape of a polarization curve depends on electrolyte solution, anode material, disk rotating speed, solution circulation, temperature, and the inter-electrode distance. Of all the factors, electrolyte is the most important. Higher limiting current (i_L) produces higher dissolution rate (R_d). Larger range (E_L) of limiting current plateau gives wider operation window for electrochemical polishing process. Therefore, a well-defined limiting current plateau (level and large E_L), and mass transport condition are desirable for electrochemical polishing.

3. Study of Copper Anodic Layers and Mass Transport Mechanisms with Electrochemical Impedance Spectroscopy

From the above discussion, we can see that anodic layers are responsible for mass transport control. Continuing anodic dissolution proceeds through either the dissolved metal ions diffusing and/or migrating through the anodic layer(s) into bulk solution, or acceptor species (anions or water) in solution get into the anodic layer(s) to form a complex with the dissolved metal ions [2, 19, 20, 27-29]. The anodic layer is usually the Nernst layer. In some system, such as Cu – HCl [26], Fe – $FeCl_2$ [30], and Ni – HCl [31], and, a salt film can form on the anode surface due to the precipitation of the dissolved metal ions when its concentration exceeds a solubility limit. Apparently, the status and property of an anodic layer has significant impact to ECP. For example, a salt film was found beneficial to the electrochemical planarization of Cu/Si wafers and for controlling the growth rate of corrosion pits and cracks in stress corrosion [30, 32, 33].

Table 1. Summary of i_L and E_L data and the trend of i_L affected by increasing the electrolytes

Solutions	i_L (mA)	Typical E_L (V)	**Effect of conc. on i_L**
Phosphoric acid (85% H_3PO_4) 13 ~ 100 %	31 ~ 284	0.7 ~ 1.7	H_3PO_4 ↓
phosphoric acid - copper oxide (85% H_3PO_4) (CuO) 24 ~ 100 % 0.5 ~ 2 M	9 ~ 206	0.8 ~ 1.7	H_3PO_4 ↓ CuO ↓
phosphoric acid - ethylene glycol (85% H_3PO_4) ($C_2H_6O_2$) 13 ~ 75 % 13 ~ 75 %	10 ~ 77	1.1 ~ 1.8	H_3PO_4 ↑ $C_2H_6O_2$ ↓
phosphoric acid - sodium tripolyphosphate (85% H_3PO_4) ($Na_5P_3O_{10}$) 10 ~ 90 % 0.1 ~ 0.5 M	16 ~ 190	0.6 ~ 1.6	H_3PO_4 ↓ $Na_5P_3O_{10}$ ↓
HEDP (60% $C_2H_8O_7P_2$) 2.5 ~ 100 %	6 ~ 81	0.5 ~ 1.4	$C_2H_8O_7P_2$ ↑↓
HEDP - hosphoric acid (60% $C_2H_8O_7P_2$) (85% H_3PO_4) 0 ~ 90 % 10 ~ 30 %	8 ~ 217	0.5 ~ 1.4	$C_2H_8O_7P_2$ ↓ 85% H_3PO_4 ↓
ethylene glycol - sodium chloride ($C_2H_6O_2$) (NaCl) 10 ~ 50 % 1 ~ 4 M	20 ~ 131	2 ~ 4	$C_2H_6O_2$ ↓ NaCl ↑
sulphuric acid - sodium nitrate (H_2SO_4) ($NaNO_3$) 1 ~ 20 % 0.5 ~ 2.5 M	63 ~ 175	2 ~ 4	H_2SO_4 ↓
sulphuric acid - potassium nitrate (H_2SO_4) (KNO_3) 1 ~ 20 % 0.5 ~ 2 M	69 ~ 422	1.6 ~ 3	H_2SO_4 ↓

Note: "↓" indicates that i_L decreases with increasing the electrolyte concentration and "↑" indicates that i_L increases with increasing the electrolyte concentration.

However, probing such anodic layers is very challenging since they may break down when the potential is switched off and the anode is taken out of the electrolyte. Several techniques have been tried to study anodic layers, such as linear sweep voltammetry (LSV) [34, 35], ellipsometry [36], holographic interferometry [37], XPS [38], and AC impedance spectroscopy. Ellipsometry and holographic interferometry can be used to measure the thickness of an anodic layer. Electrochemical impedance spectroscopy (EIS) can be used to measure the resistance of an anodic film and identify the existence of a salt film [14, 29, 30, 32, 39].

EIS is a method in which a small ac signal is applied to an electrode and the resulting current is measured. The measuring instrument processes the current vs. time and voltage vs.

time measurements to provide the impedance at different frequencies, i.e., the impedance spectrum [40]. The electrode studied can be a corroding metal (for corrosion studies) or an anode in an electrochemical cell (for characterizing the anodic layers and mass transport limiting species).

Impedance is the term used to describe the ac equivalent of dc resistance. With dc methods (such as LSV), the total resistance (polarization resistance R_p) of an electrochemical system can be measured. While an ac signal is applied, an electrochemical cell or even a simple corrosion system becomes more complicated. One has to consider polarization resistance R_p, solution resistance R_s, and double layer capacitance C_{dl}. EIS can be used to identify and quantify these components and thus characterize the anodic layers and electrochemical processes.

3.1. Equivalent Circuit of an Electrochemical Cell

To obtain the electrical properties of different components in an electrochemical cell, we have to first analyze the responses of the anode-anodic layer-electrolyte solution structure to dc and ac signals, and then find the equivalent circuit of the structure.

3.1.1. Electrode Response to ac Signal

When an ac potential is applied to an electrode (Fig. 15), the ac power supply pushes and then pulls electrons in and out of the load, at the frequency of the applied potential. There is no net current flow. During the positive half-cycle, conventional current flows into the electrode (i.e., electrons are drawn out). The electrode is then less negatively charged (more positive) than before. So it attracts the diffuse mobile layer more strongly. In result, the mobile layer moves closer. The conventional current that has flowed into the electrode has charged up the double layer capacitor.

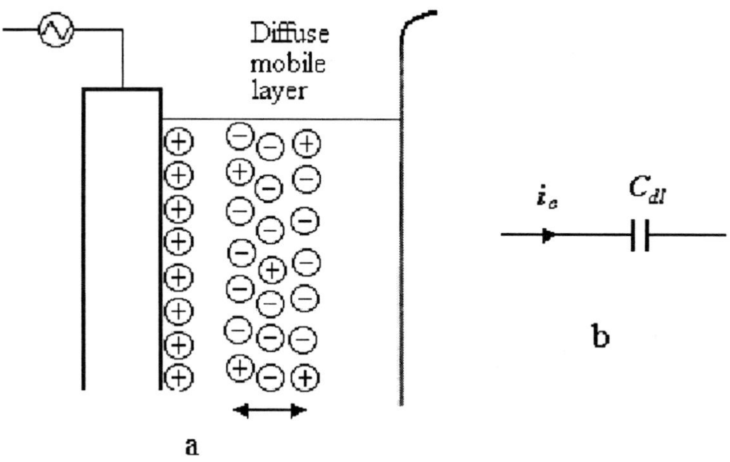

Figure 15. Illustration of the diffuse mobile layer oscillation (a) and the equivalent circuit (b).

During the negative half-cycle, conventional current flows back out of the electrode (i.e., electrons are pushed in). The electrode is now more negatively charged than before, and so repels mobile layer away. This discharges the double layer capacitor. Therefore, an alternating potential applied to the electrode makes the diffuse mobile layer oscillate at the ac frequency. And the double layer behaves like a capacitor (C_{dl}), as illustrated in Fig. 15.

3.1.2. Electrode Response to dc Signal

A direct current (dc) power supply pushes the electrons in a steady direction. As shown in Fig. 16a, a steady current flows into the anode (i.e., electrons withdrawal). This makes the electrode more positive and drives ionization of surface atoms (i.e., the anode dissolves). Meanwhile, the electrons from this ionization flow towards the dc power supply. During this process, the negative mobile layer stays at a constant distance from the electrode. This distance depends on what dc potential. The more positive the electrode, the closer is the mobile layer.

Unlike ac voltage, dc voltage makes a net current flow through the double layer, from one 'plate' to the other. The double layer capacitor 'leaks'. Therefore, with dc the double layer behaves like a resistor (R_{dl}), as shown in Fig. 16b.

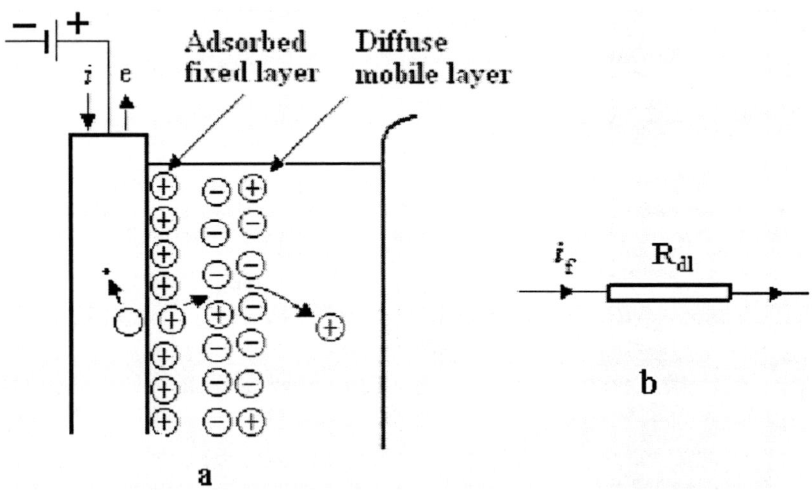

Figure 16. Current passing through the double layer under dc voltage.

3.1.3. Equivalent Circuit of an Electrochemical Cell with ac + dc Signal

In situ study of anodic layers formed in electrochemical polishing process, requires application of a dc potential and an ac signal (Fig. 17a) to the anode. In this case, the double layer is like a leaky capacitor, represented by a resistor in parallel with a capacitor (Fig. 17b). The ac component makes the diffuse mobile layer oscillate. The dc component causes chemical reaction and faradic current flow. It also controls the mid position of the oscillating mobile layer.

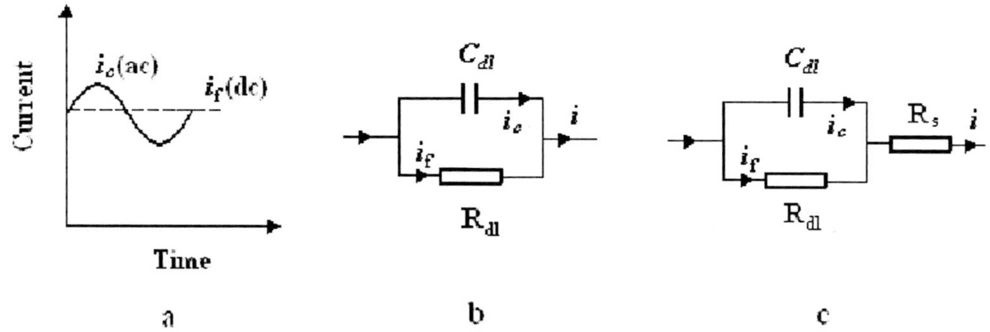

Figure 17. ac + dc signal and equivalent circuit of the double layer.

Taking into account the resistance of electrolyte solution, R_S, the equivalent circuit is therefore R_s in series with the parallel circuit R_{dl} / C_{dl} (Fig. 17c). This is valid when the applied anode potential is lower than the potential of limiting current plateau. That is, the mass transport condition is not reached yet.

3.1.4. Equivalent Circuit under Mass Transport Control

In the condition of mass transport control, a concentration gradient of the mass transport liming species is developed inside the boundary layer. The mass transport limiting species can be the anode dissolution product (ion M^{n+}) or acceptor (molecule A, or anion A^{m-}). In some cases, a salt film can form on the anode surface. Fig. 18 illustrates these anodic layer structures and the possible concentration gradient.

The concentrated layer (of mass transport liming species), or *diffusion layer*, and salt film work as diffusion and migration barriers, which limit the diffusion and migration of the dissolved metal cations (M^{n+}) from the anode surface to the bulk solution.

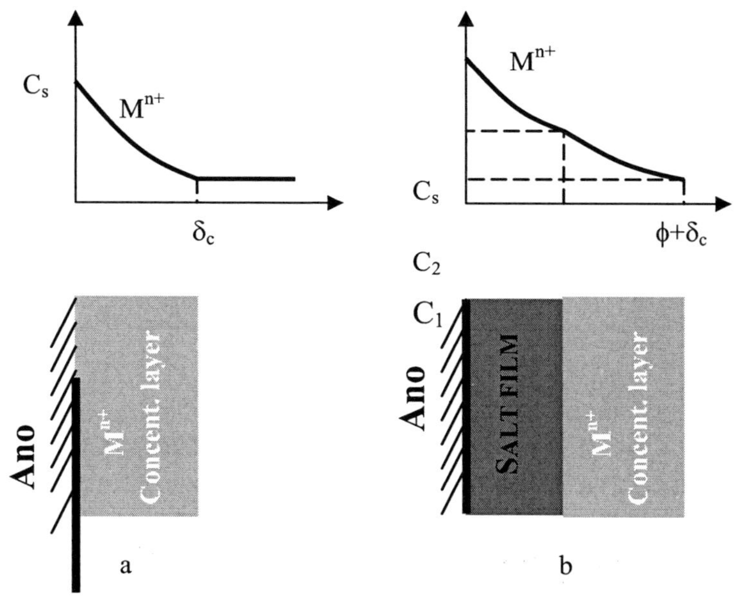

Figure 18. Concentration gradient profile in boundary layers: (a) a single M^{n+} concentrated layer, (b) a salt film and a M^{n+} concentrated layer.

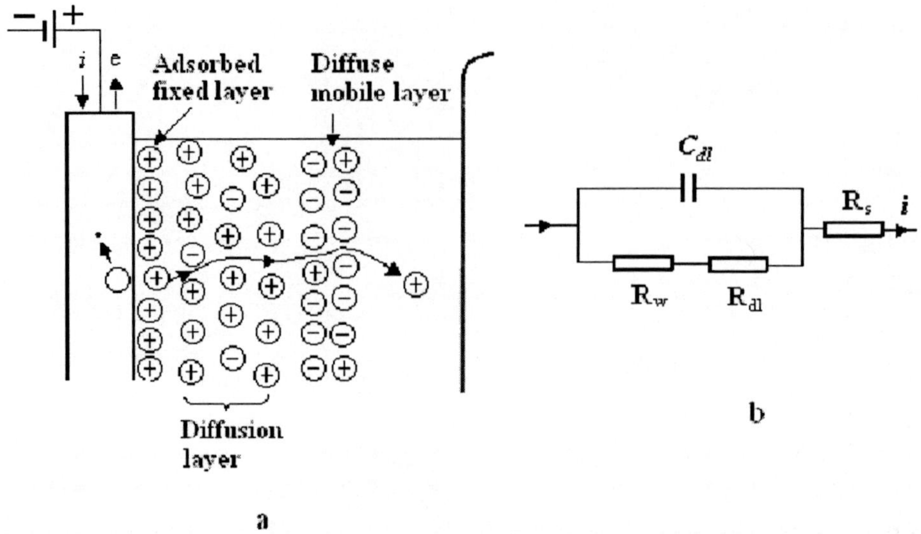

Figure 19. Desorbed ions passing through diffusion layer and diffuse mobile layer (a) and the equivalent circuit (b).

Therefore, under mass transport control, the electrode / solution interface consists of the fixed layer, diffusion layer, and the diffuse mobile layer, as illustrated in Fig. 19a. In case that M^{n+} is the mass transport control species, A M^{n+} ion in the fixed layer has to overcome two kinds of resistances to move into bulk solution: (1) the resistance from the double layer (fixed layer and diffuse mobile layer), R_{dl}; (2) the resistance from the diffusion layer, R_w, which is termed Warburg impedance [4]. The equivalent circuit in this case is shown in Fig. 19b.

In case a salt film forms on the anode surface (Fig. 20a), a desorbed ion has to overcome the resistance from the salt film, R_f, before it can pass through the double layer. The equivalent circuit for this case is shown in Fig. 20b.

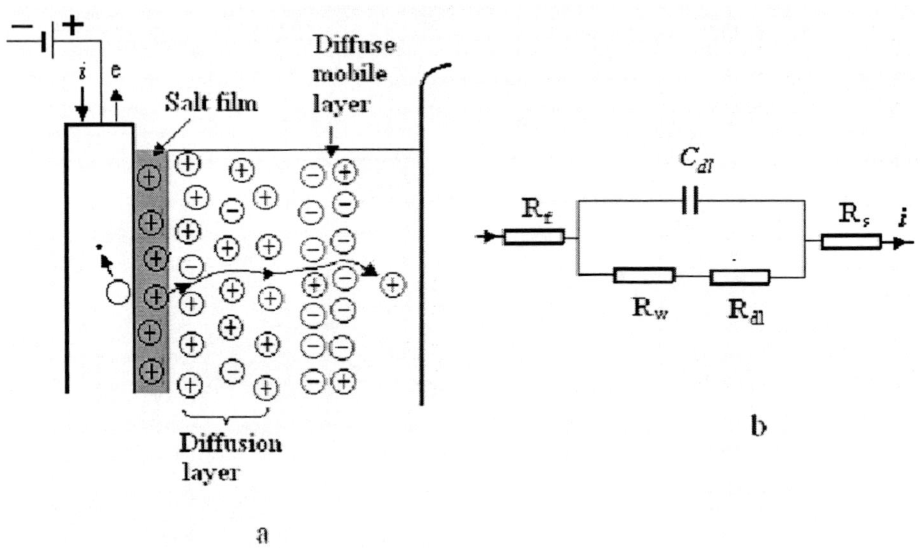

Figure 20. Desorbed ions pass through salt film, diffusion layer, and diffuse mobile layer (a) and the equivalent circuit (b).

3.2. Principle of EIS

The above equivalent circuit (Figs. 19b & 20b) may be simplified as Fig. 21, assuming

$$R_1 = R_w + R_{dl} \tag{22}$$
$$R_2 = R_f + R_s + R_m \tag{23}$$

where R_m is the resistance of the connections between electrodes and power / control units.

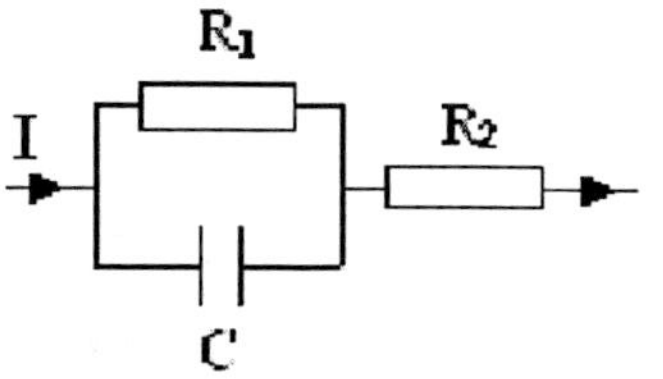

Figure 21. Simplified equivalent circuit of an electrochemical cell.

To the circuit in Fig. 21, the total impedance (Z) may be expressed as

$$Z = \frac{R_1}{1+\omega^2 C^2 R_1^{\ 2}} + R_2 - j\frac{\omega C R_1^{\ 2}}{1+\omega^2 C^2 R_1^{\ 2}} = Z' + j\,Z'' \tag{24}$$

where Z' and Z" are

$$\mathbf{Z'} = \frac{R_1}{1+\omega^2 C^2 R_1^{\ 2}} + R_2 \tag{25}$$

$$Z'' = \frac{\omega C R_1^{\ 2}}{1+\omega^2 C^2 R_1^{\ 2}} \tag{26}$$

From Eqs. (25) and (26) one may get *Nyquist plot* (*impedance spectrum*) as shown in Fig. 22. From the Nyquist plot, we may obtain R_1, R_2, and C [4, 29, 40]:

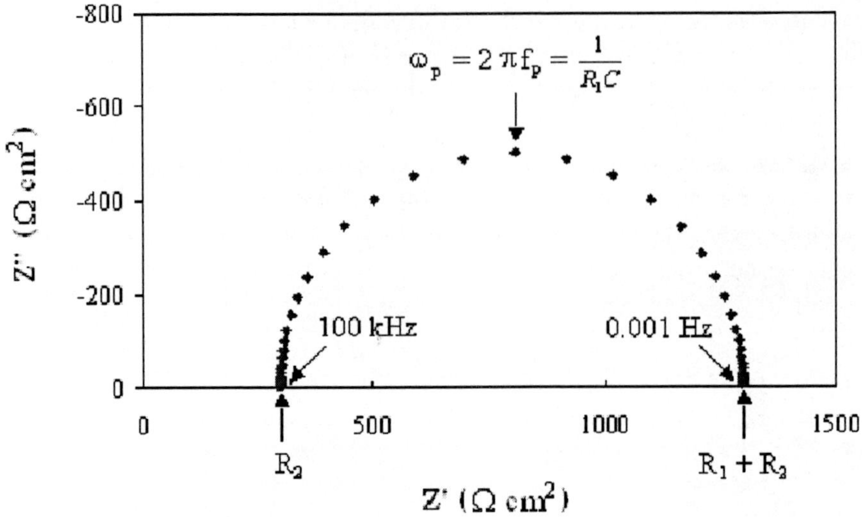

Figure 22. Nyquist plot for the circuit in Fig. 21 with $R_1 = 1000\ \Omega$, $R_2 = 300\ \Omega$, $C = 0.001$ F.

$$R_2 = Z'_{min} \tag{27}$$

$$R_1 + R_2 = Z'_{max} \tag{28}$$

$$\omega_p = 2\pi f_p = \frac{1}{R_1 C} \tag{29}$$

where ωp and fp are the angular and linear frequencies corresponding to Z"max (See Fig. 22).

3.3. EIS Measurement

EIS can be measured using an ac generator / analyzer, e.g., a frequency response analyzer (FRA) or a lock-in amplifier, which provide an ac voltage signal and analyze current and voltage outputs from a potentiostat. Two or three electrodes can be used for the measurements. According to the above discussion, both working electrode (W) and counter electrode (C) should have an equivalent circuit similar to the one shown in Fig. 21.

To avoid noise, coax cables should be used. The cables should be as short as possible. Better data can be obtained with two-electrode setup since many high frequency distortions can be associated with high impedance and Luggin-reference electrode combination [40]. However, using three-electrode setup, ac signal and output current measured are between working and reference electrodes. Therefore, the electrochemical processes occur on the counter electrode can be excluded and thus simplify impedance spectra. In either case, the shape and size of the counter electrode should be such that a uniform current distribution is provided on the working electrode.

In order to obtain a whole semicircle of the Nyquist plot, i.e., for Z" to approach 0 at high and low frequency ends, the frequency should be in range of [40]

$$\frac{100}{2\pi RC} > f > \frac{0.01}{2\pi RC} \tag{30}$$

In addition to frequency range, number of data points also needs to be determined for the measurements. More data points may have higher accuracy of data analysis. But the total time for the data collection increases. In general, 7 ~ 10 points per decade is a good compromise between accuracy and time.

The amplitude of the ac signal is another important parameter. It should be small enough so that the electrochemical process is not affected. However, if it is too small, the signal to noise ratio will be small and thus impair accuracy. Generally, 5 ~ 10 mV rms is used.

A typical system for EIS measurements is shown in Fig. 23. Using the system, the authors measured copper anodic EIS in

Solution 1: phosphoric acid,
Solution 2: phosphoric acid + copper oxide,
Solution 3: phosphoric acid + ethylene glycol,
Solution 4: HEDP, and
Solution 5: phosphoric acid + sodium tripolyphosphate.

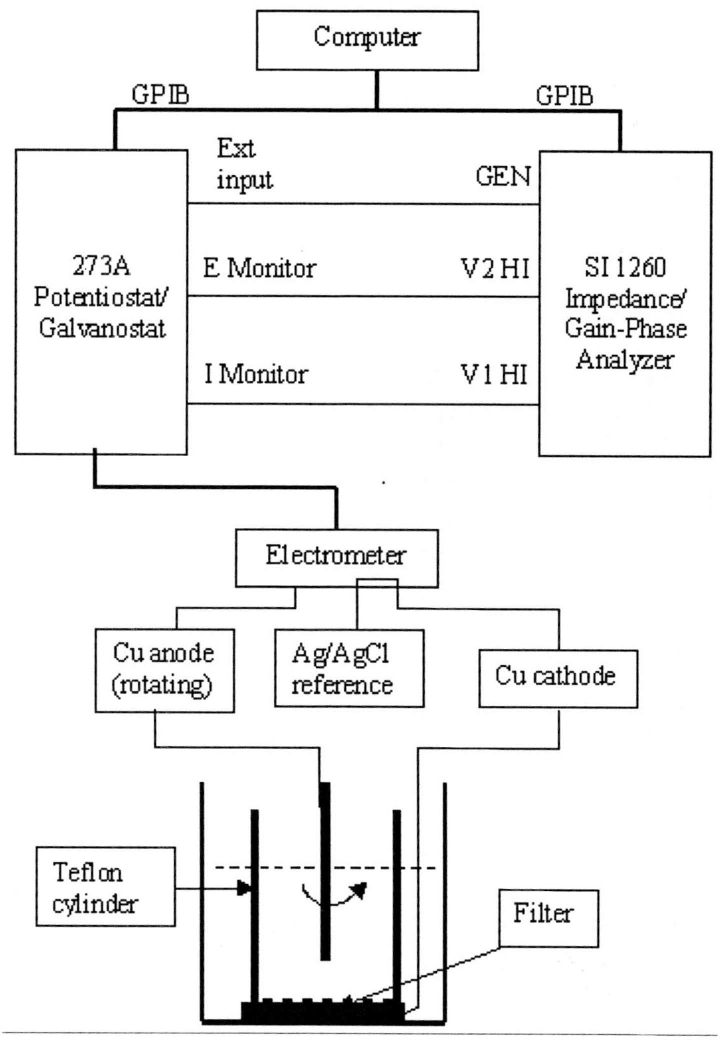

Figure 23. System for EIS measurement.

Copper disks with surface area of 1 cm^2 and 1.5 mm thickness were used as anode samples. A copper plate 12 mm thick and 43 mm in diameter served as a cathode. The cathode was chosen much larger than the anode and parallel to it so that the electrical field over the anode surface is uniform. The anode rotating speed was 100 rpm during the measurements. All the measurements were carried out at room temperature.

The measurements were performed using the software Zplot (Scribner Associate, Inc.) in "controlled E sweep frequency" mode under the following conditions: controlled dc voltage at various values: lower than limiting current plateau (e.g., 0.25 V), in the range of limiting current plateau (e.g., 1.5 V and 1.8 V); applied 10 mV ac signal with varying frequency from 100 kHz to 0.1 Hz. Before each frequency scan, a 50-second pretreatment (i.e., applied the dc voltage while ac voltage is zero) was performed. Ten points were sampled per decade. The integration time for each sampling was 5 seconds. For each solution, the scan at different potentials was conducted in batch control mode.

3.4. EIS Data and Interpretation

Two of the measured impedance spectra are shown in Fig 24. At high frequencies, a semicircle was obtained at all the dc potentials for all the solutions. At low frequencies, the EIS line shifted to left with the increase of applied dc potential for solutions 2, 3, and 5. Similar behavior was reported by Vidal and West in [29]. This phenomenon is not prominent for solution 1 and not applicable to solution 4.

R_1, R_2, and C_{dl} data were extracted from the measured impedance spectra according to their definitions in Eqs (27) ~ (29). These data are listed in Table 2.

3.4.1. Identification of Salt Film

In Table 2, all the dc potentials, except 0.25 V, are in the range of limiting current plateau. Since 0.25V is far below the potential range of the limiting current plateaus, no anodic film can form. Thus, according to the above discussion,

$$R_2\,(0.25V) = R_s + R_m \tag{31}$$

From the data in Table 2, we can see that concentrated phosphoric acid and its solutions with additives: copper oxide, ethylene glycol, and sodium tripolyphosphate, all have constant value of R_2 at all the tested potentials. Thus, from Eqs. (23) & (31) we may conclude that no resistive salt film was formed on the anode surface at potential range of limiting current plateau.

In contrast, copper-HEDP system has different values of R_2 at all the three different applied potentials. In Eqs. (23) & (31), R_m does not change with solutions and applied potential. For same solution, the change of R_s, due to anode material dissolved into solution, can be assumed negligible. So, from the R_2 data and Eqs. (23) & (31), we get the following values for Cu-HEDP system:

$R_e + R_m = 6\Omega$,

$R_f\,(1.3V) = 1\Omega$, and

$R_f\,(1.6V) = 2\Omega$.

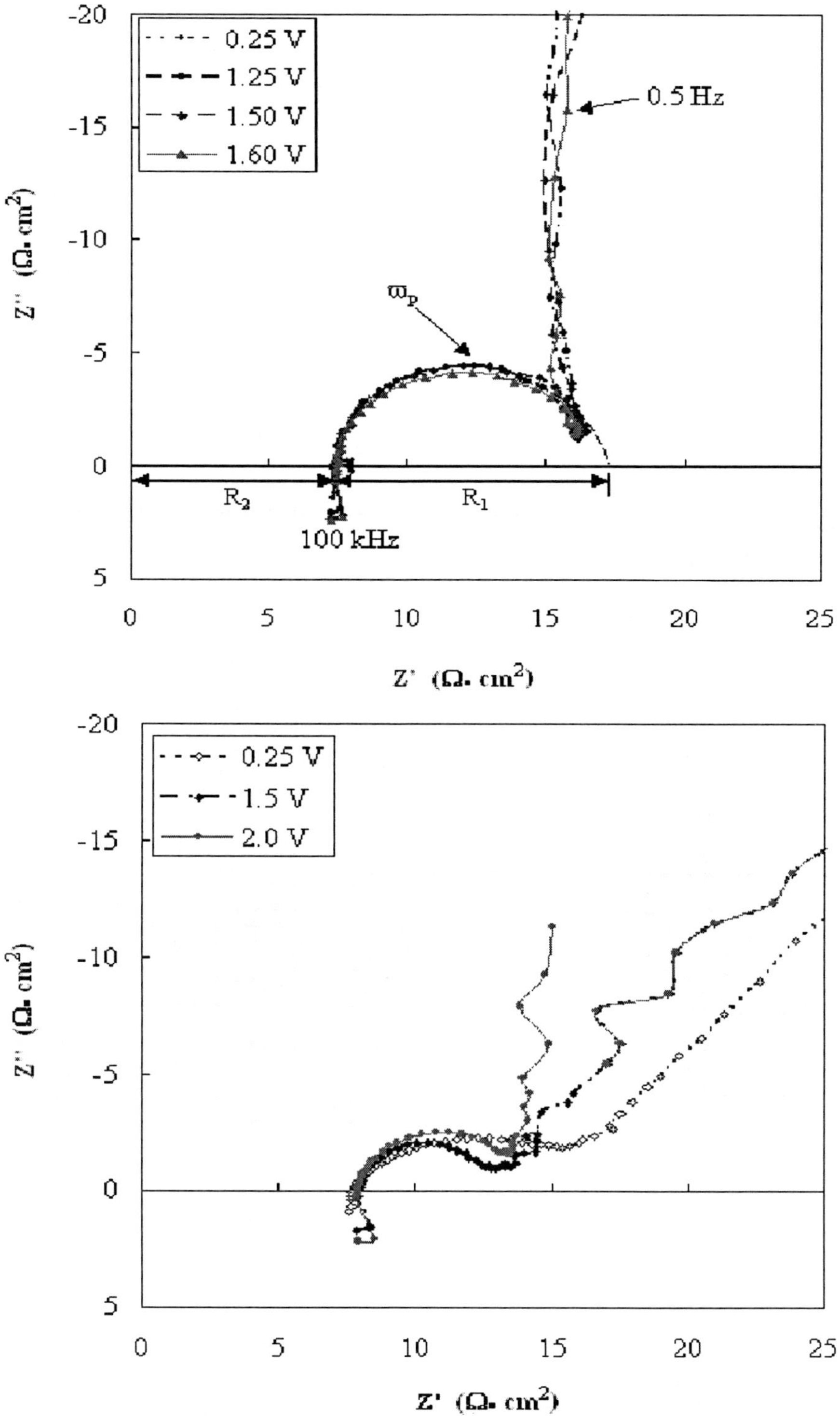

Figure 24. EIS measured from phosphoric acid (a) and phosphoric acid – sodium tripolyphosphate (b) solutions.

Table 2. Summary of impedance measurements

Solution	Potential (V)	R_2 (Ω cm^2)	R_1 (Ω cm^2)	C_{dl} ($\mu F/cm^2$)
	0.25	7.5	0.5	906
	1.25	7.5	9.0	11
	1.5	7.5	9.0	14
	1.6	7.5	9.0	18
2. 70 % phosphoric acid + 1.2 M CuO	0.25	7.6	0.4	199
	1.0	7.6	3.4	12
	1.5	7.6	3.4	74
3. 50 % phosphoric acid + 50 % ethylene glycol	0.25	29.0	-	-
	1.5	29.0	13.0	12
	1.8	29.0	15.0	11
	2.0	29.0	20.0	80
4. 70 % HEDP	0.25	6.0	14.0	180
	1.3	7.0	4.0	20
	1.6	8.0	4.5	11
5. 70 % phosphoric acid + 0.5 M sodium tripolyphosphate	0.25	8.0	9.0	88
	1.5	8.0	5.0	13
	2.0	8.0	7.0	29

Note: phosphoric acid = 85 % H_3PO_4; HEDP = 60 % $C_2H_8O_7P_2$.

This suggests the existence of an anodic salt film in Cu-HEDP system. And its thickness increases with increasing potential. This result is also agrees with the conclusion from the Levich-Koutecky plot (Fig. 11b), which indicates the existence of additional series resistance.

SEM observation further confirmed the above results. The copper anode surface after ECP in phosphoric acid was smooth (Fig. 25a). A transparent layer was observed on the surface of copper anode after ECP in HEDP solution (Fig. 25b).

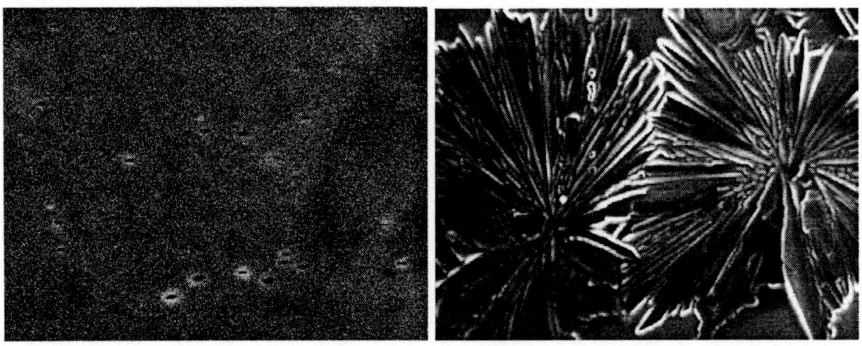

a b

Figure 25. SEM pictures of copper anode surface after ECP process in (a) phosphoric acid and (b) 70% HEDP solutions. ×1000.

3.4.2. Warburg Impedance

At a potential lower than the limiting current plateau, the process is mainly charge transfer and / or chemical reaction controlled. So R_1 (0.25V) is mainly polarization resistance R_p. While over the range of limiting current plateau, Warburg impedance (R_w) may play a much more important role. The data of R_1 in Table 2 show that in solutions 3, 4, and 5 R_1 increased with increasing potential within the range of the limiting current plateau. This suggests that diffusion impedance (R_w) increases with the potential increase.

So it is very likely that Cu^{++} is the transport limiting species. Since at higher potential, Cu atoms can be ionized to Cu^{++} more easily. That is, more Cu^{++} ions can be produced at higher potential. However, because the process was mass transport controlled, Cu^{++} ions produced due to higher potential may not necessarily reach the cathode so to complete the charge transfer and current flow cycle. Thus the limiting current did not increase. Instead, higher concentration gradient of Cu^{++} ions in the diffusion layer was developed and therefore higher diffusion / migration resistance.

In contrast, R_1 of solutions 1 and 2 remains constant at different potentials within the range of the limiting current plateau. Previous studies by Wagner [41], Vidal and West [29] have proved that Cu^{++} is not the mass transport limiting species, instead, acceptors, such as anions or water molecules are the transport limiting species in Cu-H_3PO_4 system.

3.4.3. Double-Layer Capacitance

The data in Table 2 show that all the solutions, except solution 4, have higher capacitance at higher potential in the range of limiting current plateau. Capacitance is defined as

$$C = \frac{dQ}{dt} \tag{32}$$

where Q is charge and t is time. The driving force of an electrochemical reaction is the electrical potential at the electrode surface. So higher charge transfer (e.g., Cu – 2e → Cu^{++} on anode surface) rate is expected at higher potential. That is, the adsorbed fixed layer / diffusion layer (positive ions) – diffuse mobile layer (negative ions) capacitor (See Fig. 19) has a faster charging rate at higher potential. Again, because the process is mass transport controlled, Cu^{++} ions produced due to higher potential can not reach cathode and complete charge transfer. Instead, more Cu^{++} ions are crowded in the fixed layer and diffusion layer. Correspondingly, more negative ions are attracted to the diffuse mobile layer. That is, more charge to the double layer capacitor. This can explain the observed higher capacitance at higher potential in the range of limiting current plateau.

However, for solution 4 (HEDP solution), as discussed above, a salt film was formed on the anode surface and its thickness increased with anode potential in the range of the limiting current plateau. This may inversely contribute to the capacitance, since

$$C = \frac{\varepsilon A}{r} \tag{33}$$

where A is the area, ε and r are respectively the electrical permittivity and thickness of the media between the capacitor plates. From Eq. (33) we can see that C decreases with an increase of r. Therefore, when a thicker salt film formed at higher potential, the capacitance dropped. This explains why solution 4 has a lower capacitance at 1.6V than at 1.3V.

All the solutions in Table 2 have a much higher capacitance at 0.25V than that at potentials in the range of limiting current plateau. At such a low potential, the double layer is only a couple of atomic layer thick. So the thickness (r) of the double layer capacitor, in this case, is thousands of times thinner than that with diffusion layer under mass transport condition. According to Eq. (33), the capacitance should be thousands of times higher.

Nevertheless, the charge transfer ($Cu - 2e \rightarrow Cu^{++}$) rate (dQ / dt) is low at low potential. Thus the capacitance should be small according to Eq. (32). Therefore the capacitance at 0.25V depends on solutions and is very different than it at limiting current plateau.

In summary, Cu – H_3PO_4 and Cu – (H_3PO_4 + CuO) systems have identical impedance spectra at all the tested potentials. Cu – ($H_3PO_4+C_2H_6O_2$) and Cu – ($H_3PO_4+H_5Na_5O_{10}P_3$) systems have impedance spectra with the same original points but different diameters of Nyquist semicircles at varying potentials. Cu – HEDP system has very different ac impedance spectra at different potentials.

Analysis of these spectra suggested that a resistive salt film was formed on the copper anode surface in HEDP solution while no salt film formed in the other solutions. The analysis also suggested that Cu^{++} was the mass transport limiting species in solutions of HEDP and phosphoric acid with additives ethylene glycol and sodium tripolyphosphate. Water molecules are the mass transport controlling species in solutions of phosphoric acid and phosphoric acid with copper oxide.

4. Mechanisms of Electrochemical Planarization

Why is it difficult to achieve satisfactory planarization of electroplated Cu/Si wafers by using conventional ECP solutions for bulk copper materials? What are the differences between the surface profiles of Cu/Si wafers and Cu bulk materials? What an ECP solution (condition) can be used for replacement of CMP? To answer these questions, one needs to understand ECP mechanisms.

There are three ECP mechanisms responsible for surface planarization (or leveling): *Ohmic Leveling*, driven by the difference of current density between protruding areas and receding areas of the anode due to difference in ohmic resistance to the cathode; *migration leveling*, driven by difference in electrical field strength on the anode surface; *diffusion leveling*, driven by difference in chemical potential due to the concentration gradient distribution along the anode surface. Leveling can be achieved under ohmic control or mass transport control.

4.1. Ohmic Leveling

Under ohmic control, current density is determined by the electrolyte resistance (R_e) between the electrodes (Fig. 26):

$$J = \frac{V}{R_e A} \tag{34}$$

where V (= V_1 - V_0) is the potential difference between anode and cathode. Since

$$R_e = \rho_e \frac{l}{A} \tag{35}$$

Eq. (34) may be rewritten as

$$J = \frac{V}{\rho_e l} \tag{36}$$

where ρ_e is the resistivity of the electrolyte between the cathode and anode, and l is the distance between the electrodes.

Surface removal rate (R_d, cm/s), from Eq. 15, may be expressed as

$$R_d = \frac{JM}{nFd} \tag{37}$$

where M is the atomic weight (63.5 g for Cu) and d is the density of the anode material (8.96 g/cm^3 for Cu), n (2 for Cu – 2e → Cu^{++} reaction) is the number of electrons transferred in the oxidation reaction of an anode atom, while F (= 96485 C) is Faraday constant. Substituting Eq. (36) into Eq. (37), results in

$$R_d = \frac{MV}{nFd\rho_e l} \tag{38}$$

Therefore, ohmic leveling efficiency, which is the difference of the removal rate between protruding point (P) and recessed point (Q), can be expressed as

$$\Delta R_d = R_d(P) - R_d(Q) = \frac{MV}{nFd}\frac{2b}{\rho_e(l^2 - b^2)} \tag{39}$$

Eq. (39) indicates that ohmic leveling efficiency depends on the surface profile and inter-electrode distance. Larger surface roughness and closer electrodes (i.e., smaller l) may produce better leveling effect. In addition, smaller resistivity (ρ_e) of the electrolyte solution and higher voltage (V) between the electrodes also facilitate ohmic leveling.

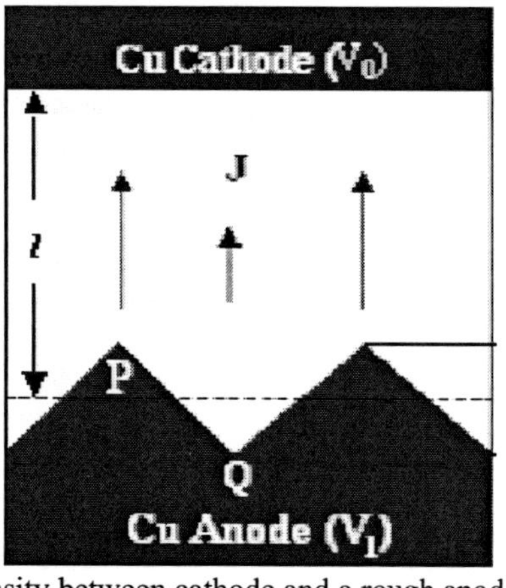

Figure 26. Schematic of current density between cathode and a rough anode.

Assuming $\rho_e = 10\ \Omega$-cm [42], b = 0.5 μm, l = 20 cm, and V = 2V, we calculated $\Delta R_d = 1.8 \times 10^{-8}$ μm/s, which is negligible compared to the average removal rate $R_d = 4.4 \times 10^{-3}$ μm/s at current density i_L = 12 mA for example (in concentrated phosphoric acid solution). This indicates that the ohmic leveling effect is negligible under normal wafer ECP conditions.

However, when the inter-electrode distance is in the order of anode surface roughness, ohmic leveling effect can be important. Fig. 27 shows the relationship of ohmic leveling effect versus inter-electrode distance with assumption: $\rho_e = 10\ \Omega$-cm, b = 0.5 μm and V = 2V. Ohmic leveling effect can be significant when inter-electrode distance is smaller that 1 mm. Note that Eq. (34) is valid only when the current density (J) is sufficiently low that mass transport limitations do not apply.

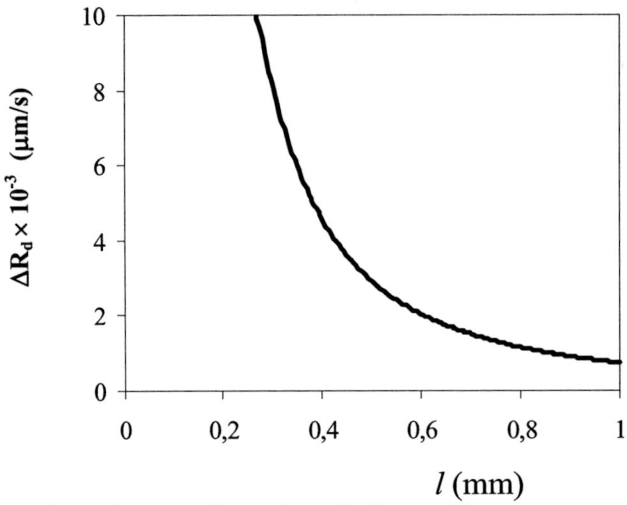

Figure 27. Ohmic leveling effect versus inter-electrode distance (l).

4.2. Migration Smoothing

Another geometrical-dependent leveling mechanism is related to ion migration, which is due to the electric field distribution on a rough surface favoring planarization. We calculated the electric field distribution along a sinusoidal surface. A sine waveform is located at the center of the anode surface in a 2D electrochemical cell, which has parameters close to an actual 300 mm wafer polishing system (See Fig. 28). This simulation was performed using a boundary element simulation software. Data obtained from the simulation (Fig. 29a) indicates that the amplitude of normal electric field (En) varies along the sinusoidal surface, similar to the shape of the surface profile except at the origin and end points of the sine waveform where a sharp geometrical change occurs. For certain amplitude (b) of the sine wave, the difference of the normal electric field $E_n(P)$ at the peak point P and $E_n(Q)$ at the recessed point Q increases as the wavelength (T) decreases (i.e., the frequency f increases), as indicated in Fig. 29b. The orientation of the normal electric field depends on the location too. At peaks and valleys, (points P and Q), E_n points perpendicular to the cathode, and deviates from this direction at other locations.

The magnitude and orientation distribution of the electric field E_n along the surface favors the migration of Cu^{++} at the peak areas where E_n has the highest value $E_n(P)$ and points towards the cathode. In contrast, Cu^{++} in the valley has the slowest migration speed. Consequently, a leveling effect can be achieved. The following discussion quantitatively shows the influence of anode surface profile on ECP due to migration leveling effect.

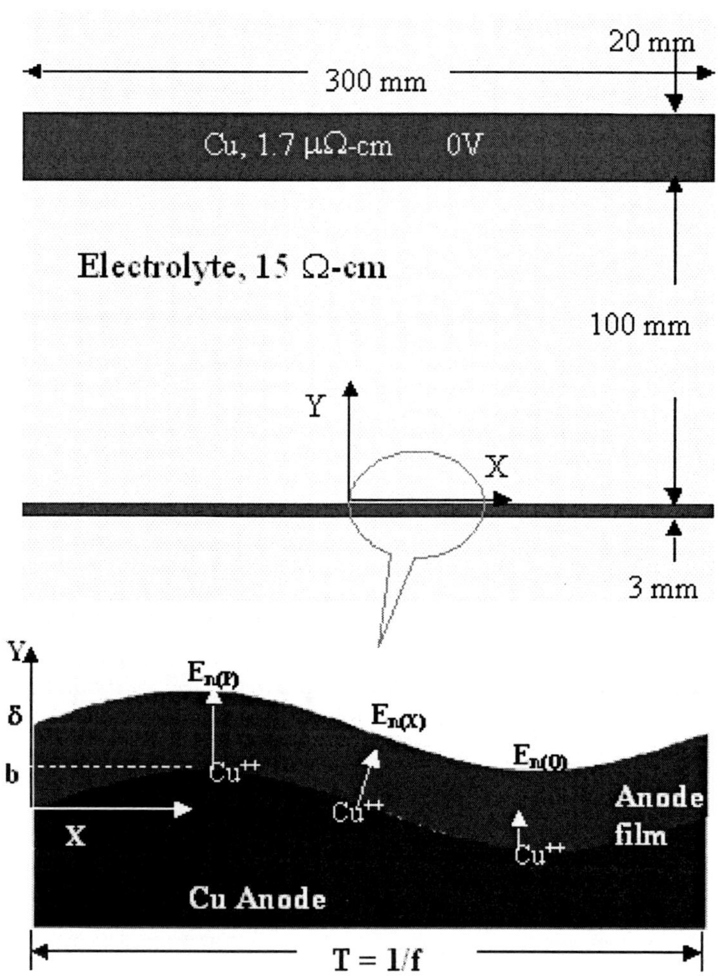

Figure 28. Schematic of an electropolishing cell and the local surface profile of the anode. The electric field (E) varies in amplitude and orientation along the sinusoidal surface.

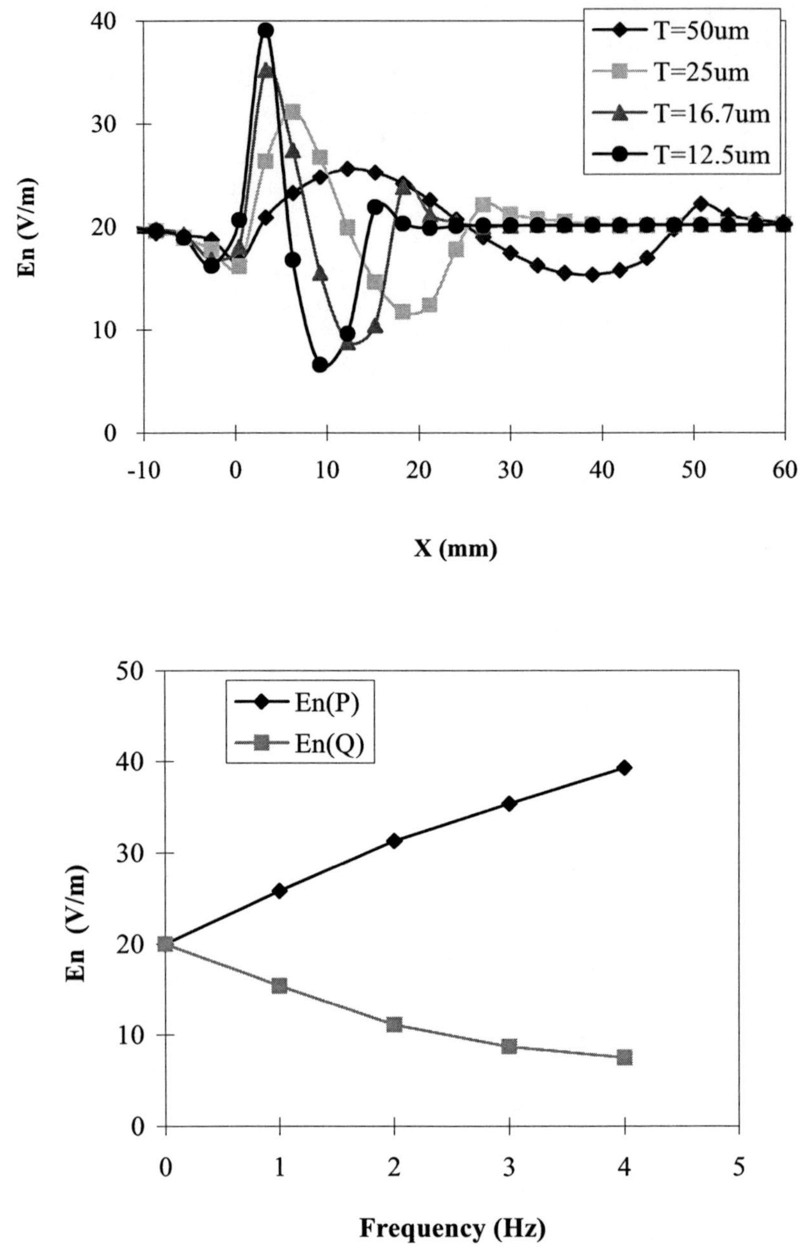

Figure 29. The amplitude of normal electric field along the sinusoidal anode surface, obtained by simulation (a); Amplitude of normal electric field at the peak and valley of the sinusoidal surface, versus the frequency of the sine wave (b).

Once a copper atom is oxidized to Cu^{++} on the anode surface, a force

$$F_E = q\,E_n \tag{40}$$

acts on the Cu^{++} resulting from the electric field E_n. As a result, Cu^{++} ions tend to move away from the anode surface towards cathode under force

$$F = F_E - F_r = q\,E_n - F_r \tag{41}$$

where q is the charge of Cu^{++} and F_r is the resistive force as Cu^{++} ions move through the anode layer(s) and the solution. In bulk solution F_E is negligible compared to the convection force (F_c) when a forced convection is applied. Inside anodic layers, F_E and concentration gradient are the only driving forces for Cu^{++} motion.

According to Newton's second law of motion, the Cu^{++} ions accelerate at a = F / m, and therefore migrate a distance

$$s = \frac{1}{2}at^2 \tag{42}$$

in time t. In the case of conformal anodic layer as shown in Fig. 28, s = 2b. It takes time t for a Cu^{2+} to migrate out of the anodic layer. From Eqs. (41) and (42), we have

$$t = 2\sqrt{\frac{mb}{qE_n - F_r}} \tag{43}$$

where m is the mass of Cu^{++}. Thus, the number of copper ions migrate into bulk solution across a unit area in unit time

$$N = N_o\,\frac{1}{t} = \frac{N_o}{2}\sqrt{\frac{qE_n - F_r}{mb}} \tag{44}$$

where N_o is the number of total Cu^{++} ions in unite area on the anode surface. N is the anode dissolution rate in atom/sec. Therefore, the difference of the removal rate between protruding point (P) and recessed point (Q) can be expressed as

$$\Delta R_d = R_d(P) - R_d(Q) = \alpha\,(N_P - N_Q) \tag{45}$$

Hence, the migration leveling effect may be obtained from Eqs.(44) and (45)

$$\Delta R_d = \frac{\alpha N_o}{2}\sqrt{\frac{q}{mb}}[\sqrt{E_n(P) - f_r} - \sqrt{E_n(Q) - f_r}\,] \tag{46}$$

assuming $F_r = qf_r$, where f_r is a constant related to the property of the anodic layer.

Similarly, for non-conformal anodic layer as shown in Fig. 30, assuming the thickness of the anodic layer at Peak and valley are $S_P = s$ and $S_Q = ks$, we may get

$$\Delta R_d = \frac{\alpha N_o}{2}\sqrt{\frac{q}{ms}}[\sqrt{E_n(P) - f_r} - \sqrt{\frac{1}{k}(E_n(Q) - f_r)}\,] \tag{47}$$

Figure 30. The structure of non-conformal anodic layer.

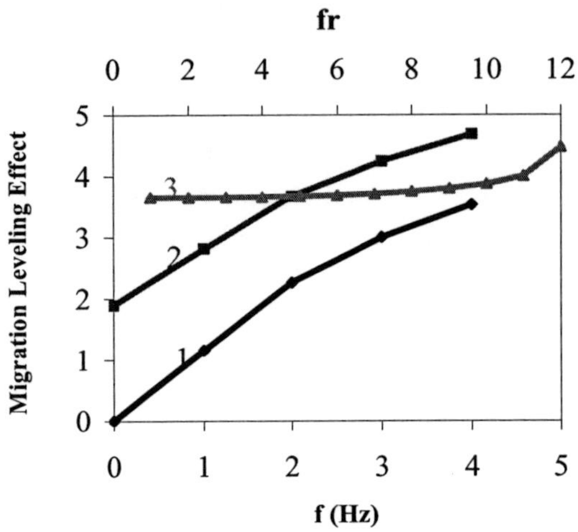

Figure 31. Migration leveling effect versus frequency of anode surface profile: 1. anodic layer is conformal to the anode surface; 2. anodic layer is non-conformal to the anode surface. Migration leveling effect versus resistance to ion migration (3).

In the above two equations, $\frac{\alpha N_o}{2}\sqrt{\frac{q}{mb}}$ and $\frac{\alpha N_o}{2}\sqrt{\frac{q}{ms}}$ are constants. So, migration leveling effect is represented by $[\sqrt{E_n(P)-f_r}-\sqrt{E_n(Q)-f_r}\,]$ for the case of conformal anodic layer, and by $[\sqrt{E_n(P)-f_r}-\sqrt{\frac{1}{k}(E_n(Q)-f_r)}]$ for the case of non-conformal anodic layer. Using the data obtained by simulation (See Fig. 29b) and assume k = 3 and f_r = 0 (since it is same for all copper ions), migration leveling effects versus the frequency of anode surface profile are plotted in Fig. 31.

Fig. 31 indicates that migration leveling effect increases with the increase of the frequency of the surface profile. In the case of conformal anodic layer, migration leveling effect tends to zero when anode surface profile frequency decreases to zero (i.e., the wavelength → ∞). In the case of non-conformal anodic layer, migration leveling effect can planarize any surface even though its surface profile tends to have infinite wavelength. In either case, shorter wavelength is planarized faster than longer wavelength.

According to Eq. (47), plot $[\sqrt{E_n(P)-f_r}-\sqrt{\frac{1}{k}(E_n(Q)-f_r)}]$ versus f_r (represents migration resistance) is shown in Fig 31, which indicates that migration leveling effect can be slightly increased by increasing migration resistance.

4.3. Diffusion Leveling

Diffusion leveling is the ECP effect due to diffusion, driven by the chemical potential results from the difference of concentration gradient. Wagner [41] calculated the dissolution speed of copper according to diffusion law (Fick's second law), under mass transport control and acceptor mechanism condition. In this case, the concentration distribution of the acceptor has a profile as shown in Fig. 32. In the figure, C_b is the bulk concentration of the acceptor. δ_c is the thickness of the diffusion layer.

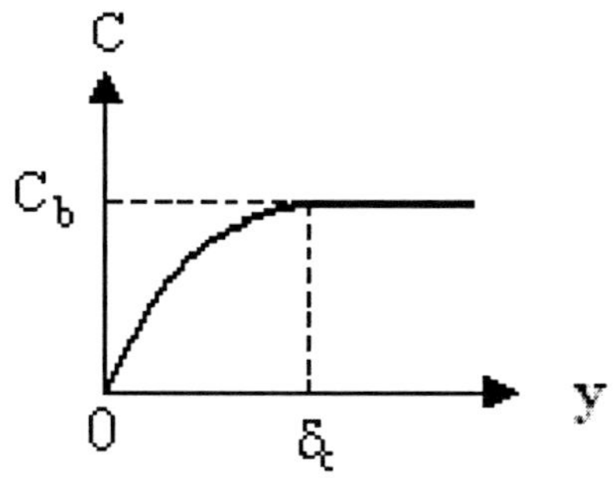

Figure 32. Acceptor concentration as a function of distance from the electrode.

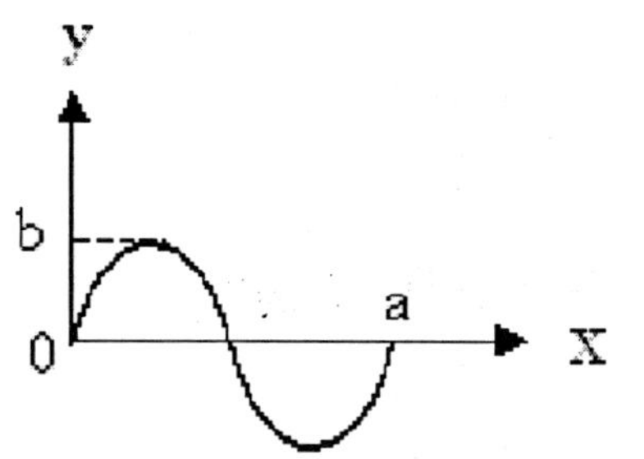

Figure 33 Anode surface profile.

Assume the anode surface (in 2D) has a sinusoidal profile, as shown in Fig. 33. The surface profile can be expressed as

$$y = b\sin(\frac{2\pi}{a}x) \tag{48}$$

where y is the distance from the average surface plane, a and b are respectively the wavelength and amplitude of the sine wave. When a and b are much smaller than δ_c, Fick's second law applies to the anode surface and adjacent area. At steady-state conditions, there is

$$\frac{\partial c}{\partial t} = D(\frac{\partial^2 C}{\partial x^2}+\frac{\partial^2 C}{\partial y^2}) = 0 \tag{49}$$

with boundary conditions in Figs. 32&33 and assumption b << a, a<< δ_c, an approximated solution of Eq. (49) can be obtained [41]

$$(\frac{\partial C}{\partial y})_{y=b\sin(\frac{2\pi}{a}x)} = B[1+2\pi\frac{b}{a}\sin(\frac{2\pi}{a}x)] \quad (50)$$

where B = $(\frac{\partial C}{\partial y})_{avg}$ is the average value of concentration gradient at the anode surface.

Since the dissolution rate (R_d) due to diffusion is proportional to the concentration gradient of the acceptor, it can be deduced from Eq. (50) that [41]

$$R_d = C(2\pi\frac{b}{a})\sin(\frac{2\pi}{a}x) \quad (51)$$

Comparing Eqs.(51) and (48), we can see that the dissolution rate versus location x has the same shape as the geometric profile of the anode surface. That is, R_d at the peak is greater than that at the valley of the anode surface. Eq. (51) also suggests that R_d increases with decreasing wavelength a. This characteristic is same as that of migration leveling effect. Similarly, if the diffusion layer has a non-conformal surface profile as shown in Fig. 30, the diffusion distance (of the mass transport limiting species) at peak points is shorter than that at the valley points and therefore planarization effect can be promoted.

In summary, Ohmic leveling effect depends on inter-electrode distance. It works best when the inter-electrode distance is comparable to the amplitude of anode roughness. Migration leveling applies when the mass transport limiting species are ions. In other words, there is no migration leveling effect when the mass transport limiting species are molecules. Both migration leveling and diffusion leveling preferentially planarize sharp undulations (i.e., the undulations that have larger b/T values). The leveling effect tends to zero when b/T → 0 (i.e., very gentle undulations) if the anodic layer is conformal to the anode surface. However, gentle undulations can be planarized by migration leveling and diffusion leveling when the anodic layer is non-conformal to the anode surface.

5. Experimental Results and Optimal Conditions of Electrochemical Planarization of Copper Bulk and Electroplated Cu/Si Wafers

5.1. Experimental Details

5.1.1. ECP with Conventional System

ECP of copper disks and Cu/Si wafer coupons was performed with computer-controlled chronoamperometry (CA). The experimental system is similar to the one used for EIS measurement (Fig. 23) but SI 1260 was disconnected.

For ECP of Cu/Si wafer coupons, a customized sample holder was made to hold 5 cm × 5 cm wafer coupons. A larger cylindrical container (about 3 liter volume) and copper cathode were used. A conical filter was also used to block hydrogen bubbles from reaching the anode surface.

The copper disks used for ECP were 1.5 mm thick, 11.28 mm in diameter (i.e., surface area = 1 cm^2), with a purity of 99.99% Cu. Before the ECP process, all the copper disks were hand polished with sand papers of 1μm grit, followed with methanol rinse and pressurized nitrogen dry. The wafer coupons for ECP were cut from 200 mm diameter trenched wafers with electroplated copper films. The copper plating on 200 mm wafers was performed using in house wafer plating system at the Electrochemistry Laboratory in OGI School of Science and Engineering, Oregon Health and Science University (Beaverton, Oregon).

The controlled potentials for ECP were the middle points of the limiting current plateau (E_L) on the polarization curves. The duration time of ECP for each sample was determined such that the original rough surface layer were totally removed. For copper disks, generally a 10 μm thick layer was removed by the ECP process. For wafer coupons, various ECP time were used to observe the surface planarization effects. The electrolyte solutions used for ECP are listed in Table 3. 100 ml solution was used for ECP of each copper disk. 1300 ml solution was used for ECP of each wafer coupon. All the experiments were performed at room temperature.

Surface profiles of the samples before and after ECP were examined with an atomic force microscope (AFM), optical and electron microscopes, as well as a surface profilometer. Mean roughness before and after ECP were determined with AFM.

Polishing (or removal) rate R_d was calculated from the weight loss and current. This verified the quantity of charge transferred in an anodic reaction.

Table 3. Electrolytes Used for ECP of copper disks

Solution No.	Chemical 1	Concentration (% Volume)	Chemical 2	Concentration (Mole / L)
1	Phosphoric acid (85% H_3PO_4)	30, 40, 70, 100. 70.	Copper oxide (CuO)	0. 0.5, 1.2, 2.0.
2	Phosphoric acid (85% H_3PO_4)	70	Ethylene glycol ($C_2H_6O_2$)	5, 15, 25 (% vol)
3	Phosphoric acid (85% H_3PO_4)	70	Sodium tripolyphosphate ($Na_5P_3O_{10}$)	0.1, 0.25, 0.5
4	Hydroxyethylidenedi phosphonic acid (HEDP)	20 ~ 80	Phosphoric acid (85% H_3PO_4)	10 ~ 30 (% vol)
5	Ethylene glycol ($C_2H_6O_2$)	20 ~ 100	Sodium chloride (NaCl)	0 ~ 2
6	Sulphuric acid (H_2SO_4)	20	Sodium / Potassium Nitrate ($NaNO_3$ / KNO_3)	1.0, 2.0.

Note: phosphoric acid = 85 % H_3PO_4; HEDP = 60 % $C_2H_8O_7P_2$.

5.1.2. ECP with Modified ECP System

1. Vertical Electrodes

To avoid hydrogen bubbles (formed on cathode) reaching anode surface and producing pits, a rectangular cell was used with vertical electrodes, as shown in Fig. 34. other conditions are same as in the above system.

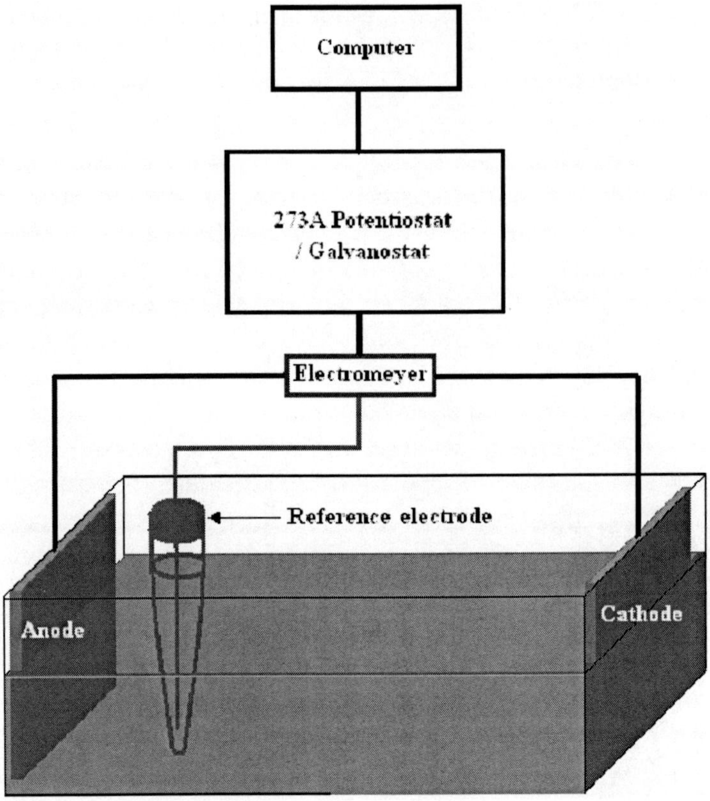

Figure 34. ECP system for reducing the influence of hydrogen bubbles on the anode surface.

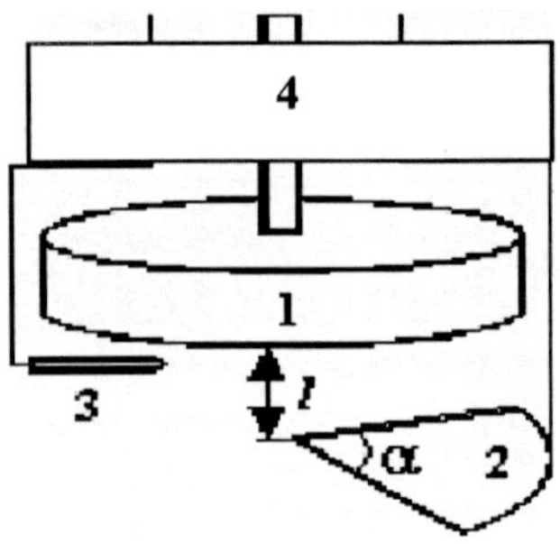

Figure 35. Schematic of a novel ECP system.

2. Sectorial cathode

For the best planarization effect on Cu/Si wafer samples, sectorial metal plates (radius 38mm) were used as cathode, as shown in Fig. 35. Copper and tantalum anodic polarization curves were measured. Planarization effect on Cu/Si wafer samples were evaluated using the system with various size of cathode and inter-electrode distance (see Table 4). Other conditions were same as in the conventional system.

Table 4. Experimental conditions for ECP using sectorial cathode

Cathode dimension (Central angle α, degrees)	10	40	90	180	360
Anode rotating speed (rpm)	100	200	300	400	500
Interelectrode distance (l, mm)	< 1	1	20	35	50

5.2. Experimental Results and Interpretations

5.2.1. ECP of Copper Disks

Typical AFM topography images of copper disk surface after ECP are shown in Fig. 36. The current density at which ECP was carried out, Cu removal rate, and mean roughness after ECP are listed in Table 5. Very good ECP effect (i.e., small mean roughness, R_a) of copper disks was achieved in solutions of phosphoric acid, HEDP, and phosphoric acid with additives copper oxide, ethylene glycol, and sodium tripolyphosphate. However, ECP effect of copper disk in ethylene glycol - sodium chloride, sulfuric acid, sulfuric acid - sodium nitrate, and sulfuric acid - potassium nitrate solutions was not so good. Table 5 lists the best ECP effects obtained from each electrolyte solution.

ECP of copper disks with horizontal and vertical electrodes setup produced same mean roughness. However, small pits were shown (Fig. 36a) on the Cu surface after ECP with horizontal electrodes setup, due to some small hydrogen bubbles passed through the filter and reached the anode surface. ECP with vertical electrodes setup produced a pit-free surface (Fig. 36b) because hydrogen bubbles could escape from the solution surface.

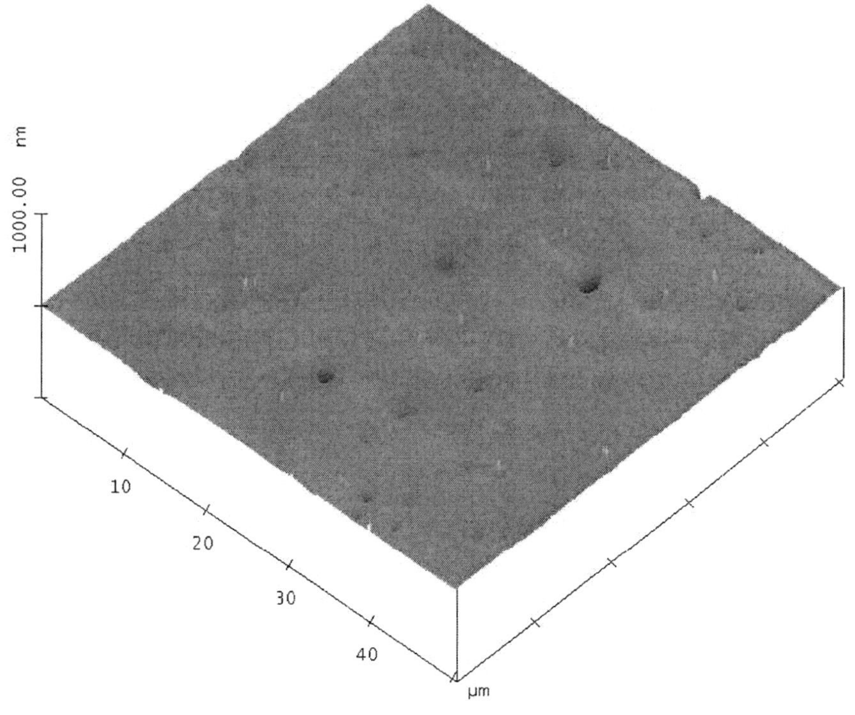

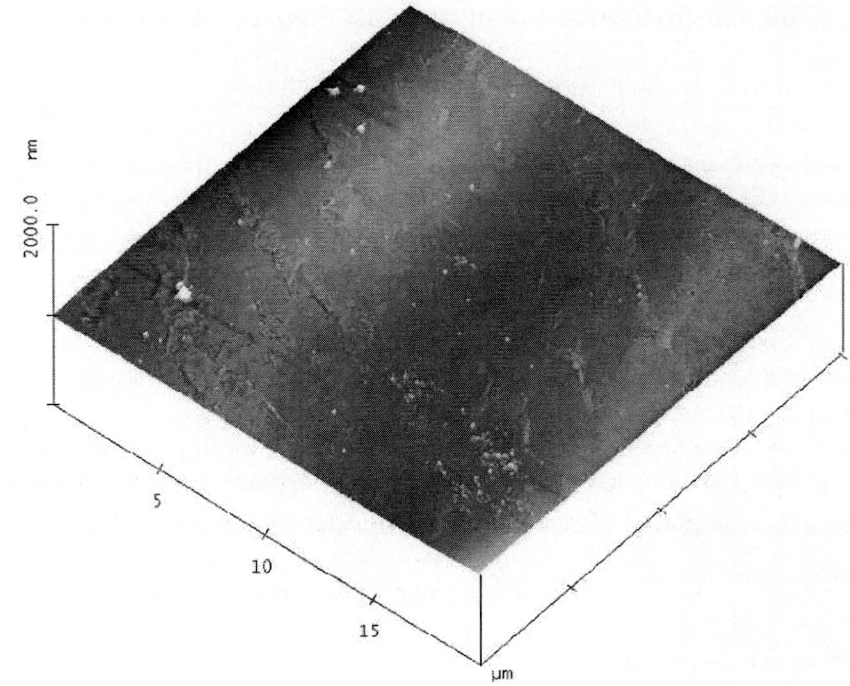

Figure 36. AFM images of copper disk surface after ECP in phosphoric acid solution with conventional ECP system (a, $R_a = 5$ nm), and after ECP in 70% phosphoric acid + 5% EG solution with vertical electrodes shown in Fig. 34 (b, $R_a = 5$ nm).

Table 5. Summary of copper disk ECP data in various solutions.

Solution	Limiting Current i_L (mA)	Removal Rate R_d (μm /min)	Roughness Ra (nm)
1. 70~100 % phosphoric acid	12~50	0.26~1.10	4~9
2. 70 % phosphoric acid + 0.5 M CuO	45	0.99	5
3. 70 % phosphoric acid + 5 % ethylene glycol	46	1.01	5
4. 70 % HEDP	30	0.66	6
5. 70 % phosphoric acid + 0.1 M sodium tripolyphosphate	40	0.88	7
6. 30 % ethylene glycol + 2 M sodium chloride	50	1.10	63
7. 20 % sulfuric acid	70	1.54	72

Note: phosphoric acid = 85 % H_3PO_4; HEDP = 60 % $C_2H_8O_7P_2$.

5.2.2. ECP of Copper Film with Conventional System

The surface profile of copper film electroplated on deep trenched wafers is characterized as flat and shiny surface with bumps and pads, which are corresponding to trenches and vias in the silicon substrate, as shown in Fig. 37. The SEM images of copper film – silicon cross-section after ECP in phosphoric acid and HEDP solutions are shown in Fig. 38.

A planar surface is obtained from ECP in HEDP solution, as shown in Fig. 38a. However, the ECP result of wafer coupons with phosphoric acid was not as good. Bumps over dense structures were left behind after ECP. As shown in Fig 38b, the area over trenches still had 0.6 μm thick of Cu film left while the Cu film over the area without trenches or vias had been totally removed.

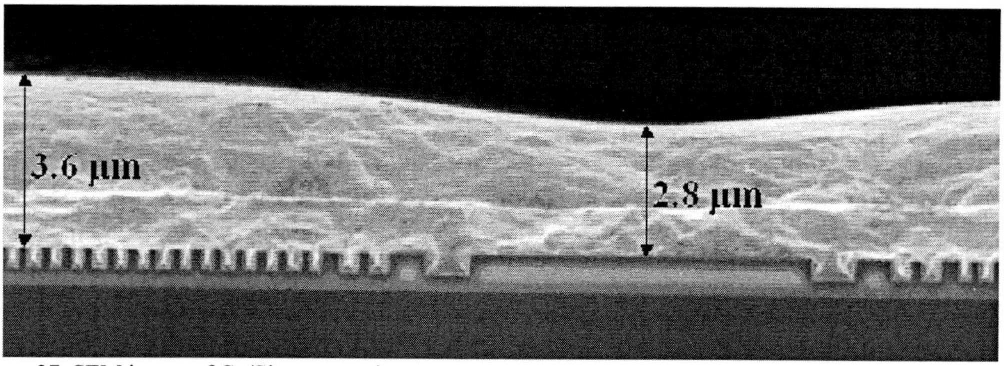

Figure 37. SEM image of Cu/Si cross section

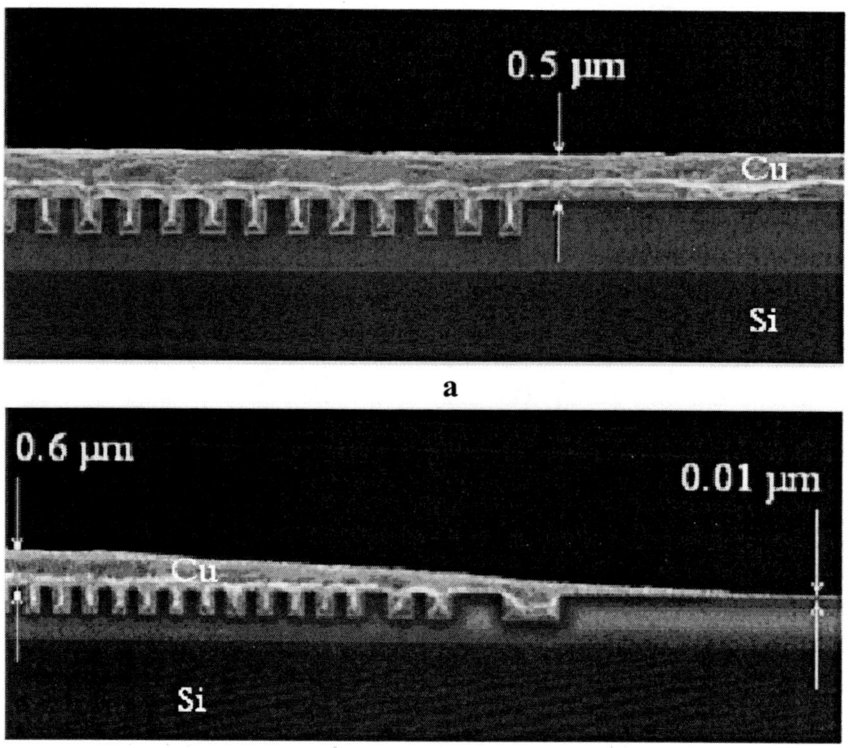

Figure 38. SEM images of copper film – silicon cross-sections after ECP in HEDP (a) and phosphoric acid solutions (b).

According the discussion about ECP mechanisms in previous section, with a conventional ECP system, the planarization effect is mainly achieved by migration and diffusion leveling mechanisms. These two mechanism favor the ECP of sharply undulated surfaces. Therefore, good ECP effects (Table 5) were obtained for Cu disks, which have sharply undulated surfaces (Fig. 1a), using phosphoric acid and several other solutions. In contrast, poor planarization effect (Fig. 38b) was obtained for electroplated Cu/Si wafer samples, which have very gently undulated surfaces (See Figs. 1c & 37), using phosphoric acid solution. Compared to this, a good planarization effect was obtained for Cu/Si wafer samples with HEDP solution. One reason is that the mass transport limiting species in Cu-phosphoric acid system is water molecules, and in Cu-HEDP is Cu^{2+} ions. So, planarization was driven by diffusion leveling only in the case of Cu-phosphoric acid while both diffusion leveling migration leveling mechanisms worked in the case of Cu-HEDP. Another reason is that the salt film formed on Cu anode surface obviously provide stronger resistance to Cu^{2+} migration and diffusion than normal diffusion layer, and thus may increase the ECP effect (Refer to section 4). Lastly, the different properties (e.g., viscosity) of normal diffusion layer and the salt film may exhibit different surface profiles, which might be also responsible for the different planarization effects.

5.2.3. ECP of Copper Film with the Novel ECP System

According the discussion in section 4, any rough surface profile can be planarized when the anodic layer is non-conformal to the anode surface profile. Therefore, good planarization effect can be expected from using very small inter-electrode distance, which may facilitate the formation of non-conformal anodic layer and ohmic leveling (in this case, migration and diffusion might dominate all over the range between cathode and anode because convection can be significantly reduced).

However, to a conventional ECP system, other problems arise when the electrodes are extremely close. For example, the hydrogen bubbles formed on cathode will reach anode easily and cause pits. These pits will cause discontinuity of Cu films. As a result, islands of Cu films will be left on the surface. Another problem is that the electrolyte circulation and thus ECP uniformity over a large wafer surface will be very poor.

To solve the above problems, a modified cathode, such as a circular cathode with holes or simply a metal bar will improve the supply of fresh solution to the anode surface and allow hydrogen bubbles to escape. Yet, caution has to be taken to make sure that there is uniform ECP driving force over the entire anode surface. That is, every point on the anode surface should have equal chance to be exposed to cathode. Discontinuity of electrical field distribution on the anode surface (such as a cathode with holes) should also be avoided. Mathematical analysis suggests that a fraction of circular plate can perfectly serve this purpose.

Fig. 39 shows two cathodes (cathode 1: a rectangular bar; cathode 2: a sectorial cathode) under a circular anode. When the anode rotating with angular speed ω, arbitrary points P & Q on anode move along tracks 1 and 2 with linear velocity $V_p = 2\pi\omega R_1$ and $V_Q = 2\pi\omega R_2$ respectively.

P & Q pass Cathode 1 over distances l_{P1} & l_{Q1} and thus face Cathode 1 for a period of time $t_{P1} = l_{P1} / V_p = l_{P1} / 2\pi\omega R_1$, $t_{Q1} = l_{Q1} / V_Q = l_{P1} / 2\pi\omega R_2$. Apparently, $t_{P1} < t_{Q1}$ since $R_1 > R_2$.

P & Q pass Cathode 2 over distances $l_{P2} = 2\pi\omega R_1$, $l_{Q2} = 2\pi\omega R_2$ and thus face Cathode 2 for a period of time $t_{P2} = l_{P2} / V_p = 2\pi\omega R_1 / 2\pi\omega R_1 = 1$, $t_{Q2} = l_{Q2} / V_Q = 2\pi\omega R_2 / 2\pi\omega R_2 = 1$. That is, $t_{P2} = t_{Q2}$. This indicates that all the points on anode have an equal chance to pass through a sectorial cathode. Any other shape of cathode does not have this property.

Fig. 40 shows the ECP results of wafer coupons with conventional apparatus and the new ECP system. Cu was preferentially removed at the edge (Fig. 41a) in the case of conventional ECP system (i.e., α=360°) and close interelectrode distance (l=1mm), due to poor solution circulation. Increasing l can improve solution circulation and thus the uniformity of Cu removal. However, the planarization effect is very poor, as shown in Fig. 40c, the Cu film over the area with trenches is much thicker than that over the area without trenches or vias. In contrast, Cu was uniformly removed over the entire surface (Fig. 40b) with the new ECP system (α = 10°, l < 1 mm). In addition, the surface was well planarized (Fig. 40d).

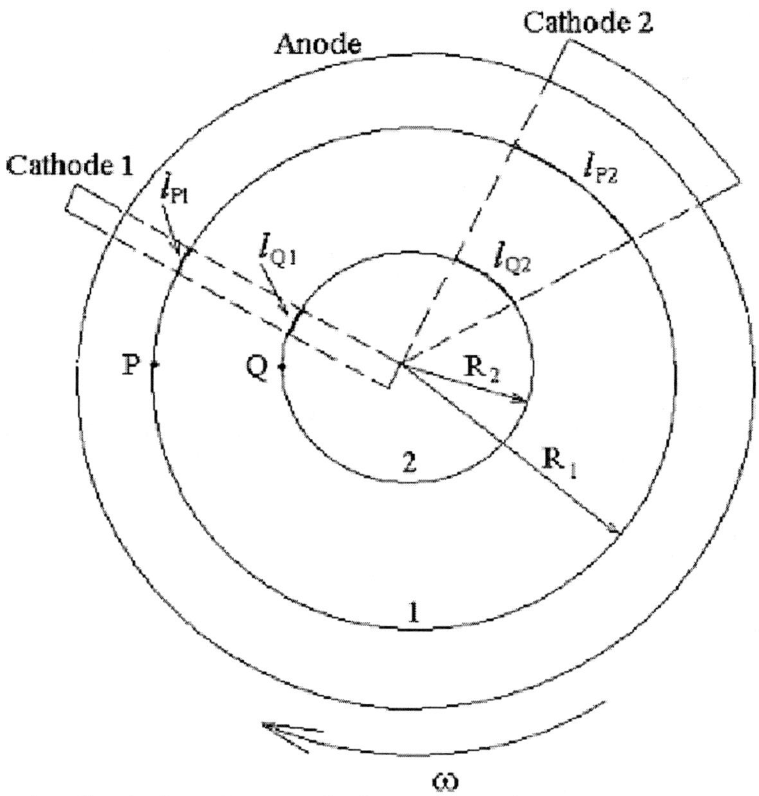

Figure 39. Illustration of tracks that points on a circular anode pass through different shape of cathodes.

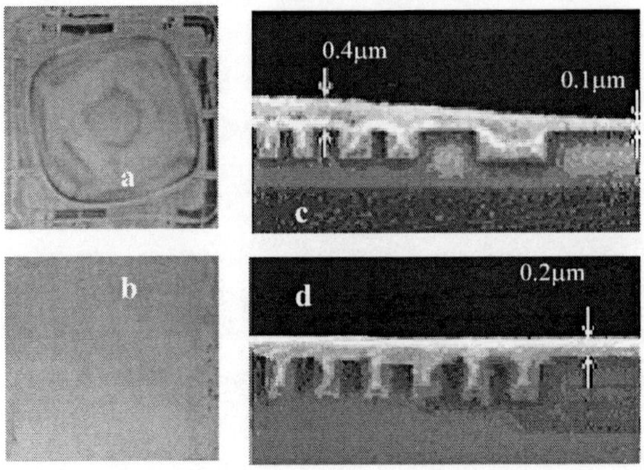

Figure 40. Topography of wafer coupons after ECP with (a) circular cathode and l=1mm, (b) sectorial cathode and l< 1mm. Cross section SEM pictures of wafer coupons after ECP with (c) circular cathode and l=35mm and (d) sectorial cathode and l< 1mm.

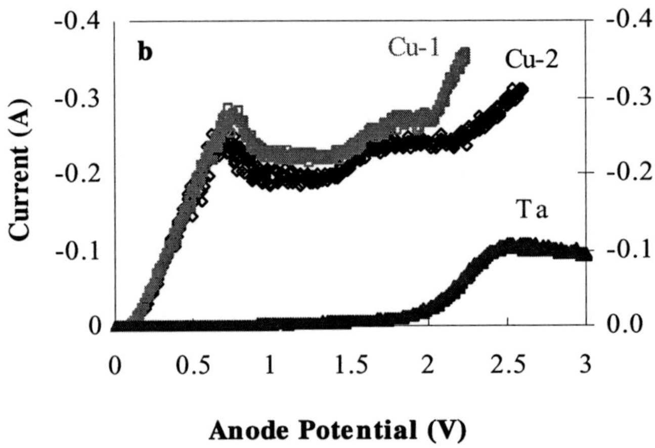

Figure 41. Cu anodic Polarization curves measured with sectorial cathode (α=10°) at different interelectrode distances (Cu-1: l=1mm, Cu-2: l=50mm) and Ta anode with circular cathode (current in mA).

Cu and Ta anodic polarization curves measured with sectorial cathodes show well-defined limiting current plateaus (Fig. 41). This and Levich-Koutecky plots measured with sectorial cathodes [43] validated mass transport condition using sectorial cathode. This is important since crystallographic etching has to be suppressed by anodic layer under mass transport condition in order to obtain a smooth surface by ECP.

From Fig. 41 we can see that in the range of Cu anodic limiting current plateaus (0.8~2V), the Ta anodic current density (right axis) is extremely small (< 0.5 mA) compared to the Cu anodic current density (about 200 mA or above). Therefore, when excess Cu is removed (i.e., Ta layer is exposed), the current density will drop significantly. This may be used to detect the endpoint of ECP process. The remaining Ta/TaN layers are usually deposited by methods such as CVD and ALD. These layers have very flat surfaces. Therefore, there are no planarization issues and they can be easily removed by ECP or etching with different solutions.

5.3. Optimal Conditions for Electropolishing Electroplated Cu/Si Wafer

The present work has been limited to copper disks and coupons cut from patterned wafers. Determination of the optimal conditions for full size wafers requires additional work. However, we can postulate that optimum results will be obtained by combining the conditions that we have found most advantageous.

5.3.1. Potential

In order to achieve good ECP effect, the process has to be mass transport controlled. Potential control has the advantage over current control. The potential range of limiting current plateau provides an operation window for potential control but not for current control.

5.3.2. Current

For a fresh anode surface, there is a transition time (about 50 ~ 100 seconds) for mass transport-limited condition to be reached [25]. During this period, a layer of copper has been dissolved under an anodic current much higher than the limiting current. Since the extra layer to be polished on a Cu/Si wafer is limited (usually about 1 μm), the ECP current should be low enough (e.g., < 30 mA) so that most of the anodic dissolution is processed under mass transport control.

5.3.3. Electrolyte Solution

Electrolytes and their concentrations in an ECP solution determine the values of limiting current and potential range of the limiting current plateau, mass transport limiting species, anodic layers, and other physical and chemical properties of the solution such as viscosity, diffusion coefficient. All these have a direct influence on ECP effects. We have observed optimal performance with thereafter exhibiting:

1. A flat and level limiting current plateau with wide potential range and low current density (e.g., <30 mA/cm^2);
2. An anodic layer with flat surface (i.e., non-conformal to the anode surface). An electrically resistive salt film (such as observed with HEDP) may further benefit the planarization;
3. Ionic mass transport limiting species, which can have migration leveling effect.

5.3.4. Solution Circulation

Solution circulation is usually performed with a pump and / or the anode disk rotating. A good circulation should produce:

1. Uniform ECP effects over a large area of anode surface; and
2. A non-conformal anodic layer(s).

Factors to be controlled include the speed, paths, and directions of the circulation.

5.3.5. Inter-Electrode Distance

Inter-electrode distance dose not affect ECP in range of 5~20 mm. However, excellent planarization effect can be obtained using sectorial cathode and very small inter-electrode distance.

5.3.6. Preventing Gas Bubbles Reaching Anode Surface

It is challenging to prevent gas bubbles from reaching the anode surface. A filter between anode and cathode may resist gas bubbles from reaching the anode surface. The finer porosity of the filter, the better filtering effect. But it also resists solution circulation. An electrochemical cell with vertical electrodes and filter close to anode can efficiently keep hydrogen bubbles from reaching the anode surface. First, larger surface area of solution allows hydrogen bubbles to get out of the solution. Secondly, hydrogen bubbles become larger when they reach the filter. So filter with larger hole size can be used, which allows better solution circulation.

5.3.7. Temperature

Temperature is not discussed in this work. Processing at higher or lower than room temperature means higher facility cost. However, temperature can affect many factors of ECP process. These factors include charge transfer, diffusion coefficient, and viscosities of solution and anodic layers. Therefore, one should be aware that temperature is an experimental variable and can possibly be used to manipulate factors such as limiting current plateau, and anode layer structure when needed.

REFERENCES

[1] Strained Silicon Yields Transistor Performance Gains, http://www.intel.com/labs /features/si12031.htm

[2] D. Landolt, Review Article Fundamental Aspects of Electropolishing, *Electrochim. Acta,* 32, 1 (1987), 1-11.

[3] K. Cui, Iron, Steel and Non-ferrous Metal Materials, Machinery Industry Press, Beijing, P. R. China, 1980. [Chinese]

[4] J. Koryta, J. Dvořák, and L. Kavan, Principles of Electrochemistry, 2nd ed. John Wiley & Sons, Inc., New York, 1993.

[5] Freitag, Creighton University, Fall 2002, CHM 203 G: General Chemestry I, Available http://freitag.creighton.edu/CHM203/notes/9.13.02.pdf [Viewed: April 14, 2003].

[6] J. O'M. Bockris and A. K.N. Reddy, Modern Electrochemistry, Vol.2, Plenum Press, New York, 1970.

[7] R. A. Robinson and R. H. Stokes, Electrolyte Solutions, 2nd ed. Butterworths Publications Limited, London, UK, 1959.

[8] Solartron Analytic, Understanding Electrochemical Cells, Technical report 17.

[9] J. Bard and L. R. Faulkner, Electrochemical Methods Fundamentals and Applications, 2nd ed. John Wiley & Sons, Inc., New York, 2001.

[10] D. T. Sawyer, A. Sobkowiak, and J. L. Roberts, Electrochemistry for Chemists, 2nd ed. John Wiley & Sons, Inc., New York, 1995.

[11] F. P. Incropera, D. P. DeWitt: Fundamentals of Heat and Mass Transfer, 4th edn, Jone Willey & Sons, Inc., New York, 1996.
[12] R. Soutebin and D. Landolt, Anodic Leveling under Secondary and Tertiary Current Distribution Conditions, *J. Electrochem. Soc.*, 129, 946-953(1982).
[13] H. Abrams and C. L. Mantell, Electrolytic Polishing of Copper and Nickel Silver, *Electrochem. Tech.*, 5, 287(1967).
[14] M. Matlosz, S. Magaino, and D. Landolt, Impedance Analysis of a Model Mechanism for Acceptor-Limited Electropolishing, *J. Electrochem. Soc.*, 141, 410-418(1994).
[15] G. Busch and H. Schade, Lectures on Solid State Physics, Pergamon Press, New York, 1975.
[16] A.J. Bard, R. Parsons, and J. Jordan: Standard Potentials in Aqueous Solution, Marcel Dekker, Inc., International Union of Pure and Applied Chemistry, New York, 1985.
[17] J. Huo, R. Solanki, and J. McAndrew, Electrochemical Polishing of Copper for Microelectronic Applications, *Surface Engineering*, 19, 11-16 (2003).
[18] V. A. Dmitriev and E. V. Rzhevskaya, Periodic Phenomena in the Anodic Dissolution of Copper in Phosphoric Acid, *Russian J. Phys. Chem.*, 35, 425-429 (1961).
[19] S. H. Glarum and J. H. Marshall, The Anodic Dissolution of Copper into Phosphoric Acid, I. Voltammetric and Oscillatory Behavior, *J. Electrochem. Soc.*, 132(1985), 2872-2878.
[20] S. H. Glarum and J. H. Marshall, The Anodic Dissolution of Copper into Phosphoric Acid, II. Impedance Behavior, *J. Electrochem. Soc.*, 132(1985), 2878-2885.
[21] M. Paunovic and M. Schlesinger, Fundamentals of Electrochemical Deposition, John Wiley & Sons, Inc., New York, 1998.
[22] I.L. Alanis and D.J. Schiffrin, The Influence of Mass Transfer on the Mechanism of Electropolishing of Nickel in Aqueous Sulphuric Acid, *Electrochim. Acta*, 27, (1982), 837-845.
[23] R. Alkire and A. Cangellari, Effect of Benzotriazole on Dissolution of Copper in Presence of Fluid Flow, *J. Electrochem. Soc.*, 136(1989), 913-919.
[24] L. J. Korb, D. L. Olson, J. R. Davis, and et al: Metals Handbook, 9th ed., Vol.13, ASM International, Metals Park, Ohio, 1978.
[25] J. Huo, *Electrochemical planarization of copper for microelectronic application,* doctoral dissertation, Oregon Health and Science University, Oregon, 2004.
[26] F.K. Crundwell, Model for the ac-impedance of an electrode coated by a precipitated salt film, *Electrochimica Acta*, 36, 7(1991), 1183.
[27] K. Kojima and C. W. Tobias, Solution-Side Transport Processes in the Electropolishing of Copper in Phosphoric Acid, *J. Electrochem. Soc.*, 120 (1973), 1026-1033.
[28] J. Edwards, The Mechanism of Electropolishing of Copper in Phosphoric Acid Solutions, *J. Electrochem. Soc.*, 100, 8(1953), 223C-230C.
[29] R. Vidal and A. C. West, Copper Electropolishing in Concentrated Phosphoric Acid, I. Experimental Findings, II. Theoretical Interpretation, *J. Electrochem. Soc.*, 142(1995), 2682-2694.
[30] R. D. Grimm, A. C. West, and D. Landolt, AC Impedance Study of Anodically Formed Salt films on Iron in Chloride Solution, *J. Electrochem. Soc.*, 139, 6 (1992), 1622-1629.
[31] N. Sridhar, D.S. Dunn, In situ study of salt film stability in simulated pits of nickel by Raman and electrochemical impedance spectroscopies, *J. Electrochem. Soc.*, 144, 12 (1997), 4243-4253.

[32] J. Huo, R. Solanki, and J. McAndrew, Electrochemical Planarization of Patterned Copper Films for Microelectronic Applications, *Proceedings of the 22nd Heat Treating Society Conference and the 2nd International Surface Engineering Congress*, 15-17 September 2003, Indianapolis, Indiana, USA, 389-397.

[33] E.P. Simonen, R.H. Jones, M.J. Danielson, Predicted salt film formation effects on stage I stress corrosion cracking of anodically polarized nickel *Corrosion Science*, 34, 6 (1993), 899-914.

[34] P. Neufeld and T. Zervas, The Relationship Between Polarization Characteristics and the Electropolishing Behavior of Copper, *Surf. Tech.*, 8, (1979), 129-135.

[35] A. S. Mansour and T. H. El Sherify, Electropolishing of Horizontal Copper Ring Disks in H_3PO_4 Solutions, *Surf. Tech.*, 19 (1983), 355-361.

[36] M. Novak, A. K. N. Reddy, and H. Wroblowa, An Ellipsometric Study of Surface Films on Copper Electrodes Undergoing Electropolishing, *J. Electrochem. Soc.*, 117, 6 (1970), 733-737.

[37] R. J. Schaefer and J. A. Blodgett, Holographic Study of Electropolishing, *J. Electrochem. Soc.*, 123, 11 (1976), 1701-1705.

[38] J. L. Fang and N. J. Wu, Determination of the Composition of Viscous Liquid Film on Electropolishing Copper Surface by XPS and AES, *J. Electrochem. Soc.*, 136, 12 (1989), 3800-3803.

[39] S. Magaino, M. Matlosz, and D. Landolt, An Impedance Study of Stainless Steel Electropolishing, *J. Electrochem. Soc.*, 140(1993), 1365-1373.

[40] R. Cottis, S. Turgoose, and B. C. Syrett, Corrosion Test Made Easy Electrochemical Impedance and Noise, NACE International, USA, 1999, 19-33.

[41] C. Wagner, Contributions to the Theory of Electropolishing, *J. Electrochem. Soc.*, 101(1954), 225-228.

[42] K. D ianatkhah, 'Passivation Behavior of Copper Thin Films for Electrochemical Planarization Application', MS thesis, Oregon Graduate Institute of Science and Technology, Beaverton, Oregon, 1999.

[43] J. Huo, R. Solanki, and J. McAndrew, A New Electropolishing System for Replacement of CMP, *Electrochemical and Solide-State Letters*, submitted.

In: Focus on Electrochemistry Research
Editor: Magdalena Nunez, pp. 147-168
ISBN: 1-59454-545-6

Chapter 5

ELECTROCATALYTIC HYDROGENATION OF UNSATURATED ORGANIC MOLECULES: SOME FEATURES AND PREPARATIVE OPPORTUNITIES

L. M. Korotaeva and T. Ya. Rubinskaya
N. D. Zelinsky Institute of Organic Chemistry, Russian Academy of Sciences, Moscow, Russian Federation

ABSTRACT

Influence of the nature and surface condition of metallic cathodes (Cu, Ni, Zn, Hg) on the direction of citral electrochemical reduction has been studied. Electrocatalytic hydrogenation (ECH) on modified Cu and Ni cathodes has been effectively implemented. X-ray photoelectron and Auger-electron spectroscopy data have confirmed that the Cu cathode surface layer composition was modified by preliminary annealing, what is likely to be responsible for a considerable buildup in electrocatalytic activity.

Effects of homogenous factors, i.e. the nature and concentration of solvent and acid components, on the composition of the products of ECH of citral have been closely examined.

The identified regularities were taken in consideration in ECH of *trans*-2-allyl-6-R-1,2,3,6-tetrahydropyridines (R = Me, All, Ph). The products of complete hydrogenation of all double bonds on the Ni cathode have been obtained in quantitative yields. Selective hydrogenation of terminal double bonds of allyl substituents on the Cu cathode with retention of the intracycle double bond has been successfully realized.

The proposed ECH concept that takes into account the influence of both heterogeneous and homogenous factors on ECH efficiency may appear useful in selecting hydrogenation conditions for polyfunctional unsaturated organic molecules.

INTRODUCTION

A direction of electrochemical reduction (ER) of organic compounds is governed by their structure, nature and the number of functional groups. A distinctive feature of the process is cathode functioning as a co-reagent. The electrode nature in many instants has a substantial effect on the direction and mechanism of organic molecule transformations. A principle of metallic electrode separation frequently employed in the electrosynthesis was formulated by Antropov [1] back in 1950. It bases on the Met–H bond energy and bears a methodological character. Two basic types of metallic cathodes were proposed: with a high (Pt, Pd, Ni) and low or zero hydrogen sorption capacity (Hg, Pb, Cd). This approach logically leads to two types of the ER mechanism [2]. A direct electron transfer to the organic substrate occurs preferably on metals with high hydrogen overvoltage (Hg, Pb, Cd) producing respective anion radicals. Their further transformations are highly dependent on the electrolysis conditions. Electrocatalytic hydrogenation (ECH) is implemented on metals with low hydrogen overvoltage (Pt, Pd). The ECH mechanism [2] actually consists of addition of atomic hydrogen generated electrochemically and adsorbed on the electrode surface to multiple bonds of an unsaturated organic molecule. An unsaturated compound can be hydrogenated either in the adsorbed state or in the close-to-electrode layer of solution [3] depending on the electrode material, its surface condition and composition of the environment. ECH often proceeds with high selectivity [4] and may result in hydrogenation of one or a few groups in the polyfunctional molecule.

Such mechanism is realized in heterogeneous catalytic processes [5, 6] where organic compound adsorption on the catalyst is determinant for the hydrogenation direction. For this reason, a lot of regularities revealed in catalytic reactions appear applicable to ECH processes.

It should be pinpointed that implementation of ER and ECH mechanisms often leads to different products. The process sometimes follows a combined mechanism resulting in a complex mixture of products. Each mechanism has individual favorable conditions in relation to the composition of the reaction medium [4]. Feasibility assessment for either of the mechanisms in each particular case may substantially facilitate selection of the conditions beneficial for obtaining the target product.

The extensive experimental material summarized in the reviews [7, 8] allows a conclusion that a boundary between metals of 1st and 2nd groups is highly provisional and that many metals can exhibit catalytic activity in proton-donating media during electrolysis.

Modification of the process mechanism and kinetic parameters resulting from cathode pretreatment was communicated in a number of publications [7, 9], e.g. a decrease of the hydrogen evolution potential (E_H) on metals with the loose surface. Because of that it is advisable to carry out hydrogenation reactions on metals with the developed surface [2, 7, 9]. Expansion of the electrode working surface leads to a rise in the adsorbed hydrogen concentration and hence to higher hydrogenation efficiency.

In our opinion, most attractive and promising for ECH are metals characterized by rather low electrocatalytic activity with respect to the hydrogen evolution reaction (Ni, Cu, Zn). They are not so costly as Pt and Pd, available and thus more suitable for the routine electrosynthesis. These metals have close atomic radii though they differ in the electron structure and formally belong to different groups. The Met–H energy declines in transition from Ni to Zn [8]. However, with its decrease the metal-substrate interaction, i.e. adsorption, becomes more and more meaningful. That is likely to reason special selectivity of individual functional groups during ECH.

It seemed interesting to trace how the direction and selectivity of ECH of an unsaturated organic compound change depending on a electrode material and its surface condition (pretreatment) and to evaluate the influence of a homogenous factor, i.e. the medium composition, using Ni, Cu and Zn cathodes.

ELECTROREDUCTION OF CITRAL

Citral, an available chemical frequently used as a synthon in the synthesis of valuable fragrance substances and pharmaceuticals [10], was chosen as a model compound. Unlike the majority of monofunctional models traditionally employed in electrocatalytic researches, the citral molecule has three multiple bonds with different polarity: C=O conjugated with C=C and isolated, sterically hindered C=C in the bulky alkene radical.

It is known that the main products of ER of α,β-unsaturated carbonyl compounds on the Hg cathode are dimers with different structure [11]. However, monomolecular products having different hydrogenation degree as well as the destruction products were produced at ECH of acrolein and crotonaldehyde [12] on the platinated Pt cathode. Pinacol and an oxytetrahydrofuran derivative were synthesized in the total yield of 50 – 70% during the citral electrolysis on the Hg cathode in the water-ethanol buffer medium [13]. A mixture of dimeric and oligomeric products was obtained in anhydrous or 10% aqueous DMF with supporting tetrabutylammonium perchlorate on the Hg cathode at $E = -2.0 \div -2.2$ V relative to the saturated calomel electrode (s.c.e.) [14]. The same products were generated by ER of citral in 40% aqueous DMF on the Cu cathode ($E = -1.8$ V).

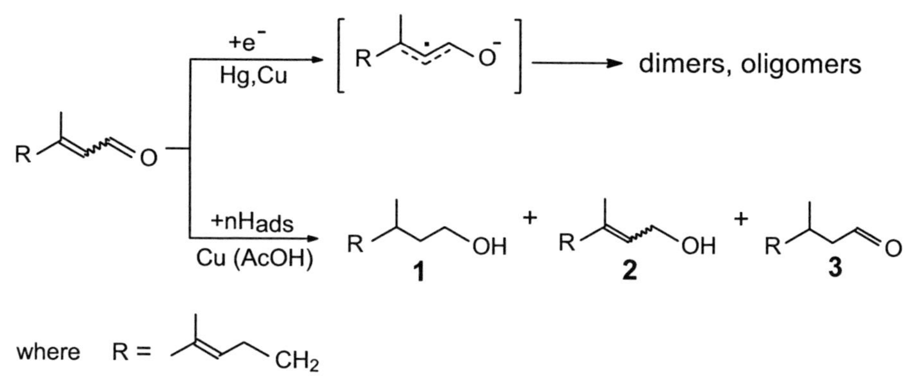

Scheme 1

However, when the process is conducted in excessive AcOH ($E = -1.4 \div -1.5$ V) it is predominated by monomeric products of double bonds hydrogenation (Scheme 1) - citronellol (**1**), isomeric nerol and geraniol (**2**) (the proportion of nerol and geraniol agrees well with the neral:geranial ratio in the starting citral in all test runs). In a few cases, an insignificant amount of citronellal (**3**) is present in the product mixture. Presumably, in the absence of AcOH the direct ER mechanism is realized on Hg and Cu cathodes, whereas in the presence of AcOH the ECH mechanism is implemented on the Cu cathode.

This assumption was confirmed in the study [15] of the citral electrochemical behavior on Hg (a hanging drop technique), Au, Pt, Cu and glass carbon microcathodes in DMF (supporting salt – Bu_4NPF_6) using a cyclic voltamperometry technique. In the absence of proton donors (AcOH, H_2O), a peak of the reduction potential falls within the range of −2.03 ÷ −2.05 V (s.c.e.) regardless of the electrode material. With AcOH present, hydrogen evolution on Cu and particularly on the Pt electrode occurs at considerably less negative potentials ($E_H < -1.0$ V for the Cu cathode). Hence the monomeric products of hydrogenation of double bonds **1**, **2** and **3** on the Cu cathode in the presence of AcOH under the preparative electrolysis conditions are formed *via* the ECH mechanism.

Evidently, the ECH mechanism in its pure form can be realized on cathodes with medium hydrogen overvoltage only when the direct ER potential is much more negative than E_H, otherwise the mechanism may be of a mixed character. Both direct ER and ECH processes are realizable on metallic cathodes, such as Cu, Ni and Zn [7, 9]. This offers high opportunities to the studies on the influence of various factors, both heterogeneous and homogenous, on the electrochemical process direction and selectivity. Of special concern here is electrode pretreatment. Conventional methods of electrode activation are [9]: mechanical or chemical treatments, application of asymmetric alternating current (cathode-anode polarization) and addition of metal salts to electrolysis solution. Annealing of a Cu cathode was employed in some cases for surface treatment.

Electrocatalytic Hydrogenation of Citral on the Cu Cathode

Table 1 summarizes the results of ECH of citral (detailed in [16]) obtained using different Cu cathode pretreatment alternatives. They demonstrate noticeable changes in both the total yield and in the ratio of alcohols 1 and 2 in the product mixture (Tests 1 - 3). Unsaturated alcohols 2 predominated when traditional cathode pretreatment techniques were used (Tests 1 and 2). More efficient appeared to be Cu cathode annealing in the glassblowing burner flame (Cu_{ann}) followed by copper oxide elimination by dilute (1:10) HNO_3. The electrolysis on this cathode leads to a rise both in the total yield of alcohols and in the portion of target citronellol 1 in the mixture (Test 3).

Such strong effect of the heterogeneous factor on the composition of the ECH products is indicative of a classical mechanism of the process - a reaction between the citral molecule adsorbed on the cathode surface and active hydrogen. Adsorption of organic compounds on the Cu cathode reaches its maximum [17] within the range −1.0 ÷ −1.2 V (s.c.e.). Consequently, on the Cu_{ann} cathode the E_H decline to below −1.2 V is beneficial for citral adsorption efficiency on its surface. Addition of $CuCl_2$ to the electrolyzed solution results in a decrease of both the product yield and ECH selectivity (Test 5). Copper precipitation on the Cu_{ann} cathode during the electrolysis apparently disturbs the surface structure formed under annealing, negatively affecting the citral adsorption conditions.

Optimization of ECH conditions (composition of electrolytic solution) [14, 18] allowed production of citronellol 1 in rather high yields (Tests 4 and 6). It should be noted that the sterically hindered C=C bond distanced from the C=C-C=O grouping remains unaffected.

Table1. Total yield and ratio of the products of electrocatalytic hydrogenation of citral on the Cu cathode relative to a pretreatment technique [a]

No.	Cu cathode pretreatment	E, V	DMF,%	C_{AcOH}/C_{cit}	Total yield of 1+2 (%)	Ratio 1/2
1	mechanical treatment	−1.4	55	20	54	0.57
2	HNO_3 (dilute)	−1.5	55	20	40	0.48
3	a) annealing b) HNO_3 (dilute)	−1.2	40	20	86	2.3
4	a) annealing b) HNO_3 (dilute)	−1.2	40	10	85	4.2
5	a) annealing b) HNO_3 (dilute) c) in the presence of 1% $CuCl_2$	−1.2	40	10	40	0.3
6	a) annealing b) HNO_3 (dilute)	−1.2	40	5	90	5.0

[a] The electrolysis was carried out in a cell with a glass diaphragm at constant potential, anode – Pt, supporting salt - 0.1 *M* KCl, C_{cit} = 0.024 *M*.

EXAMINATION OF THE CU CATHODE SURFACE BY XPS AND XAES

It is known [19] that metallic copper lacking vacant d-orbitals in the ground state exhibits catalytic activity in hydrogenation only after its activation due to the appearance of excited valent states. We assume that the catalytic activity buildup in ECH of citral is most likely associated with the modification of its surface state caused by Cu cathode annealing. It seemed interesting to examine the annealed cathode surface by physicochemical methods. A popular and efficient method to study the surface massive metal electrodes is Auger spectroscopy [20, 21]. Using XPS (X-ray photoelectron spectroscopy) and XAES (X-ray Auger electron spectroscopy) [22] we analyzed the surface of the copper samples: a) non-annealed and b) annealed in the gas burner flame and then washed with dilute $HN0_3$.

According to the XPES and XAES data, annealing of the copper plate in the above conditions leads to the surface layer enrichment with oxygen both as a Cu_2O component and in the adsorbed and interstitial forms. It should be pinpointed that deep penetration of oxygen into the crystalline lattice of smooth platinum to produce dissolved oxygen during oxygen-hydrogen thermal treatment is regarded as an ascertained fact [23].

A detailed discussion of the influence of the structure and state of the metallic catalyst surface on its adsorption and electrocatalytic properties was provided in [20, 21]. As a rule, thermal treatment of electrodes results in surface smoothening and formation of more ordered and stable structures [23]. The data on the physicochemical properties of copper allowed us to assume [16] that the increase in catalytic activity of the annealed Cu cathode was associated with recrystallization processes. According to Bochvar rule [24], the copper recrystallization temperature is about 370° C. At this temperature, crystallite grains in the deformed pure copper rapidly grow in size from $2 \cdot 10^{-5}$ to $2.5 \cdot 10^{-4}$ cm during a short interval at the beginning, their ultimate size being independent from a deformation degree of the starting sample. After three-minute annealing of the Cu cathode in the burner flame, the temperature [22] stops rising at the point of 340° – 360° C (measurements by thermocouple).

Two parallel processes are likely to occur during annealing: emergence of more defects in the crystalline lattice as affected by interstitial oxygen and Cu_2O and recrystallization or crystal grain coarsening, i.e. formation of smooth regions. The removal of interstitial oxygen under the ECH conditions should lead to the appearance of new adsorption sites. Thus two types of active sites different in structure can be formed on the Cu_{ann} cathode surface.

A comparison of the adsorption and kinetic data obtained in ER of a few ethylene derivatives on smooth platinum let assume [25] that atomic hydrogen and molecules of organic substances adsorbed on different adsorption sites. An interaction of olefin with atomic hydrogen proceeds at the boundaries of these regions and determines the rate of the entire process.

A comparative study on the behavior of organic compounds from different classes evidences to a dominating trend to π-adsorption [26]. The adsorption rate of organic compounds (e.g. benzene and phenylacetic acid) on smooth Pt is higher than on the platinated Pt [27], and a special arrangement of adsorption sites and their structural consistency with adsorbing molecules are necessary for efficient ECH of complex organic compounds.

In the case of the annealed Cu cathode, the potential of hydrogen evolution from the water-organic medium decreases (by 200 - 300 mV) and the crystalline copper lattice defects may perform a function of adsorption sites for the evolving hydrogen. The smooth regions formed on the Cu_{ann} cathode surface due to recrystallization are most likely beneficial for π-adsorption of large organic molecules. Thus there should not be any competition for an adsorption site between the organic molecule and atomic hydrogen during ECH. That, in the long run, may be a reason for a sharp rise in electrocatalytic activity of the copper electrode resulting from annealing.

When a massive metal is used as an electrode-catalyst, of high concern is its history, i.e. a fabrication technology (forging, molding, punching, etc.) [9, 21], since it determines the structure of a metal and variations in the concentration of crystalline lattice defects. For ECH of citral [16], in addition to a cylindrical Cu cathode (refined copper), we employed a spiral copper wire cathode. The spiral cathode was subjected to 5 annealing runs under the standard conditions. Yields of alcohols 1 and 2 and their ratio after ECH of citral were similar to those obtained on the annealed cylindrical Cu cathode. Hence the technological defects in metal are leveled during annealing and the electrocatalytic properties of copper become standardized.

After two years of experimenting with the cylindrical Cu cathode, we have found that its efficiency in ECH of citral somewhat decreases whereas E_H shifts towards positive values. The Cu cathode surface becomes darker and more friable. Basing on the XPS and XAES data, we arrived at a conclusion that the decline of structure ordering and reduction of the crystalline lattice defect concentration occurred due to the prolonged use of the multiply annealed Cu cathode. It is known [28] that the penetration of supporting electrolyte cations (alkaline metals) into the cathode is inherent to the electrolytic process. The results reported in [29] show that Cl^- anions are adsorbed and incorporated into the crystalline lattice of metallic copper as evidenced by characteristic coloration of the flame during annealing. The adsorbed organic molecules and chlorine are removed from the cathode during annealing whilst the crystalline lattice defects accumulate. A thorough mechanical treatment of the cylindrical Cu cathode surface to gain the lustrous quality and its subsequent annealing (3 - 5 times) resulted in complete recovery of electrocatalytic activity.

The XPS and XAES data indicate that in-air storage of the samples after standard treatment leads to 100% oxidation of surface copper atoms to Cu^{2+}. This change in the surface condition after exposure to air oxygen implies that the annealed cathode should be used in the electrochemical process immediately after its pretreatment and storage in solution under argon. Of importance is that the described method of electrode preparation ensures high reproducibility of EHC results. Moreover, special experiments on ECH of citral proved applicability of this electrode for the multi-hour electrolysis.

Thus, both the literature and relevant XPS and XAES data allow a conclusion that the copper surface layer enrichment by oxygen in the form of Cu_2O and atomic adsorbed oxygen as well as the structural changes caused by the elevated temperature are responsible for the rise in electrocatalytic activity of the Cu electrode. Annealing of the Cu cathode in the burner flame is a timesaving and practical electrode preparation alternative for ECH as it provides well-reproducible results.

Influence of the Solvent Composition

A detailed study of the influence of homogenous factors, though in a heterogeneous process, was possible owing to high sensitivity of ECH of citral on the Cu_{ann} cathode manifested in product composition changing even at minor variations of the composition of the medium (Table 1, Tests 3, 4, 6).

As reported in [6], selectivity of catalytic hydrogenation of citral varies with changes in the solvent nature. The composition of the medium should provide maximum efficiency of the organic substrate interaction with the cathode surface. However components of the medium contacting with the electrode may block the surface [3, 30], thereby deteriorating the substrate adsorption conditions. In the instant of citral, there are additional difficulties related to the presence of a bulky alkene radical. These are lower solubility in water, hindered diffusion toward the electrode and desorption of the reaction products from its surface. The influence of the medium composition and solvent nature was closely examined in [14, 31] where the total product yield was shown to vary from 0 to 90% in terms of the organic phase nature and content.

Water is known to possess an ordered structure [32]. Water clusters do not obstacle interactions of small polar molecules with the electrode surface though they hinder adsorption of bulk organic molecules. That is likely to account for low yields of the ECH products in aqueous solutions. For this reason and because of low citral solubility in water, we used mixed water-organic solutions. Figure 1 shows the results of citral ECH on the Cu_{ann} cathode with different compositions of a mixed DMF-water solvent (for details see [14]). According to [33], a DMF concentration increase in aqueous solution involves non-monotonic changing of some physical properties of the system. Small DMF additions (17.5 volume % DMF) to water stabilize the water structure while further DMF content increasing leads to its destruction promoting organic substrate adsorption on the electrode surface. The mixed solution acquires maximum viscosity at ~70% DMF and its value is higher than that of pure ingredients. This is related to the emergence of DMF-H_2O associates differing in composition.

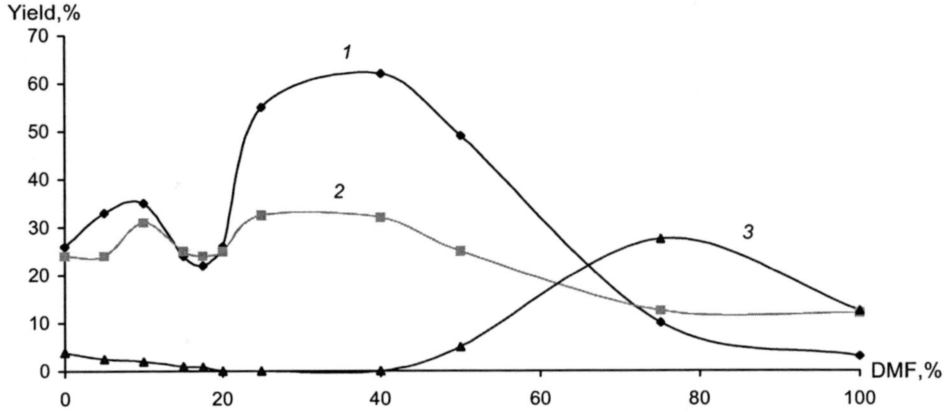

Figure 1. The depedence of yields of the products of ECH of citral from the DMF proportion in solvent (supporting salt – 0.1 *M* KCl, C_{cit} = 0.024 *M*, C_{AcOH}/C_{cit} = 10): *1* – citronelol (1), *2* – isomeric nerol and geraniol (2) (overall yield), *3* – citronellal (3).

Indeed, when the DMF concentration in solution reaches ~20%, there is a descending bend on the curve of the ECH product yield dependence. The maximum yield of alcohols 1 и 2 is achieved in 40% aqueous DMF. As the DMF concentration grows further, solubility of citral increases, what naturally [34] involves a decline of its adsorption on the electrode surface and hence lower alcohol yields.

The situation changes when EtOH–H_2O mixtures are used as solvents (Fig. 2). It appeared that in ECH of citral carried out in solution with <30% EtOH alcohols 1 and 2 are produced in approximately equal amounts and their total yield does not exceed 50% (the process is accompanied by resinification) [31]. A further EtOH concentration buildup leads to a marked decrease of the alcohols yield and at EtOH >70% no ECH products are produced. Evidently, that does not merely result from a rise in organic substrate solubility. The dependence established between the ECH product yield and the solvent composition reflects a non-linear behavior of changes in the physical properties of EtOH–H_2O mixtures. EtOH is known to have a high association degree [35]. Intermolecular hydrogen bonds link EtOH molecules forming long chains of associates (up to 50 molecules, coordination number ~2). The structure of solution with more than 70% alcohol generally corresponds to the structure of pure alcohol [36]. As the water content in the mixture increases, gradual destruction of the alcohol structure occurs accompanied by generation of associates with different composition. In our opinion, it is exactly the destruction of long-chain associates at EtOH <70% that is responsible for better citral adsorption on the electrode surface and hence for ECH implementation. The ECH results reflect the water structure stabilizing effect [37] at EtOH <20%: yields of alcohols 1 and 2 decrease (the curves feature a clearly seen descending bend in this region, Fig. 2).

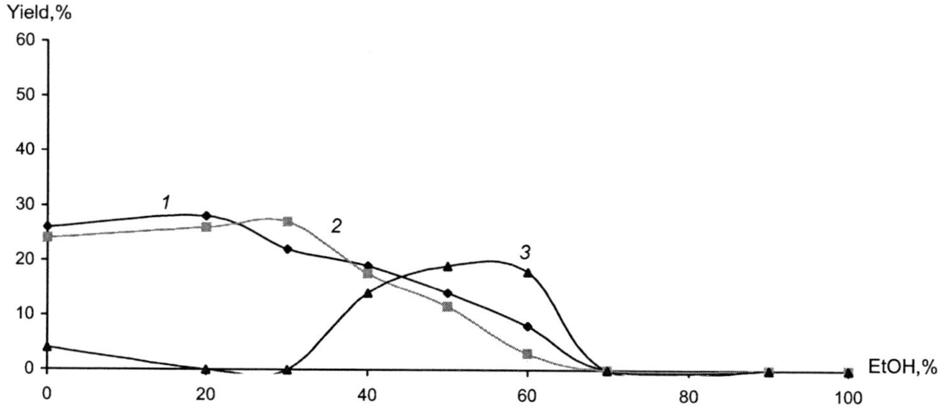

Figure 2. The depedence of yields of the products of ECH of citral from the EtOH proportion in solvent (supporting salt – 0.1 *M* KCl, C_{cit} = 0.024 *M*, C_{AcOH}/C_{cit} = 10): *1* – citronelol (1), *2* – nerol and geraniol (2) (overall yield), *3* – citronellal (3).

It is well-known that viscosity of H_2O–EtOH solution is maximal at EtOH ~40% and almost 3-fold higher than viscosity of each individual component. Note that an appreciable amount of aldehyde 3 was detected in the ECH products when the solvent had maximal viscosity, meanwhile the yield of products 1, 2 and 3 was about the same [31]. Of concern is that in ECH of citral proceeding in DMF-H_2O solutions (Fig. 1) the yield of aldehyde 3 is essentially higher that the yield of alcohols 1 and 2 in 70–80% DMF, i.e. under maximal viscosity conditions. The dependences obtained (Fig. 1 and 2) reflect the non-monotonic changes in the physicochemical properties and structure of water-organic solutions, which, in turn, are likely to affect the citral adsorption behavior.

Therefore the analysis of the influence of the water-organic solvent composition (H_2O–DMF and H_2O–EtOH) shows that proton-donating solvents, specifically lower alcohols, considerably reduce ECH efficiency and selectivity. 40% aqueous DMF appeared the best choice for ECH of citral. Consideration of the solvent structure influence (a homogenous factor) is obligatory in conducting ECH of bulk organic molecules, even though the process may be of a pronounced heterogeneous character.

Influence of the Acidic Component

The acidic component nature and concentration noticeably influence the citral ECH efficiency [18]. The E_H value is much lower when strong acids are used than in the presence of AcOH: $E_H = -0.6 \div -0.7$ V for HCl and $E_H = -0.7 \div -1.0$ V for oxalic acid (refer to Table 1). At the same time, both the total yield of alcohols 1 and 2 and the ratio 1/2 tend to decreasing (down to 30 – 40% and 0.3, respectively); in addition, dimeric by-products are generated in considerable amounts. It is known [38] that, in the presence of acids, compounds having a system of conjugated C=C and C=O bonds may undergo chemical transformations, such as vinyl polymerization, condensation, etc. Under such conditions, ECH of citral is likely to be accompanied by its resinification in bulk solution.

On the other hand, hydrogen evolution proceeds with a less negative potential if compared to the potential of maximal adsorption of organic compounds (E = –1.0 ÷ –1.2 V [17]). It is evidently unbeneficial for citral adsorption on the cathode surface because the process of hydrogen evolution and recombination prevails.

When AcOH is used, ECH selectivity is rather high and varies over a broad range (from 4 to 10) of the molar ratio of AcOH to citral, the total yield of the alcohols being 75 – 90%. Maximum selectivity is achieved when acid is present in the amount needed for complete hydrogenation of the system of conjugated C=C and C=O bonds form the stoichiometric standpoint ($C_{AcOH}/C_{cit} = 4$).

In ECH AcOH obviously acts as a depolarizer - a hydrogen source for hydrogenation. At the same time, the polarography data [18] evidences to the AcOH engagement in generating stable associates with participation of multiple citral bonds. Here, adsorption of associates most likely proceeds on the copper electrode surface and is accompanied by a hydrogen release from them. Such close mutual positioning of the reaction sites should lead to a rise in hydrogenation selectivity. However the higher is the AcOH concentration, the lower is the ECH efficiency - probably because of citral being forced from the electrode surface by the adsorbed AcOH molecules. A higher excess of acid may cause electrode surface blocking to result in a decline of both the alcohols yield and process selectivity. (At the ratio C_{AcOH}/C_{cit} = 40, the total yield of the alcohols – 53% and the relevant ratio 1/2 – 0.68.)

Thus the detailed research on the influence of the acidic component nature and concentration allows a conclusion [15] on the practicability of weak organic acids (e.g. AcOH) in ECH provided they are used in the amounts needed for hydrogenating conjugated double bonds from the stand point of stoichiometry.

The optimal composition of the medium (40% aqueous DMF, $C_{AcOH}/C_{cit} = 4$) identified as a result of our close examination of ECH of citral may be useful for raising efficiency of ECH of citral conducted on other cathodes (Ni and Zn) as well.

Electrocatalytic Hydrogenation of Citral on the Ni Cathode

It is known that, unlike copper, nickel refers to the Pt-group metals characterized by lower overvoltage of hydrogen. However electrocatalytic activity of nickel largely depends on an electrode pretreatment technique. For the carbonyl series, it was shown [7] that reduction on a smooth nickel electrode occurred *via* the direct ER mechanism, whereas this process on Raney nickel appeared to be electrocatalytic. ECH of a few α,β-unsaturated ketones on Raney nickel fixed on the cathode by a magnet made it possible [39] to perform selective hydrogenation of the conjugated C=C bond with retention of the carbonyl group and to obtain relevant saturated ketones in high yields.

As reported in the reviews [21, 40] discussing the role of structural factors in the electrocatalysis, high-dispersed metals differ from massive metals in the adsorption and electrocatalytic properties. It is also should be noted that the crystallographic orientation and defectiveness of the metal-electrocatalyst surface affect the electrocatalytic parameters to a less degree than their dispersity. Although ECH of unsaturated organic compounds on different catalysts was examined in numerous researches, the electrode material and electrochemical conditions have been so far selected empirically in each particular case [3, 23].

The authors [41] studied changes of the electrocatalytic properties of a nickel cathode on the example of ECH of citral using different pretreatment alternatives. A primary objective was to find a convenient and effective technique for activating the nickel cathode. It appeared (Table 2) that on the polished Ni cathode hydrogen evolved at a surprisingly high cathode potential (E_H −1.7 V). Presumably, a direct electron transfer to the citral molecule may occur as well. Indeed, the electrolysis products (Test 1) comprised the hydrogenation products (alcohols 1, 2 and aldehyde 3) along with dimers and oligomers. Such makeup of the products is indicative of parallel implementation of two mechanisms - ECH and direct ER.

Table 2. Results of electrocatalytic hydrogenation of citral on the nickel cathode[a]

No.	Type of Ni cathode	Pretreatment conditions	E, V	Yield, (%)		
				1	2	3
1	polished	a) mechanical treatment b) HNO_3 (dilute)	−1.7	17	29	5
2	polished	a) in-flame annealing b) HNO_3 (dilute)	−1.9	0	0	0
3	polished	a) mechanical treatment b) cathode-anode activation	−1.0	51	17	17
4	Polycrystalline Ni/Ni	a) mechanical treatment b) bath (chloride-borate)	−1.3	54	11	0
5	dispersed Ni/Ni	a) mechanical treatment b) precipitation from $NiCl_2$	−1.0	95	0	0

[a] ECH conditions: 40% aqueous DMF, $C_{AcOH}/C_{cit} = 4$, supporting salt – 0.1 M KCl.

After Ni cathode annealing in the gas burner flame, we managed to electrolize citral at E = −1.9 V (Test 2), though it yielded only dimeric and oligomeric products. Thus heat treatment of the Ni cathode, opposite to the Cu cathode, slows down the hydrogen evolution reaction. The Ni cathode loses its electrocatalytic properties and behaves as a metal with high hydrogen overvoltage. This pretreatment technique extends opportunities of the Ni cathode use in implementing direct ER due to a larger region of operational potential.

Cathode-anode activation with a periodic electrode polarity alternation is another method often recommended [9] for the electrode surface preparation. In addition to the surface cleaning from contaminants, such activation leads to a change in the surface structure, i.e. its loosening, due to the formation of oxides and their subsequent reduction [20]. Anyhow, this activation technique proved low-efficient for ECH of citral. Although E_H decreases markedly (to -1.0 V), the selectivity is low: the mixture contains a whole set of hydrogenation products (Test 3) and a considerable amount of the unreacted citral.

Electrodeposition of metals on electrode surface is a conventional way to increase catalytic activity of electrodes. Unlike skeleton catalysts, electrolytic precipitates can be prepared directly prior to use avoiding labor-consuming preliminary stages [42]. The precipitate structure may dramatically vary against the preparation conditions [23, 43], such as the composition of the medium, presence of additives, concentration of the components, temperature, current density and support material. The available comprehensive publications on electrocrystallization of metals mainly focus on design of smooth solid coatings with anticorrosive and decorative properties. Nonetheless the developed procedures allow obtaining surfaces with the pre-given structure, e.g. a precipitate produced by nickel electrodeposition from the chloride-borate bath [43] is characterized as polycrystalline and flat, with pronounced grains. We used the same conditions to perform nickel electrodeposition on the electrode subjected to thermal treatment. A satisfactory yield of alcohol 1 (77%) was attained only in the first run of thermally pretreated Ni cathode electrodeposition on the surface. The total product yield and ECH selectivity were considerably lower in the reruns after nickel electrodeposition on the mechanically treated surface (Test 4).

The structure of the polycrystalline precipitate is known [23] to be largely dependent on the nature and state of the support metal surface. Therefore to obtain the electrode surface characterized by catalytic activity and good reproducibility it is needed not only to strictly observe the electrodeposition regime but also to control the state of the support.

A nickel electrode with the dispersed nickel surface (Ni_{disp}/Ni) electrodeposited from aqueous solution of $NiCl_2$ appeared to be more efficient for ECH of citral. Its surface was renewed before each electrolysis run. The single product was obtained on this electrode in a series of test runs - alcohol 1 in a high yield (Test 5) (products 2 and 3 were detected only in trace quantities). Hydrogen evolution potential decreases gradually from test to test during ECH of citral. Presumably, multiple electrodeposition of dispersed nickel entails relevant structuring of the metal surface. Reduction of nickel precipitation time (from two hours to thirty minutes) in the next test runs led to a minor decrease in the alcohol yield. This effect is similar to the "memory effect" described in [20], i.e. ECH results tend to improvement in a series of electrolysis runs conducted with the same sample and the same electrode pretreatment.

Therefore a feasible procedure of nickel electrodeposition under the above-described conditions is useful for achieving high reproducibility of ECH results and high yields of alcohol 1. It should be highlighted that, like in the experiments with the Cu_{ann} cathode, the conjugated C=C-C=O system is hydrogenated completely and the isolated C=C bond remains unaffected provided there are three multiple bonds in the molecule.

ECH on the Ni_{disp}/Ni cathode is likely to proceed by a mechanism different from the ECH mechanism on the Cu_{ann} cathode. In the instant of massive metals [8], the energy of the Cu-H bond is much lower than of Ni–H, i.e. a compact Cu electrode has a considerably lower capacity to adsorb hydrogen, yet adsorbability of organic compounds on copper from the aqueous medium is higher than on nickel [17]. Moreover, a layer of adsorbed water dipoles forms on the surface of disperse nickel in aqueous solutions [44]. It is also known [45] that Ni (skeleton, electrodeposited) adsorbs a relatively large amount of hydrogen, a considerable fraction of which is present on the electrode surface in the active form. Apparently, this hinders adsorption of an organic substrate on a cathode from a water-organic solvent. It is also known [3, 21, 27] that adsorption of organic compounds on electrodes with a loosened surface is lower than on smooth ones. Consequently, on the Ni_{disp}/Ni cathode more probable is hydrogenation of the molecule arriving from solution than of the adsorbed unsaturated organic molecule. Such ECH mechanism was earlier examined on the Ni_{disp}/Ni cathode in the aqueous environment for ethylene [44] and 1-(2-propionitrile)pyrrolidine [46].

Of interest is that in ECH of olefins on a polished Ni cathode (similar to a Pt cathode) their destruction is possible [45], yet no destruction of citral occurs on the Ni_{disp}/Ni cathode under the conditions we proposed. The obtained results (Table 2, Test 5) prove that the environment composition conditions selected earlier for ECH of citral on the Cu_{ann} cathode are also optimal for ECH on the Ni_{disp}/Ni cathode, though another mechanism is realized for the latter.

Electroreduction of Citral on the Zn Cathode

Zinc is a metal characterized by high hydrogen overvoltage and a preferable direct ER mechanism of electrochemical processes. However it is also capable of a specific interaction with a carbonyl group in reduction of streptomycin [7]. The catalytic action of electrolytic powders and skeleton Zn in hydrogenation of a triple bond to a double bond in dimethylethynylcarbinol was described in [39]. We assumed that the specific interaction of a carbonyl group with a metal could occur under ECH of citral on the Zn cathode, yielding isomeric unsaturated alcohols 2 as main reaction products.

Intensive hydrogen evolution is observed on a mechanically treated (polished) Zn cathode in 40% aqueous DMF and a fourfold molar excess of AcOH at $E = -1.7$ V. The total yield of alcohols 1 and 2, hydrogenation products, does not however exceed 12%. Preliminary precipitation of dispersed Zn from aqueous solution of $ZnCl_2$ on the Zn cathode surface merely results in an insignificant reduction of E_H at the very beginning of the electrolysis and then the potential rises to $E = -1.8$ V. That should be probably related to specific adsorption of the carbonyl molecule slowing down the hydrogen evolution reaction. The total alcohol yield increases only up to 26% (the alcohol ratio 1/2 ~ 0.8). Basically, two parallel processes are likely to proceed on the Zn cathode: a) H evolution and its rapid recombination, and a) direct ER of citral resulting in noticeable resinification, similar to the Hg cathode. A contribution of the ECH reaction into the electrochemical process is insignificant.

For comparison, on the Hg cathode ($E= -1.7$ V) with the same composition of the medium (40% aqueous DMF, $C_{AcOH}/C_{cit} = 4$), a complex mixture of dimeric and oligomeric products was obtained; as for aldehyde 3, its content was as low as ~10%, i.e. direct ER of citral predominates.

Thus, unlike Cu and Ni cathodes, electrocatalytic activity of which alters dramatically in terms of the surface condition, the Zn cathode acts mostly as a metal with high hydrogen overvoltage, similar to the Hg cathode. Modification of its surface by dispersed Zn does not affect the electrochemical process direction.

Electrocatalytic Hydrogenation of Citral on Modified Cathodes

As mentioned above, electrolytic powders or so-called "spongy" electrodes are recommended for hydrogenation purposes [7, 9]. Of concern however are a material and condition of the cathode surface where on the dispersed layer is deposited [23].

The data in Table 3 allow comparison of performance of various spongy cathodes in ECH of citral. Electrodeposition on the surface of mechanically treated cathodes (except for Cu_{ann}) was carried out from 0.5 *M* aqueous solution of corresponding salts $CuCl_2$, $NiCl_2$ and $ZnSO_4$ according to the procedure similar to that described in [42].

Table 3. Results of electrocatalytic hydrogenation of citral on spongy cathodes [a]

No.	Cathode	*E*, V	Total yield of 1+2 (%)	Ratio 1/2
1	Zn_{disp}/Zn	–1.8	26	0.8
2	Ni_{disp}/Zn	–1.2	44	>30
3	Ni_{disp}/Cu_{ann}	–1.1	74	9.0
4	Cu_{disp}/Cu_{ann}	–1.0	68	2.3
5	Ni_{disp}/Ni	–1.0	95	>30

[a] ECH conditions: 40% aqueous DMF, $C_{AcOH}/C_{cit} = 4$, supporting salt – 0.1 *M* KCl.

From Table 3 it follows that the Zn cathode coated with dispersed nickel is more efficient in ECH of citral than the spongy Zn cathode (Tests 1 and 2). Actually only alcohol **1** in a yield higher than the total yield of alcohols 1 and 2 in Test 1 is produced on the Ni_{disp}/Zn cathode at a much lower E_H. The use of the Ni_{disp}/Cu_{ann} cathode leads to higher product yields, though the selectivity is dramatically lower than in Test 2.

Discontinuous dispersed precipitates are typical for electrodeposition of metals [23]. In our case, discontinuous coating of the cathode surface (support) may be responsible for the unexpected results of ECH of citral performed with electrodeposition of Ni_{disp} on Zn and Cu cathodes. On the Ni_{disp}/Zn cathode, ECH presumably occurs only on the section coated with dispersed nickel leading actually to one product - alcohol **1** though in a relatively low yield (Test 2). On the Ni_{disp}/Cu_{ann} cathode (Test 3), ECH proceeds on the uncoated sections as well, yet by a different mechanism: citral is hydrogenated in the adsorbed state. Parallel implementation of the two mechanisms leads to a selectivity decrease if compared to Tests 2 and 5. It should be noted that selectivity of ECH of citral is higher on the Cu_{ann} cathode modified by dispersed nickel whereas the product yield is somewhat lower than on the non-modified Cu_{ann} cathode (refer to Test 6, Table 1). When the Cu_{ann} surface is modified by the dispersed copper precipitate (Test 4, Table 3) there is a drop in ECH efficiency as well as in the product yield and process selectivity. Alcohol **1** in a high yield is produced only in the case of Ni_{disp}/Ni (Test 5).

Hence our results refute the proposed thesis that all spongy metals are more efficient in the electrolysis [9].

Electroreduction of Trans-2-Allyl-6-R-1,2,3,6-Tetrahydropyridines

It seemed interesting to verify a possibility of using the proposed ECH method on Cu_{ann} and Ni_{disp}/Ni cathodes for compounds with another structure. Recently synthesized [47] *trans*-2-allyl-6-R-1,2,3,6-tetrahydropyridines (R = Me (4a), All (4b), Ph (4c)) are heterocyclic compounds with terminal and intracycle double bonds differing in reactivity.

A lot of plant and insect species produce asymmetrically 2,6-disubstituted piperidine alkaloids that play a vital role in their life. Many alkaloids exhibit biological activity [48] and most likely function as pest antagonists.

According to the preliminary polarographic measurements in non-aqueous DMF with 0.1 *M* Bu_4NClO_4 as a supporting electrolyte [49], allylpiperidines 4a and 4b are inactive up to the supporting electrolyte discharge potentials (E_{se} = -2.9 V). For piperidine 4c, the beginning of the reduction wave is however observed at a potential more positive than E_{se} by approximately 30 mV probably because of intracycle double bond activation caused by the phenyl substituent in position 6 of the heterocycle.

Direct electroreduction of 4c on a mercury pool cathode in unaqueous DMF with 0.1 *M* Bu_4NClO_4 as a supporting electrolyte at E = -2.74 V gives the reduction products of the heterocycle double bond - a mixture of *cis*- and *trans*-2-allyl-6-phenylpiperidine (5c) in a 5 : 7 ratio, with the yield by substance and the current efficiency each approaching 100%; the allyl double bond remains unaffected (Scheme 2). The ER mechanism was detailed in [50].

The polarograms of allylpiperideines 4a – 4c in H_2O-DMF solutions show a wave at E = –2.3 V [49]. The height of this wave depends on the concentration of the depolarizing agent and drastically rises in mixtures containing water in 40% excess. On this basis we assume that this wave is similar to catalytic waves of hydrogen evolution in non-buffered water-alcohol solutions of pyridine and its homologs [51].

In the electrolysis of 4c in 40% aqueous DMF (Scheme 2), the process follows another route: catalytic evolution of hydrogen (E = -2.3 V) occurs probably due to reduction of an amine associate with water. Current does not drop even under impact of a large charge (Q = 15 F/mol) and free-radical conversions lead to practically complete destruction of amine 4c [49].

Thus a change in the state of nitrous base in the presence of a proton donor (water) radically alters the process direction. A possibility of implementing ECH of allylpiperideines 4a – 4c on Cu_{ann} and Ni_{disp}/Ni is conditioned by less negative E_H values than those of the discussed electrochemical process.

Electrocatalytic Hydrogenation of Trans-2-Allyl-6-R-1,2,3,6-Tetrahydropyridines

ECH of allylpiperideines 4a – 4c on the Ni_{disp}/Ni cathode in a tenfold molar excess of AcOH leads to total hydrogenation of all C=C bonds. Piperidines 6a – 6c were obtained in almost quantitative yields by substance (Scheme 2). The same products are generated by catalytic hydrogenation in acetic acid over Raney Ni in the autoclave (100° C, 100 atm. H_2) [47, 50]. It should be noted that ECH is conducted under much milder conditions (room temperature and atmospheric pressure) than heterogeneous catalytic hydrogenation. Moreover, preliminary activation of a cathode by nickel electrodeposition is much simpler that the catalyst preparation procedure.

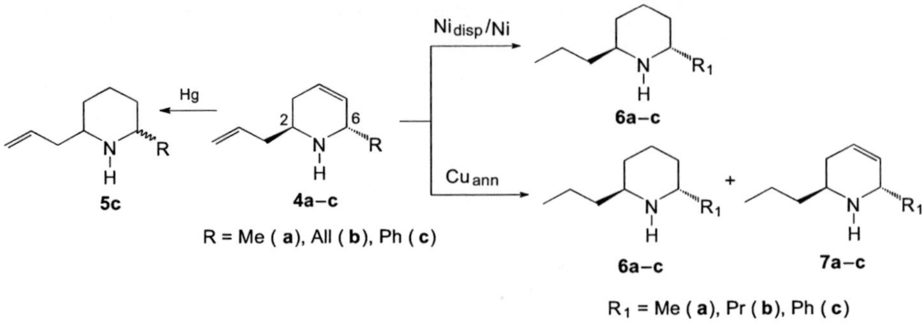

Scheme 2

In the case of ECH on Ni_{disp}/Ni, the character of the substituent in position 6 does not significantly affect the yield of ECH products. The process proceeds with retention of substituents *trans*-positioning relative to the piperidine ring.

Mixed products both of hydrogenation of only allyl groups with unchanged intracycle double bonds in 7a - 7c and of complete hydrogenation of compounds 6a - 6c were obtained on the Cu_{ann} cathode under the same conditions ($C_{AcOH}/C_{4a\text{-}c}$ = 10). Note that all generated products also have a *trans*-configuration. However, changing of the nature of the substituent in position 6 of the piperideine ring discernibly affects both the yield and the proportion of the ECH products when Cu_{ann} is used. The overall hydrogenation product yield goes down from 90 to 50% in the piperideine series 4a, 4b, 4c. The products of hydrogenation of solely terminal double bonds 7a - 7c predominate in the post-electrolysis mixtures. A portion of the complete hydrogenation products reduces from 40 to 25% in the same sequence.

For ECH on Cu_{ann} of importance is the character of 4a - 4c adsorption on the electrode surface as the product yield and composition are defined by the influence of both steric and electronic factors. For example, the presence of a bulky Ph substituent in position 6 of the piperideine ring (4c) probably hinders the interaction between the cathode surface and π-electrons of both the allyl group and the intracycle double bond. It is precisely the factor that accounts for the decrease in the total yield of hydrogenation products as well as in the portion of 6c - a product of complete hydrogenation.

The effect of AcOH on selectivity of hydrogenation on Cu_{ann} and Ni_{disp}/Ni was examined on the example of *trans*-2,6-diallyl-1,2,3.6-tetrahydropyridine 4b (Scheme 3).

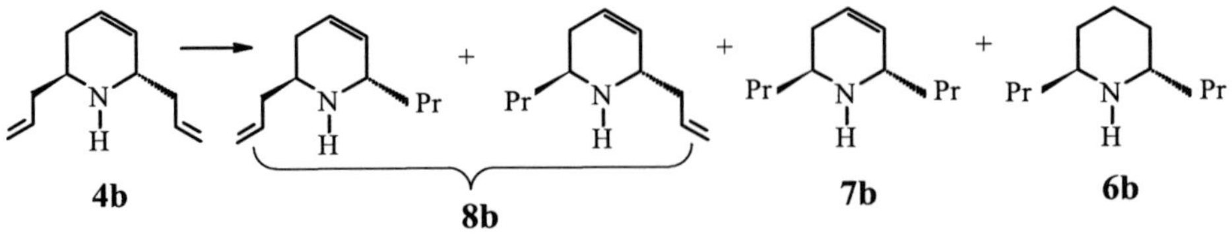

Scheme 3

Figure 3 illustrates how the yield of all products of ECH of diallylpiperideine 4b on Cu_{ann} varies against the molar ratio of AcOH and 4b. The character of the curves indicates a special role played by AcOH in the ECH process, which if likely to be related with the organic substrate nature: 4b is a cyclic amine with an unshared electron pair at the nitrogen atom. Although π-adsorption is normal for adsorption of organic compounds, an adsorption type characterized by the heteroatom involvement is more preferable for heterocyclic compounds. An interaction of 4b and the electrode in the absence of AcOH is likely to occur due to the free electron pair of nitrogen in the heterocycle. The hydrogenation products (piperideine 7b and isomers 8b) are produced in almost equal amounts, with the overall yield by substance reaching nearly 70%. In the presence of an equimolar AcOH amount, an associate between amine 4b and acid should be formed in bulk solution. The unshared electron pair in this case is blocked and hampers the interaction between the organic molecule and the electrode leading to some decline in hydrogenation performance (Fig. 3).

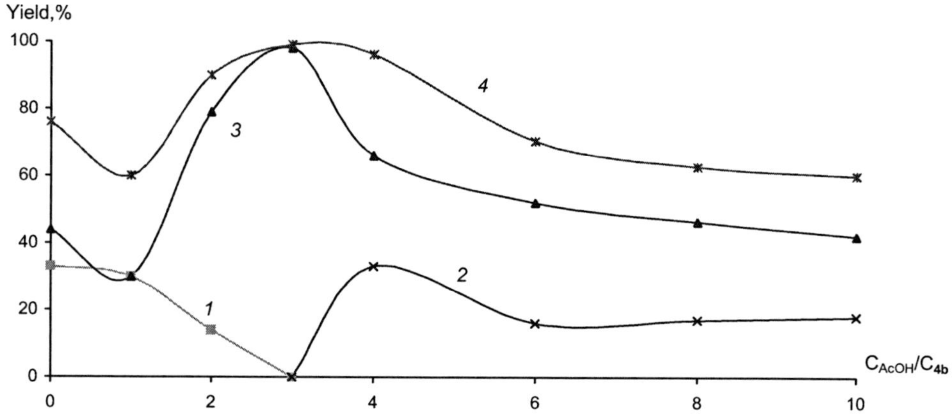

Figure 3. Yield of products of ECH of *trans*-2,6-diallyl-1,2,3,6-tetrahydropyridine (4b) on the Cu_{ann} cathode at different values of C_{AcOH}/C_{4b} in solution during electrolysis (supporting electrolyte – 0.1 *M* KCl in 40% aqueous DMF, C_{4b} = 0.015 *M*): *1* – isomers 8b; *2* – piperidine 6b; *3* – dipropylpiperideine 7b; *4* – total 6b, 7b, 8b in the mixture.

In the case of a threefold excess of AcOH, its association with the C=C bonds of 4b becomes possible, which probably facilitates the interaction between the organic substrate and the negatively charged electrode. As follows from Fig. 3 (Curve 4), at C_{AcOH}/C_{4b} = 3, less sterically hindered C=C bonds of allyl substituents are hydrogenated selectively (in a yield close to quantitative) and produce piperideine 7b. With a growing AcOH excess, the complete hydrogenation product 6b is generated whereas the total yield of 7b and 6b gradually decreases. At the same time, the ratio between these products almost does not change with a higher AcOH excess. The drop of the overall hydrogenation product yield may be explained by competing AcOH adsorption and forcing of 4b from the cathode surface.

In our opinion, the principal function of AcOH is to finely regulate adsorption of 4b on the Cu_{ann} cathode. The threefold AcOH excess is optimal for selective generation of product 7b. Selectivity in this case is conditioned by the steric factor. Adsorption of the organic molecule is hindered by the intracycle double bond that is why terminal double bonds are more readily, if compared to the intracycle bond, available for adsorption and subsequent hydrogenation.

Figure 4 shows a dependence between the yield of the products of electrocatalytic hydrogenation of diallylpiperideine 4b on Ni_{disp}/Ni and the molar ratio of AcOH and 4b. It is obvious that AcOH in a fourfold excess is optimal for obtaining a product of complete hydrogenation of 6b (Fig. 4, Curve 2). At $C_{AcOH}/C_{4b} = 4$, blocking of the nitrogen free electron pair occurs as well as activation of C=C bonds owing to the formation of associates of the π-electron system of bonds with AcOH - actually one AcOH molecule is needed for each hydrogenated double bond. A mixture of 7b, 6b and 8b was obtained at $C_{AcOH}/C_{4b} <4$. Although the product of hydrogenation of only terminal double bonds 7b predominates in the mixture (at $C_{AcOH}/C_{4b} = 3$) we failed to attain it selectively on the Ni_{disp}/Ni cathode [52]. A higher AcOH excess actually does not affect the yield of 6b and ECH selectivity. Both this fact and the absence of influence of the nature of the substituent in position 6 of the heterocycle on the yield and composition of the ECH products confirm the assumed mechanism of hydrogenation on the Ni_{disp}/Ni cathode with participation of the piperideine molecule arriving from bulk solution rather than of the adsorbed one.

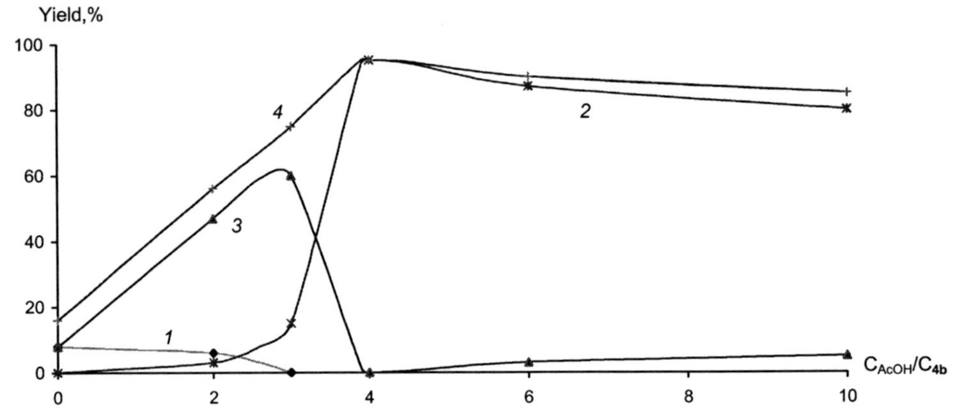

Figure 4. Yield of products of ECH of *trans*-2,6-diallyl-1,2,3,6-tetrahydropyridine (4b) on the Ni_{disp}/Ni cathode at different values of C_{AcOH}/C_{4b} in solution during electrolysis (supporting electrolyte – 0.1 *M* KCl in 40% aqueous DMF, $C_{4b} = 0.015$ *M*): *1* – isomers 8b; *2* – piperidine 6b; *3* – dipropylpiperideine 7b; *4* – total 6b, 7b, 8b in the mixture.

Thus, in both cases, despite different nature of the cathodes used, at maximum ECH selectivity, $C_{AcOH}/C_{4b} = n+1$ where n - the number of hydrogenated double bonds. A necessary and sufficient condition for complete hydrogenation of all double bonds of diallylpiperideine 4b on Ni_{disp}/Ni ($n = 3$) is the use of AcOH in a fourfold excess.

Selective hydrogenation of allyl double bonds in piperideines 4a – 4c with retention of the intracycle double bond may be realized on the Cu_{ann} cathode, provided a certain optimal ratio between concentrations of acetic acid and initial piperideine is rigorously observed. In this case $C_{AcOH}/C_{4a\text{-}4c}$ also equals to $n+1$, where n - the number of terminal double bonds. The applicability of the proposed formula for ECH of other piperideines was demonstrated on the example of *trans*-2-allyl-6-phenyl-1,2,3,6-tetrahydropyridine (4c) ($n = 1$). On the Cu_{ann} cathode and at $C_{AcOH}/C_{4c} = 2$, the product of ECH of the allyl group was obtained [49] in a quantitative yield (as calculated for the reacted piperideine (4c)) with retention of the intracycle double bond 7c.

Thus the conditions identified earlier for ECH of citral - electrode material and pretreatment (Cu_{ann} and Ni_{disp}/Ni), composition of the solvent, and nature of the acidic component - proved to be optimal for ECH of a few *trans*-allyl-tetrahydropyridines, the structure of which principally differs from citral. Different ECH results were obtained on Cu_{ann} and Ni_{disp}/Ni whereas the solvent composition was the same in all the cases. That was dictated by the nature of these electrodes, i.e. the electron structure and surface condition, and vividly demonstrates distinctions in their catalytic activity and relevant mechanisms.

The ECH process on a Cu_{ann} cathode is more sensitive to variations in both the molecule structure and the environment composition. This cathode shows more promise in design of selective hydrogenation alternatives for double bonds in polyunsaturated compounds.

CONCLUSION

This case study demonstrates that efficiency, selectivity and even direction of ECH of citral may dramatically vary in relation to the cathode surface condition (resulting from different pretreatments). The metal nature specifics should be taken into account in selecting an alternative for metallic cathode activation. To intensify the electrocatalytic properties of metals with medium overvoltage of hydrogen (Cu and Ni) the following pretreatment is recommended: annealing for Cu and dispersed Ni electrodeposition for Ni. On the other hand, these metals, upon appropriate pretreatment, may be utilized as cathodes with high hydrogen overvoltage.

Of high significance for ECH of unsaturated organic compounds is the environment composition during the electrolysis, which is associated with changes in substrate adsorption on the electrode surface.

The experiments with allylpiperideines 4a – 4c have shown that the use of cathodes with medium hydrogen overvoltage paves the way to both selective (Cu_{ann}) ECH of unsaturated bonds on allyl substituents and complete hydrogenation (Ni_{disp}/Ni) of all double bonds on the condition of precise $C_{AcOH}/C_{4a\text{–}4c}$ maintenance with account of the number of hydrogenated bonds.

It has been established that the phenyl substituent as well as non-activated, isolated and sterically hindered double bonds do not undergo hydrogenation during ECH on Cu_{ann} and Ni_{disp}/Ni cathodes under the proposed conditions.

REFERENCES

[1] Antrohov, L. I. (1950). Theory of electrolytic reduction of organic compounds. *Zh. Fiz. Khim., 24,* 1428 - 1435 (Russ).

[2] Tomilov, A. P. (1993). Role of the cathode material in electrochemical synthesis of organic compounds. *Zh. Org. Khim., 29,* 657 - 665 (Russ).

[3] Bagotskii, V. S. (1980). Main problems of electrocatalysis. In V. S. Bagotskii, *Problems of electrocatalysis* (5 - 23). Moscow: Nauka (Russ).

[4] Couper, A. M., Pletcher, D. & Walsh, F. C. (1990). Electrode materials for electrosynthesis. *Chem. Rev, 90,* 837 - 865.

[5] Krylov, O. V. & Matyshak, V. A. (1996). *Intermediates in heterogeneous catalysis.* Moscow: Nauka (Russ).

[6] Pak, A. M. & Sokol'skii, D. V. (1983). *Selective hydrogenation of unsaturated oxo compounds.* Alma-Ata: Nauka (Russ).

[7] Tomilov, A. P., Khomutov, N. E. & Fedorova, L. A. (1973). Selection of optimum conditions for cathodic electrosynthesis of organic compounds. In A. N. Frumkin, *Electrosynthesis and mechanism of organic reactions* (90 - 121). Moscow: Nauka (Russ).

[8] Korovin, N. V. (1991). Relationship between the electrocatalytic activity of metals in the hydrogen evolution reaction and their properties. *Elektrokhimiya, 27,* 1629 - 1634 (Russ).

[9] Tomilov, A. P., Mairanovskii, S. G., Fioshin, M. Ya. & Smirnov, V. A. (1972). *Electrochemistry of organic compounds.* New York: Wiley (Transl. from Russ).

[10] Grigor'eva, N. Ya., Veselovskii, V. V. & Moiseenkov, A. M. (1987). Synthesis of polyprenols, membrane-active participants in the biosynthesis of carbohydrate-containing biopolymers in the cellular wall. *Khim.-Farm. Zh., 21,* 845 - 854 (Russ).

[11] Baizer, M. M. (1973). The α,β-unsaturated carbonyls. In M. M. Baizer, *Organic electrochemistry: an introduction and a guide* (399 - 411). New York: Marcel Dekker.

[12] Horanyi, G. & Torcos, K. (1982). Formation of hydrocarbons from acrolein and crotonaldehyde. *J. Electroanal. Chem., 136,* 301 - 309.

[13] Johnston, J. C., Faulkner, J. D., Mandell, L. & Day, R. A., Ir. (1976). Electrochemical reduction geranial, farnesal and crotonaldehyde. *J. Org. Chem., 41,* 2611 - 2614.

[14] Rubinskaya, T. Ya., Korotaeva, L. M. & Gultyai, V. P. (1993). Electrochemical hydrogenation of citral. 2. The effect of the components of the medium on the process of electrochemical reduction at a copper cathode. *Russ. Chem. Bull.,42,* 1835 - 1838.

[15] Gultyai, V. P. & Strelets, V. V. (1999). Electrochemical hydrogenation of citral. 6. Voltammetric measurements in DMF solutions of citral and acetic acid. *Russ. Chem. Bull., 48,* 202 - 204.

[16] Korotaeva, L. M., Rubinskaya, T. Ya. & Gultyai, V. P. (1993). Electrochemical hydrogenation of citral. 1. The role of the copper cathode in the electroreduction of citral. *Russ. Chem. Bull., 42,* 1830 - 1834.

[17] Bockris, J. O'M. & Swinkels, D. A. J. (1964). Adsorption of *n*-decylamine on solid metal electrodes. *J. Electrochem. Soc., 111,* 736 - 743.

[18] Korotaeva, L. M., Rubinskaya, T. Ya. & Gultyai, V. P. (1997). Electrochemical hydrogenation of citral. 4. Role of the acid component in electrochemical hydrogenation. *Russ. Chem. Bull., 46,* 459 - 462.

[19] Sokol'skii, D. V. & Zabotin, P. I. (1979). *Electrochemical aspects of heterogeneous catalysis of redox reactions.* Alma-Ata: Nauka Kaz. SSR (Russ).

[20] Podlovchenko, B. I., Pshenichnikov, A. G. & Scundin, A. M. (1993). Metal and metal-modified electrocatalysts. *Elektrokhimiya, 29,* 422 - 432 (Russ).

[21] Skundin, A. M. & Kudryashov, I. V. (1980). Structural factors in electrocatalysis. In V. S. Bagotskii, *Problems of electrocatalysis* (76 - 99). Moscow: Nauka (Russ).

[22] Korotaeva, L. M., Tkachenko, O. P., Rubinskaya, T. Ya. & Gultyai, V. P. (2003). State of the Cu electrode surface studied by X-ray photoelectron and Auger spectroscopies. *Russ. Chem. Bull., 52,* 2346 - 2351.

[23] Korovin, N.V., & Podlovchenko, B. I. (1980). Metallic catalysts in electrochemical systems. In V. S. Bagotskii, *Problems of electrocatalysis* (100 - 127). Moscow: Nauka (Russ).

[24] Ivanov, V. I., & Osipov, K. A. (1964). *Recovery* and *recrystallization in metals during fast heating*. Moscow: Nauka (Russ).

[25] Kita, H., Nakamura, T., Iton, H., & Kano, H. (1978). Electroreduction of dimethyl maleate, dimethyl fumarate and α,β-dimethyl maleic acid on a platinum electrode in sulfuric acid aqueous solution. *Electrochim. Acta, 23,* 405 - 411.

[26] Damaskin, B. B. & Petrii, O. A. (1993). Electrochemistry and surface science. *Ross. Khim. Zh., 37,* 16 - 26 (Russ).

[27] Kuliev, S. A., Vasil'ev, Yu. B. & Bagotskii, V. S. (1987). Influence of the roughness factor of a platinum electrode on the adsorption of benzene and phenylacetic acid. *Elektrokhimiya, 23,* 553 - 556 (Russ).

[28] Tomashova, N. N. & Chekavtsev, A. V. (1991). Electrochemical incorporation into copper of alkali metals from aqueous solutions. *Elektrokhimiya, 27,* 937 - 939 (Russ).

[29] Batrakov, V. V., Dittrikh, Yu. & Popov, A. N. (1972). Effect of copper surface structure on hydrogen generation overvoltage in acid solution. *Elektrokhimiya, 8,* 640 - 643 (Russ).

[30] Horanyi, G. & Torkos, K. (1980). Effect of chemisorbed methanol on the rate and selectivity of electroreduction processes at a platinized platinum electrode. *J. Electroanal. Chem., 111,* 279 - 286.

[31] Rubinskaya, T. Ya., Korotaeva, L. M. & Gultyai, V. P. (1996). Electrochemical hydrogenation of citral. 3. Effect of the nature of the solvent on electrocatalytic hydrogenation at a copper cathode. *Russ. Chem. Bull., 45,* 74 - 78.

[32] Samoilov, O. Ya. (1957). *Structure of aqueous solution of electrolytes and hydration of ions.* Moscow: Izd. Akad. Nauk (Russ).

[33] Mishustin, A. J. & Kessler, Yu. M. (1974). Liquid-phase interactions between water and dimethylformamide. *Zh. Strukt. Khim., 15,* 205 - 209 (Russ).

[34] Blomgren, E., Bockris, J. O'M. & Jesch, C. (1961). The adsorption of butyl, phenyl and naphthyl compounds at the interface mercury-aqueous acid solution. *J. Phys. Chem., 65,* 2000 - 2010.

[35] Komarenko, V. G., Skryshewskii, A. F., Dorosh, A. K. & Manzhelii, V. G. (1968). Structure of alcohols and their kinetic properties at low temperatures. *Sb. Nauch. Tr. Fiz-Tekn. Inst. Nizk. Temp.,* 44 - 64 (Russ).

[36] Belousov, V. P. & Morachevskii, A. G. (1964). Thermodynamic properties of binary propyl alcohols-water system. *Zh. Fiz. Khim., 38,* 184 - 189 (Russ).

[37] Krestov, G. A. & Krestov, Al. G. (1993). Thermodynamic characterization of interparticle interactions in solutions. *Sb. Nauch. Tr., Thermodynamics of electrolyte solutions* (3 - 14). Ivanovo: Ros. Akad. Nauk (Russ).

[38] Nesmeyanov, A. N. & Nesmeyanov, N. A. (1969). *Principles of organic chemistry,* book 1. Moscow: Khimiya (Russ).

[39] Tomilov, A. P. & Kirilyus, J. V. (1982). *Cathodic syntheses of organic compounds.* Alma-Ata: Nauka (Russ).

[40] Scundin, A. M. (1984) Structural factors in electrocatalysis. *Itogi Nauki Tekn., ser. Elektrokhim., 18,* 228 - 263 (Russ).

[41] Korotaeva, L. M., Rubinskaya, T. Ya. & Gultyai, V. P. (1998). Electrochemical hydrogenation of citral. 5. The use of a nickel electrode in the electrocatalytic process. *Russ. Chem. Bull., 47,* 1487 - 1490.

[42] Filimonova, V. J., Bekenova, U. B., Zakarin, S. Z. & Kirilyus, I. V. (1996). Electrocatalytic reduction of acetylene derivatives. *Russ. J. Electrochem., 32,* 141 - 142.

[43] Brown, H. & Knapp, B. B. (1974). Nickel. In F. A. Lowenheim, *Modern electroplating* (3rd ed., 287 - 341). New York: Wiley.

[44] Ratiani, N. P., Kakhadze, A. E., Michri, A. A. & Pshenichnikov, A. G. (1978). Ethylene hydrogenation on a nickel electrode. Relation between ethylene hydrogenation kinetics and hydrogen discharge ionization. *Elektrokhimiya, 14,* 95 - 98 (Russ).

[45] Sokol'skii, D. V. & Sokol'skaya, A. M. (1970). *Metal catalysts for hydrogenation.* Alma-Ata: Nauka Kaz. SSR (Russ).

[46] Zerina, T. A., Ivanovskaya, L. N., Michri, A. A. & Pshenichnikov, A. G. (1992). Effect of cathode material on the electroreduction of 1-(2-propionitryl)pyrrolidine. *Elektrokhimiya, 28,* 349 - 359 (Russ).

[47] Bubnov, Yu. N., Klimkina, E. V. & Ignatenko, A. V. (1998). Preparation of *trans*- and *cis*-2-allyl-6-alkyl(aryl)-1,2,3,6-tetrahydropyridines based on the reductive *trans*-2,6-dialkylation of pyridine. Synthesis of (±)epidihydropinidine and (±)dihydropinidine. *Russ. Chem. Bull., 47,* 451 - 458.

[48] Tawara, J. N., Blokhin, A., Foderaro, T. A., Stermitz, F. R. & Hope, H. (1993). Toxic piperidine alkaloids from Pine (Pinus) and Spruce (Picea) trees. New structures and a biosynthetic hypothesis. *J. Org. Chem., 58,* 4813 - 4818.

[49] Korotaeva, L. M., Rubinskaya, T. Ya., Klimkina, T. V., Gultyai, V. P. & Bubnov, Yu. N. (2000). Peculiarities of electroreduction of *trans*-2-allyl-6-methyl(allyl,phenyl)-1,2,3,6-tetrahydropyridines. *Russ. Chem. Bull., 49,* 2037 - 2044.

[50] Korotaeva, L. M., Rubinskaya, T. Ya., Klimkina, T. V., Gultyai, V. P. & Bubnov, Yu. N. (1999). Preparative electroreduction of *trans*-2-allyl-6-methyl(phenyl)-1,2,3.6-tetrahydropyridine. *Russ. Chem. Bull., 48,* 754 - 757.

[51] Mairanovskii, S. G. & Gul'tyai, V. P. (1965). Polarografic study of the effect of ethanol concentration on the protonation rate of pyridine and 2,6-lutidine by water. *Elektrokhimiya, 1,* 460 - 466 (Russ).

[52] Korotaeva, L. M., Rubinskaya, T. Ya. & Gultyai, V. P. (2003). Selective electrocatalytic hydrogenation of *trans*-2-allyl-6-methyl(allyl,phenyl)-1,2,3,6-tetrahydropyridines on the nickel and copper cathodes. *Russ. J. Electrochem. 39,* 1203 - 1210.

ACKNOWLEDGMENTS

The authors greatly appreciate valuable comments and recommendations of Prof. V.P. Gultyai as well as highly efficient technical assistance of E.R. Abdrakhimova.

In: Focus on Electrochemistry Research
Editor: Magdalena Nunez, pp. 169-213
ISBN: 1-59454-545-6

Chapter 6

DEVELOPMENT OF CO TOLERANT ELECTROCATALYSTS FOR REFORMATE HYDROGEN AND METHANOL OXIDATIONS ON PROTON EXCHANGE MEMBRANE FUEL CELLS

Elisabete I. Santiago, William H. Lizcano-Valbuena, Edson A. Ticianelli and Ernesto R. Gonzalez
Instituto de Química de São Carlos – USP, Av. Trabalhador Sãocarlense 400,
13560-970 São Carlos, SP, Brazil

ABSTRACT

The search for CO-tolerant catalysts has been a paramount task for the successful development of more efficient polymer electrolyte (PEMFC) and direct methanol fuel cell (DMFC) systems. This work discusses different aspects of the electrocatalysis of the hydrogen oxidation reaction in the presence of CO and of the methanol electro-oxidation on Pt-based materials, supported on high surface area carbon, from the fundamental and practical points of view. The emphasis is given to bimetallic materials like Pt-Ru/C, Pt-Os/C, and Pt-Mo/C, regarding the following aspects: i) the preparation and characterization methods and the effect of thermal treatments, and ii) the influence of the particle morphology and the crystal structure on the electrochemical performance. The characterization of the prepared materials was made using several techniques, such as Energy Dispersive X-rays (EDX) analysis, Electron Back-scattering (EBS), X-ray Diffraction (XRD), Transmission Electron Microscopy (TEM), High Resolution Scanning Electron Microscopy (SEM), and X-ray Absorption Spectroscopy (XAS). The electrochemical evaluation of the several catalysts was made by voltammetric techniques, chronoampetrometry, and potential vs. current density curves obtained in a single fuel cell for anodes fed with pure hydrogen, reformate hydrogen and methanol. The results are discussed on the basis of the responses of smooth and single crystal electrodes, in comparison to those of the carbon supported materials.

1. INTRODUCTION

In recent years there has been a steadily growing concern with the degradation of the environment and the effect of pollutants on human health [1]. High levels of pollutants are produced by internal combustion engines, particularly those running on diesel, in large urban centers. Today, the use of clean energy sources is considered an urgent necessity and, among the alternatives, fuel cells are attracting much interest [2] due to its clean, silent operation and high energy efficiency (45-55 %). In particular, polymer electrolyte membrane fuel cells (PEMFC), are considered good candidates for transportation and portable applications because they are capable of delivering high power densities and can start operating at room temperature [3].

In spite of the efforts to develop the direct methanol fuel cell (DMFC), which has the advantage of using a liquid fuel, the most efficient low temperature fuel cells still use hydrogen as fuel. The cheapest way of producing hydrogen is by vapor reforming of fossil fuels or low molecular weight alcohols [3]. This process produces 6-7 % CO, which can be reduced to 1-2 % by a shift reaction and to levels below 100 ppm by partial oxidation [4]. This CO adsorbs strongly on the Pt catalyst of the fuel cell electrode inhibiting the anodic reaction. The DMFC is not free of this problem, because the oxidation of methanol on Pt produces CO as an intermediate [5]. The possibility of using polymer electrolyte membrane fuel cells (PEMFC) in transportation applications is a strong reason to search for solutions of the CO poisoning problem. For the DMFC a solution of this problem would allow to obtain power densities above 0.3 Wcm^{-2}, in order to make the fuel cell powered car competitive.

When impure hydrogen is used as a fuel, several approaches have been tested to make PEMFC's run with up to 100 ppm CO [6-20]. These involve: i) adding small amounts of oxygen (up to 5%) to the hydrogen entering the anode [6,7], ii) performing a catalytic oxidation using Co oxide [8], iii) using in the anode Pt alloys with less noble metals like Ru [9-13], Sn [14], Rh [15], Mo [16-18], Re [19] or some noble metals like Au [20], and iv) using Pt-Ru alloys and adding small amounts of hydrogen peroxide, as an oxygen carrier, to the water hydrating the membrane [21,22].

1.1. Mechanism of CO Electro-Oxidation

Although some reports point to the Pt-Mo alloy as being the more effective [17,18], Pt-Ru alloys have been more extensively investigated. The use of Pt alloys is based on the fact that the less noble metal forms the hydrated oxides necessary to oxidize CO at lower potentials. Thus, the CO adsorbed on Pt is oxidized by the second metal through the so-called bifunctional mechanism proposed initially by Watanabe et al. [23]:

$$\text{Pt-CO} + 2\text{Ru-OH} \rightarrow CO_2 + \text{Pt} + 2\text{Ru} + H_2O \qquad (1)$$

In order to understand better the oxidation of CO to CO_2 several studies have been carried out on the adsorption of CO on Pt, using techniques as Fourier transform infrared spectroscopy (FTIR) [24] and differential electrochemical mass spectrometry (DEMS) [25]. It is observed that CO adsorbs on Pt in two forms: bridge and linear bonding [26,27]. Morimoto et al. [28] used *in situ* FTIR and cyclic voltammetry (CV) to study the characteristics of the adsorption of CO on surfaces with different morphologies. They found that on smooth surfaces the bridge form amounts to 12%, while for rough surfaces this figure increases to 29%.

Ianniello et al. [29] studied the oxidation of CO on smooth Pt and Pt-Ru alloys by *in situ* FTIR identifying the OH species that participate in the process. The lowest oxidation potential was found for the 50:50 Pt-Ru alloy. A similar result was presented by Arico et al. [30] on Pt-Ru alloys supported on carbon. On smooth rotating disk electrodes (RDE) Gasteiger et al. [31] proposed that the oxidation of CO on Pt-Ru follows a Langmuir-Hinshelwood mechanism because of the shift to lower potentials of the CO oxidation peak. More recently, Azevedo et al. [32] showed by electrochemical impedance spectroscopy (EIS) that the rate of CO oxidation is dependent on the energy of adsorption of OH, which is larger for Ru if compared with Pt, making the CO oxidation on Pt-Ru slower that on Pt. This result is in agreement with a proposal by Gasteiger et al. [31]. McBreen et al. [33] used X-ray absorption spectroscopy (XAS) to show that Ru modifies the electronic structure of Pt, and that this effect has to be included in the bifunctional mechanism. Buatier de Mongeot et al. [34] carried out a comparative scanning tunneling microscopy (STM) and temperature programmed calorimetry (TPD/TPR) study on the adsorption and oxidation of CO on different surfaces of Pt, Ru and Pt-Ru and showed that the tolerance of Pt-Ru materials to CO can be at least partly explained by the reduced adsorption energy on these surfaces, caused by electronic modifications of the Pt surface atoms due to the interaction with neighboring Ru atoms.

1.2. Mechanism of Methanol Electrooxidation

The case of methanol electro-oxidation is different. On Pt there are two pathways: the direct oxidation and via intermediates [35-37], while on Pt-Ru catalysts it is accepted that methanol is oxidized only via intermediates. The oxidation of methanol has been studied on several electrocatalysts [38-45]. Pt is the best material for this reaction but, unfortunately, the formation of intermediates, like CO [45,46], formic acid and methylformate, identified by differential electrochemical mass spectrometry (DEMS) and Fourier transform infrared spectroscopy (FTIR) [46], which poison the electrode surface, inhibiting the oxidation reaction.

The alloying of Pt with other metals like Ru [39-46], Mo [47,48], Sn [49,50], W [51,52], Os [53,54] has been tested as a way of minimizing the effect of poisoning intermediates. The second metal is chosen so that it has a larger tendency to form oxides than Pt, so oxygenated species are formed at lower potentials, promoting the conversion of CO to CO_2 on Pt, through the so-called bifunctional mechanism as was discussed above. This step is related to the decisive stages of the oxidation of methanol when this reaction takes place through the formation of intermediates [38]. Ruthenium is the most common second metal used in binary Pt-based materials. Ru, deposited by different methods [44] on Pt, forms islands of different sizes, but only the Ru on the borders of Pt particles participates in the bifunctional mechanism [45]. In addition, the size of the islands has much relevance in the activity of the Pt-Ru material [44]. Studies on single-crystal Pt showed that the plane (111) is the more active for the oxidation of methanol [43], a result that can be of interest in practical systems using supported catalysts, because it is known that nanoparticles of Pt (2-4 nm) used in the production of gas diffusion electrodes for fuel cells present preferentially the plane (111) [55,56]. Ternary [53,54] and quaternary [54] alloys containing Pt and Ru have been tested without much improvement over binary Pt-Ru alloys. Unfortunately, until today Pt alloys capable of dealing efficiently with the problem of Pt poisoning with CO, have not been found.

With respect to the most widely used Pt-Ru catalysts, one controversial point is the optimum Pt-Ru atomic relation. Using smooth electrodes Gasteiger et al. [57] proposed, on the basis of the need of having three sites on Pt for methanol adsorption, an atomic relation near 10 at. % of Ru. Recently, Iwasita et al. [44] showed that the activity of smooth Pt-Ru electrodes with Ru contents between 10 and 40 at. % was similar in the potential region between 0.35-0.6 V vs ERH. On supported catalysts, a recent report by Coutanceau et al. [58] indicates that Pt-Ru 80:20/C would be the best catalyst.

The optimum particle size for methanol oxidation has been investigated and the values vary with the preparation procedure, the conditions of the reaction and the metals present on the catalyst. Attwood et al. [59] found an optimum specific activity for Pt with particle sizes of 3 nm. Other reports [60,61] show that the specific activity decreases for sizes below 4.5 nm. Particles below 5.0 nm have enough coordination sites to produce strong Pt-OH bonds leading to high coverages with this species. The role of geometric parameters and changes in the electronic structure due to specific adsorbates, such as hydrogen and oxygenated species (OH, CO) on methanol oxidation, were studied by Mukerjee and McBreen [62]. They found by XAS studies that as the particle size of carbon supported Pt clusters is reduced below 5.0 nm there is a strong adsorption of H, OH and C1 compounds such as CO. On small particles, H adsorption is sufficiently strong to induce restructuring and morphological changes in Pt clusters at negative potentials. The strong adsorption of OH for potentials more positive than 0.8 V on small particles (<5.0 nm) inhibits the reduction of oxygen and the combined effect of strongly adsorbed CO and OH on these small particles inhibits the oxidation of methanol. The reports of several groups that work with DMFC show that the performance depends on the method of preparation of the supported catalysts [63], the membrane and electrodes assembly (MEA) preparation procedure [64] and the methanol feed system [65-67].

Another aspect of consideration is that in direct methanol fuel cells (DMFC) that use Nafion® as polymer electrolyte, the crossover of methanol to the cathode through the membrane is a serious problem that causes a mixed potential and the poisoning of the cathode, decreasing the performance of the cell [68,69]. Some approaches have been tested to solve this problem, as i) the development of proton exchange polymeric membranes less permeable to methanol, ii) modification of the Nafion® membrane and iii) the development of electrocatalysts for the oxygen reduction reaction that are inactive for methanol electrooxidation, but so far without definitive results. Today, Nafion® is still the most commonly used membrane [70].

In spite of the considerable number of published works directed to the understanding of the mechanism of the oxidation of methanol on smooth [39] and single crystal Pt and Pt alloys [39,42], the extrapolation of the results to practical electrodes is open to question. Christensen et al. [71] discussed the difficulty and the risks of extrapolating results obtained on smooth and rough electrodes to porous diffusion electrodes.

In summary, an important although not unique aspect of the catalysis of methanol oxidation is related to the catalysis of CO oxidations and the search of CO-tolerant catalysts has been a paramount talk for the successful development of more efficient PEMFC systems.

For practical reasons, this chapter is divided into three sections. The first one is dedicated to the description of the methods to prepare and characterize Pt-based carbon-supported electrocatalysts. The second and third are devoted to discuss the catalysis of the hydrogen oxidation reaction in the presence of CO and the methanol oxidation processes, focusing in the improvements reached in the understanding of the reaction mechanisms for different materials.

2. Development of Supported Electrocatalysts

2.1. Treatment of The Carbon Supports

Silva et al. [72] analized different supports submitted to thermal treatments in several atmospheres. Fullerene soot originally presents a high level of surface impurities than the most used support (carbon Vulcan XC-72, Cabot). Heat treatments with reduzing or oxidant agents lead to a general decrease of the total amount of impurities and also to an increase in the surface concentration of oxygenated groups. On the other hand, treatments with hydrogen promote a decrease in the amount of impurities like C=O, COH, COOH, but still leave groups more difficult to reduce/oxidize such as aromatic esthers. After the thermal treatment, fullerene soot presents a higher surface area than Vulcan. A value close to 380 m^2g^{-1} represents a high real surface area of this carbon when compared with 240 m^2g^{-1} of the Vulcan carbon. Cyclic voltammetry indicates that fullerene soot presents a very large charge storage capacity, being a very promising material to be used as substrate for ion intercalation electrodes. For supported catalysts the Pt active areas are similar when Vulcan or fullerene soot are used as substrates. Pinheiro et al. [73] determined by the changes in the double layer region that thermal treatments at 850 °C for 5 hours in Ar atmosphere results in a two-fold increase in the electroactive areas. These results are in agreement with the BET areas for the same samples. A very recent approach was that of M. do Carmo et al. [74], who used carbon nanotubes as support for Pt and Pt-Ru catalysts. The results in the DMFC show an enhancement in the electrochemical performance compared with supported catalysts on high surface area carbon like Vulcan XC-72.

2.2. Preparation Methods

Several approaches have been used for the production of catalysts, both supported and unsupported. Some are the following:

1) Reduction with formic acid (formic acid method, FAM) : A dispersion of carbon is prepared in a formic acid solution and the precursors are added to this carbon dispersion [75].

2) Modified formic acid method [76]: this is similar to the original method but uses lower concentrations of formic acid.

3) Reduction with bisulfite (Watanabe´s method, WM): Watanabe et al. [77] proposed a method to prepare catalysts based on the use of sodium bisulphite to convert $PtCl_6^{2-}$ ions to a sulfite complex, which is converted to Pt oxides by the addition of hydrogen peroxide. Ru is added as $RuCl_3$, and then the carbon powder, followed by H_2 bubbling.

4) Modification of Watanabe′s method [78] (MWM): This consists in applying the above procedure simultaneously to the platinum and ruthenium salts, as suggested by Radmilovic et al. [79].

5) Reduction with $NaBH_4$ [80] (BM): In this case carbon powder is suspended in a solution containing Pt (IV) and Ru (III) or other ions. The borohydride solution is added slowly.

6) Reduction with formadehyde [81]: A solution containing the precursors of Pt and Ru or other metals is added to a stirred suspension of carbon. While the mixture is heated to 80 °C, NaOH is added until pH 8.5, after which a solution of formaldehyde is added.

7) Reduction with hydrazine [82]: Precursors of Pt and the other metal, neutralized with an appropriate amount of sodium carbonate, are added to a carbon suspension and refluxed for 2 h. After addition of hydrazine, the mixture is stirred for 1 h, filtered, and washed.

8) Impregnation and reduction with formaldehyde [83]: The carbon support is impregnated in a H_2PtCl_6 solution in water and reduced with formaldehyde.

9) Colloidal procedures: Two preparation procedures have been employed. a) Coloidal Pt was prepared by the method of Turkevich and co-workers [84] in which Pt^{4+} is reduced with sodium citrate under reflux until a black sol is formed. b) Colloidal Pt on a carbon support (carbon as received or in oxidized form) has been prepared by the Bonnemann method [85]. $PtCl_2$ is suspended in tetrahydrofurane (THF) and reduced with ($N(octyl)_4Bet_3H$) where the ammonium salt functions as a stabilizer for the colloidal metal particles.

Reduction of the metallic ions from their salts by different methods [75-85] has been proposed, as described above, resulting on catalysts with such diverse physical characteristics that direct comparison of the electrochemical performance of materials is a very tentative task. In the case of bimetallic catalysts, the reduction of the second metal can be done simultaneously with the first one or over a metal supported first, as was shown by Colmati et al. [86] for Pt-Ru/C and Shukla et al. [87] for Pt-Co/C.

2.3. Determination of the Metal Composition of the Catalysts

The metal composition of the materials has been investigated with EDX analyses. For Pt-Ru, the emitted radiation energies are $L\alpha$ = 9.441 kV and M = 2.048 kV for Pt and $L\alpha$ = 2.558 kV and M = 0.461 kV for Ru. Thus, the value of the L energy for Ru is close to the value of the M energy for Pt, but the resolution of the used electron microscope (0.15 kV) is enough to avoid overlapping errors. Table 1 shows the atomic composition of the Pt-Ru/C catalysts with different Pt:Ru atomic ratios (prepared by the original formic acid (FAM) procedure [75,88]) As it can be observed it was not possible to anchor more than 25 % of Ru. Previous results also show that when the formic acid method is used to prepare Pt-Ru/C catalysts it is very difficult to obtain alloy compositions with more than 30 % Ru [16,76,88]. On the other hand, other methods allow the preparation of alloys with an atomic Ru content up to 50 % [78,79].

Table 1. Nominal and EDX compositions, mean particle sizes (XRD and TEM) and lattice parameters (XRD) for several Pt-based catalysts

Catalyst Nominal atomic ratios (%)	Pt:M EDX Atomic ratio (%) (M=Ru,Os,Mo)	Mean particle size (nm) XRD	TEM	Lattice parameter *a* (nm)	Interatomic distances R (nm)
Pt/C (E-TEK)	100	2.9	3.6±1.0	0.39244	0.277
Pt-Ru/C (E-TEK) 50:50	50:50	-	3.2±0.7	--	--
Pt-Ru/C (F) 80:20	84:16	2.6	4.4±1.3	0.39121	0.276
Pt-Ru/C (F) TT1 80:20	84:16	2.9	6.6±2.3	0.39216	0.276
Pt-Ru/C (F) 70:30	75:25	3.4	5.1±2.1	0.39107	0.276
Pt-Ru/C (F) TT1 70:30	75:25	4.5	5.6±1.6	0.39188	0.277
Pt-Os/C (B) 60:40	77:23	--	--	--	--
Pt-Os/C (B) TT2 60:40	77:23	8.0	--	--	--
Pt-Os/C (F) 60:40	91:9	2.3	--	--	--
Pt-Os/C (B) TT2 60:40	91:9	5.0	--	--	--
Pt-Os/C (W) 60:40	88:12	1.8	--	--	--
Pt-Os/C (W) TT2 60:40	88:12	3.3	--	--	
Pt-Mo/C (F) 60:40	72:28	2.4	--	0.3917	0.277

TT1 = thermal treatment in H_2 (300 °C, 1h).
TT2 = thermal treatment in H_2 (200 °C, air 1h/ 500 °C H_2 1h).

Similar analyses were carried out for Pt-Os/C catalysts prepared using the formic acid, Borohydride and Watanabe's methods. It is seen that there is a clear difference between the molar compositions of the precursor solutions (Table 1) and of the resulting Pt-Os/C catalysts, with the BM giving the higher Os content. These differences were attributed to partial reduction of the osmium ions or the difficulty of anchoring onto the carbon substrate, which depends on the preparation conditions and the strength of the reducing agent. In contrast to Pt-Os/C and Pt-Ru/C, EDX analyses of Pt-Mo/C materials have shown a high efficiency in the reduction of Pt and Mo ions onto carbon by using the formic acid method, with the catalyst composition ratio resulting close to that of the precursors in solution.

2.4. Characterization of the Catalyst Particle Size and Bond Distance

In general, as shown in Figure 1 for the case of Pt-Ru/C catalysts prepared by the FAM and for different atomic Pt:Ru ratios, only peaks at approximately $2\theta = 40$, 47, 67 and 82 are observed. These are associated to the (111), (200), (220) and (311) planes, respectively, of the fcc structure of platinum, or a Pt-Ru phase rich in platinum that retains the fcc structure. It was not observed any peak due to metallic ruthenium, or to materials rich in Ru with hexagonal structure. These structures would produce a peak at $2\theta = 44$, due to a reflection of the (101) plane, the most intense peak of hcp Ru that could indicate their presence. However, it is known that for equal quantities of Pt and Ru the intensities of Ru peaks are lower than those of Pt peaks and for this reason metallic Ru may not be detected in the carbon support [89]. Also, it has to be taken into account that usually Ru is present in less quantities than Pt and the peaks may be superimposed because of the broadening due to small particle sizes and the strain lattice effect. Ru, present as amorphous Ru oxides, has not been observed.

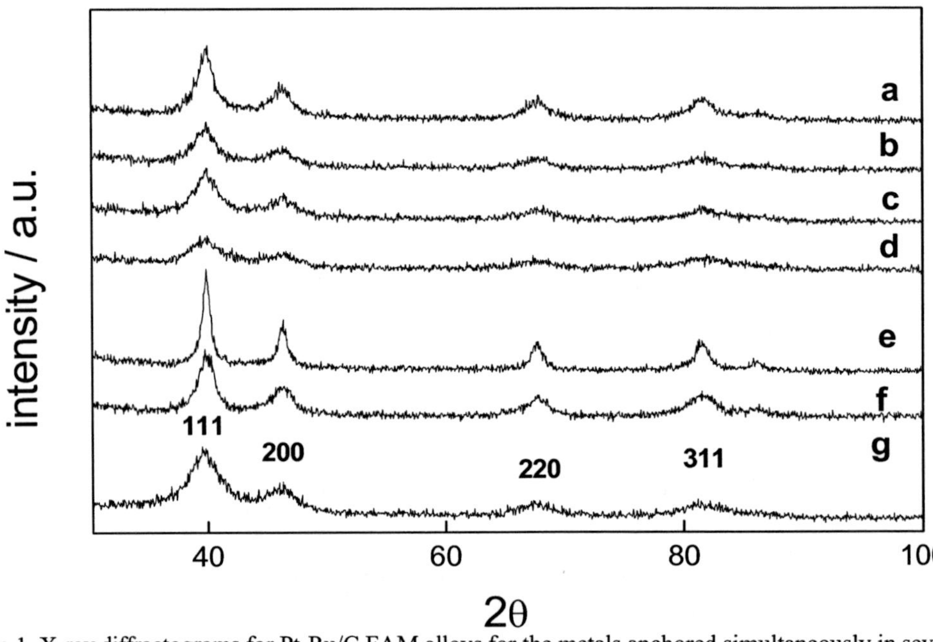

Figure 1. X-ray diffractograms for Pt-Ru/C FAM alloys for the metals anchored simultaneously in several atomic relations. a) Pt-Ru/C 75:25 T. T.; b) Pt-Ru/C 75:25; c) Pt-Ru/C 84:16 T. T.; d) Pt-Ru/C 84:16; e) Pt-Ru/C 92:08 T. T.; f) Pt-Ru/C 92:08; g) Pt/C E-TEK. T.T.: thermal treatment.

After several heat treatments [90,91], it is observed that the peaks in the diffractograms become narrower, which could be due to i) an increase in the crystallinity, ii) a lowering of the internal stress in the net of the alloy caused by a possible migration of Ru (as a substitution defect) or iii) an increase of the particle size. For Pt-Ru/C catalysts with different Ru contents, the mean particle size can be determined from the X-ray diffractograms using Scherrer's equation [92], considering a spherical form for the particles (this condition is implied by the value of 0.9 for the K factor in the equation). In this case, the (220) peak of fcc platinum is chosen because it appears in a region where there are no contributions from the carbon substrate and is possible to fit the peaks using gaussian or lorentzian functions to eliminate the background contributions. In order to obtain more precise measurements, the patterns for particle size determinations can be obtained for 2θ between 60 and 80 ° with a step scan mode, in general with a rate of 0.02 ° s^{-1} and using a step time between 1 and 10 seconds.

The mean particle sizes for all the Pt-Ru/C and Pt-Os/C catalysts are collected in Table 1. It is observed that the mean particle size of the Pt-Ru/C catalysts vary between 2.6 and 4.5 nm, being comparable to those resulting from other preparation methods [55,78,79,85,91]. In the case of Pt-Os/C, it is noted that the Pt-Os/C (FAM) and Pt-Os/C (WM) materials present smaller average particle sizes than Pt-Os/C (BM). Finally, Table 1 also shows that in all cases the particle size increases with the heat treatment, probably due to sintering and agglomeration of the particles.

The morphological characteristics of the catalysts can be studied by electron backscattering transmission electron microscopy (TEM) micrographs. Figure 2 shows these results for a Pt-Ru/C catalyst with and without thermal treatment (1h, 300 °C, H_2). For all the catalysts it is observed the formation of agglomerates with sizes varying in the range 100 – 200 nm. Sometimes, images that suggested agglomerates with higher dimensions were observed. Because this means a loss of active area, the formation of these large agglomerates should affect the performance of the catalysts. The phenomenon is also seen for commercial E-TEK catalysts, Pt/C and Pt-Ru/C 50:50.

a

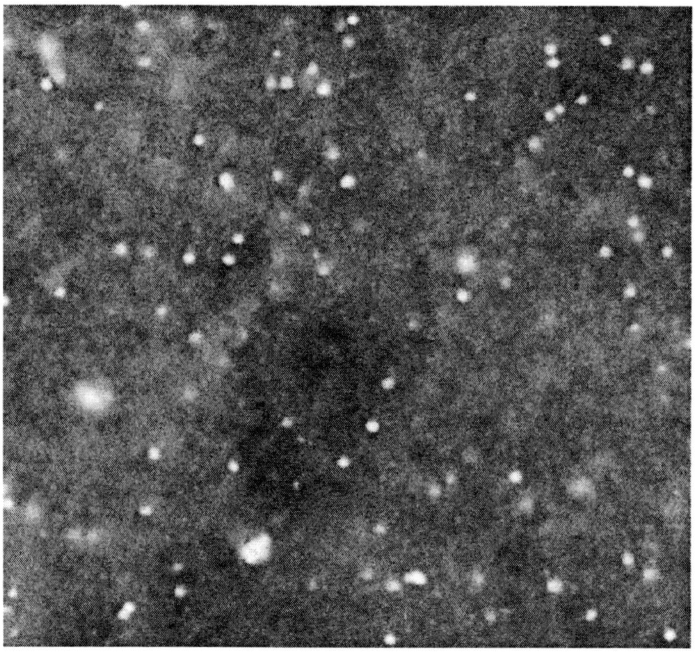

b

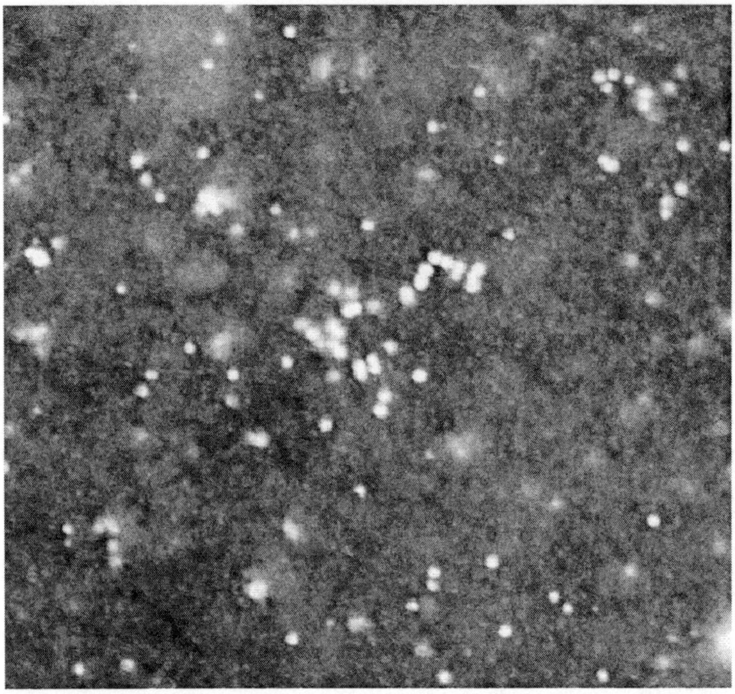

Figure 2. Back-scattering micrographies of catalysts prepared by the FAM: a) as-prepared Pt-Ru/C 75:25 and b) thermally treated Pt-Ru/C 75:25 TT.

Standard TEM micrographs for a Pt-Ru/C 75:25 catalyst are presented in Figure 3. Important characteristics of the morphology of the catalysts are shown: i) there are some portions of the carbon support that do not show the presence of catalyst, an aspect that must be directly related to the preparation procedure and ii) it is apparent the presence of agglomerates of metal particles with sizes ranging from 15 to 25 nm on the carbon support (the large agglomerates observed by backscattering cannot be observed in the TEM micrographs because of absorption of the electron beam). The formation of agglomerates had been reported before for Pt-Ru/C [67,78,79,91] and, although the reasons for their formation are not well known, the phenomenon might be related to the nature of the support and/or the method of preparation.

a

b

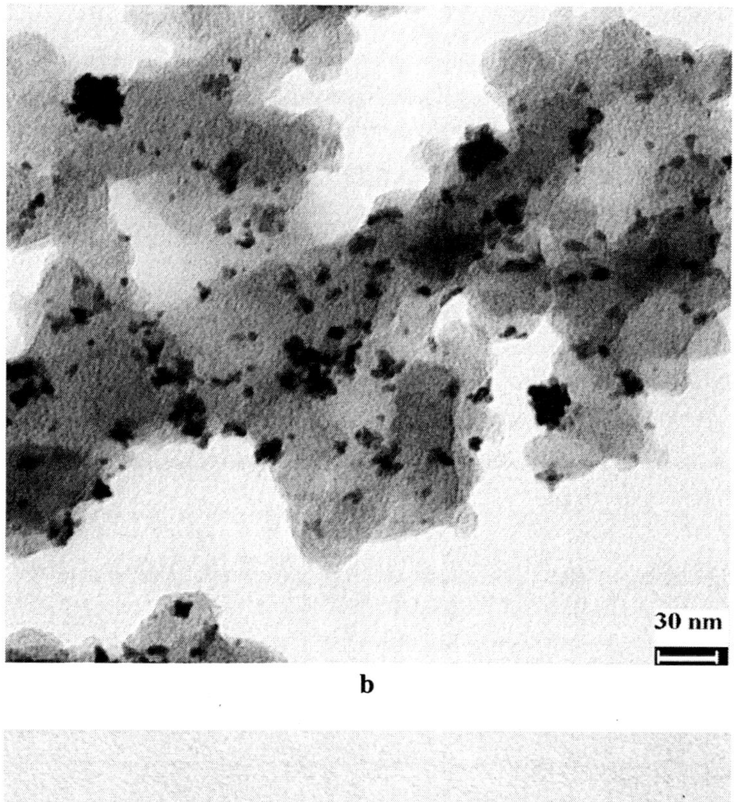

Figure 3. TEM micrographies of catalysts prepared by the FAM: a) as-prepared Pt-Ru/C 75:25 and b) thermally treated Pt-Ru/C 75:25 TT.

In Figure 3B it is shown that the thermal treatment contributes to increase the particle size of the catalysts. This phenomenon must be influenced by the previous presence of metal agglomerates in the catalysts, as shown in Figure 3A. There is no question that TEM is a better technique than XRD to determine changes in particle size because it allows the construction of histograms that reveal the distribution of particle sizes, while XRD gives the crystallyte size, which is smaller than the particle size. The histograms obtained for these materials from the TEM micrographs for the different catalysts prepared by the FAM were built by counting particles within 0.5 nm using several micrographs of the same catalyst. Some of these results are shown in Figure 4. The histograms of the distribution of particle sizes for the Pt-Ru/C catalyst were quite broad and, in the majority of cases, the particles present diameters between 3 and 6 nm. Slightly larger fractions of smaller particles (2-3 nm) are found for the catalysts prepared at 53 °C.

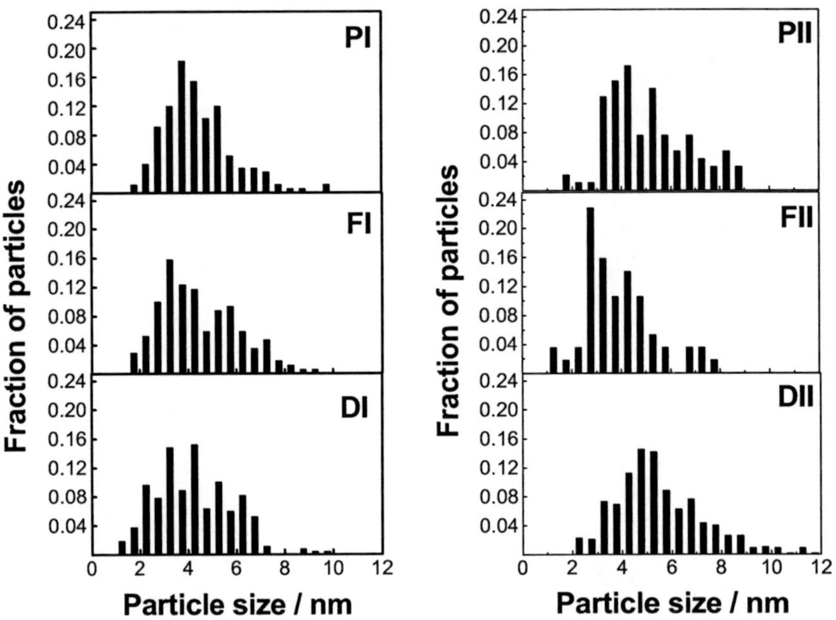

Figure 4. Histograms of particle size distribution for the Pt-Ru/C materials.

X-ray diffractograms have been also employed to access the unitary cell parameter *a* by using the angular relation (as reported by Radmilovic et al. [79]) for the supported catalysts. The values of *a* parameter for PtRu/C and Pt/C are shown in Table 1. In all cases the values of *a* are smaller than those found for pure Pt (0.3923 nm, JCPDS card 4-802), which indicates the presence of Ru substituting Pt in the fcc lattice. This fact is observed as a shift to higher 2θ values at the collected diffractogram in very controlled conditions as mentioned by Arico et al. [93]. It must be noted that the value of *a* is slightly larger for the materials submitted to thermal treatment indicating some segregation of the bimetallic material. The values of *a* for materials prepared with the formic acid are larger when compared with those of the catalysts obtained by other methods. Arc melted Pt-Ru alloys show $a = 0.3904$ nm for Pt-Ru 80:20 and

0.3864 nm for Pt-Ru 50:50, while this last composition has a value of 0.38965 nm when prepared by chemical reduction with sodium borohydride on a carbon support. Radmilovic et al. [79] found 0.3898 nm for the 75:25 and 0.3884 nm for the 50:50 Pt-Ru/C alloys. Thus, this method seems to favor the formation of solid solutions more than the FAM.

Recently, it was reported by Takasu et al. [83] that the Ru precursor used to produce the catalysts has an effect on the structure of the material. Thus, by using $RuCl_3$ to produce a Pt-Ru/C 50:50 alloy the value of *a* was 0.391 nm, which is similar to that mentioned above for the BM. With $Ru_3(CO)_{12}$ the values of *a* were 0.3882 and 0.390 nm for Pt-Ru/C 75:25 and 50:50 respectively, and with $RuNO(NO_3)_x$ the values were even smaller: 0.3862 and 0.3865 nm, respectively. The possible explanation for the low incorporation of Ru when $RuCl_3$ is used, is the formation of $[Ru(CO)(H_2O)Cl_4]^{-2}$ from $RuCl_3$ and HCOOH in solution when the reaction time is increased. To reduce the metallic ions other methods use organic compounds as formaldehyde (0.3908(5) nm) and hydrazine (0.3916(4)) [94]. The lattice parameter values are smaller than the value for Pt/C (0.3923 nm, JCPDS, card 4-802), suggesting the insertion of Ru on the Pt fcc lattice.

More detailed analyses of the catalyst structure have been made using X-ray absorption spectroscopy, particulary the extended X-rays absorption fine structure (EXAFS) data obtained at the L_3 Pt absorption edge. Figure 5 shows the Fourier transform (FT) of the EXAFS signal for Pt-Ru/C (as prepared), Pt-Os/C, and Pt-Mo/C catalysts. The similarities of the FT features indicate that all catalysts retain the fcc structure of Pt. For Pt-Ru/C and Pt-Mo/C, the absence of a strong peak at around 1.5 Å, the Pt-O distance [95-98], indicates that no strong Pt oxidation is apparent, in contrast to Pt-Os/C treated under air at 200 °C for which the strong peak at around 1.5 Å is related to the presence of the Pt-O coordination in a metal oxide phase. The high peak centered at 2.5 Å represents the contribution of the first metal-metal coordination shell to the EXAFS oscillation. It is seen that the magnitude of these peaks are smaller for the dispersed catalyst compared to bulk Pt, and this is just due to the very small particle size, as observed in previous studies for Pt/C [95,96], Pt-Ru/C [33,96,99] and other Pt-alloy supported catalysts. A single peak is apparent for all materials, suggesting that they are not true alloys. Otherwise, more complicated spectra than those presented in Figure 10 would be expected. This is illustrated by the results in Figure 6 obtained for the Pt-Ru/C catalyst after some heat treatments (200 °C under air, followed by 300 °C or 700 °C under hydrogen, here designated as Pt-Ru/C 200/300 or Pt-Ru/C 200/700). These results show that there is an increase in the conversion of ruthenium into an alloyed state and an increase of particle size with increase of heat treatment temperature. In agreement with these results, other works have shown that there is considerable segregation of the Ru phase in the Pt-Ru materials, and even for a 50:50 atomic composition ratio, only *ca.* 10% of Ru is present in the first coordination shell of Pt [33,99].

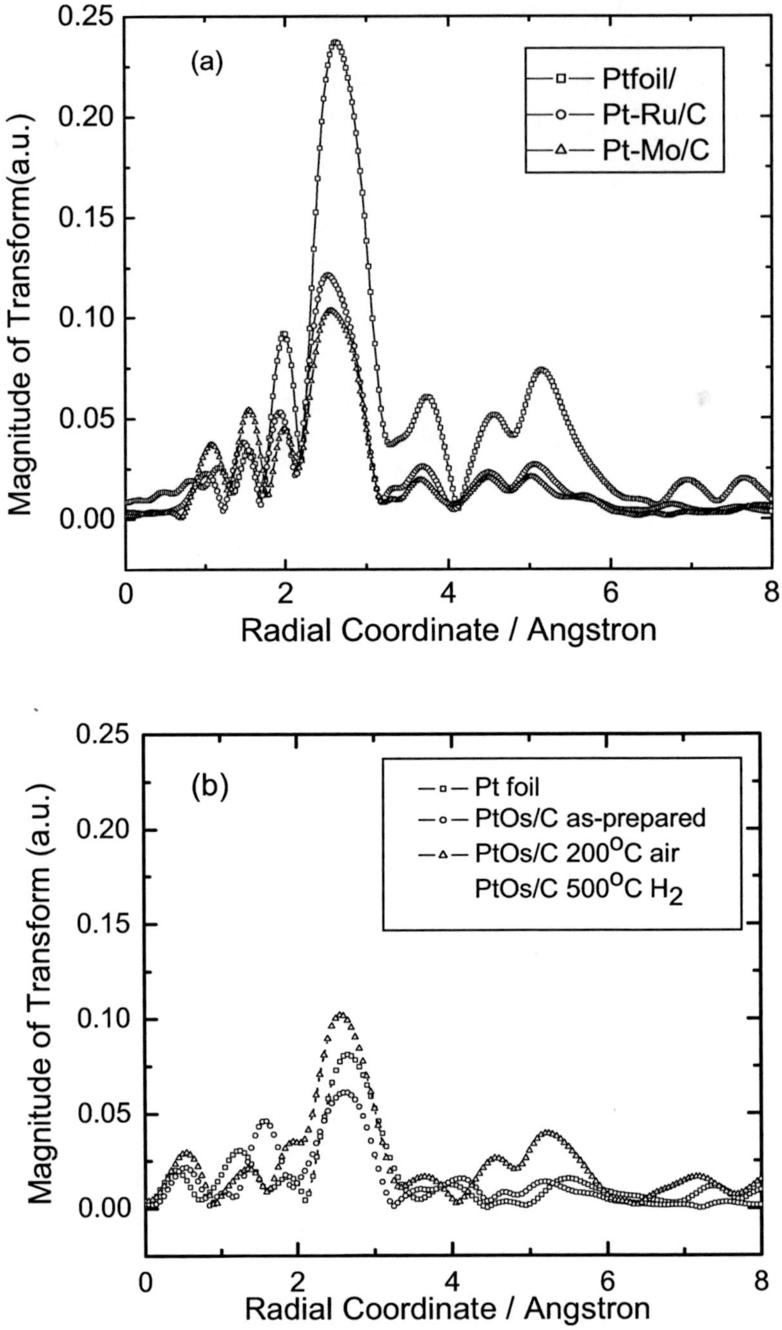

Figure 5. Fourier transform (k^3 weighted) of the EXAFS oscillations at the Pt L_3 edge for: a) PtRu/C, and PtMo/C catalysts and b) PtOs/C heat-treated at different atmospheres and temperatures; $\Delta r = 3.6\text{-}12.6$ Å^{-1}.

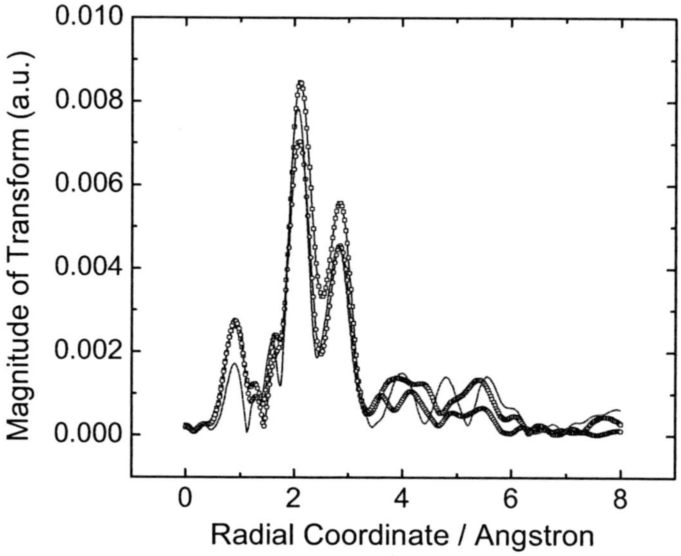

Figure 6. Fourier transform (k^3 weighted) of the EXAFS spectra for: (solid line) as received PtRu/C E-Tek; (O) PtRu/C 200/300; (□) PtRu/C 200/700. Catalysts were analyzed at 0.080 V vs. RHE.

Further analyses of the Pt-Ru/C EXAFS results were made from the Pt-Pt and Pt-Ru first coordination shell signals in the Fourier transform. Phase and amplitude data as a function of the radial coordinate (r) were fitted to those calculated using the FEFF program [100] to obtain the coordination number (N), the bond distance (R), the Debye-Waller factor ($\Delta\sigma^2$), and the edge energy shift (ΔE_o). Table 2 presents a summary of the physical parameters for all samples, at 0.08V and 0.80 V vs. RHE. In all the theoretical calculations a S_o^2 value of 0.72, as obtained by fitting the results for the bulk foil, was used. It has to be noted that the two-shell fit (Pt-Pt and Pt-Ru) was only satisfactory for the less disordered materials constituting real alloys.

Table 2. Results of *in situ* EXAFS analysis on the carbon supported Pt electrocatalysts at 0.080 vs. RHE

Electrocatalyst	Coordination shell	N	R (Å)	$\Delta\sigma^2$ ($Å^2$)	ΔE_o (eV)
Pt/C E-Tek	Pt-Pt	10.4	2.74	0.0061	4.55
PtRu/C E-Tek	Pt-Pt	7.2	2.70	0.0066	0.36
	Pt-Ru	0.8	2.66	0.0010	0.36
PtRu/C 200/300	Pt-Pt	7.2	2.72	0.0067	3.18
	Pt-Ru	1.8	2.69	0.0031	3.18
PtRu/C 200/700	Pt-Pt	7.2	2.72	0.0056	4.02
	Pt-Ru	2.1	2.70	0.0025	4.02

The values of coordination numbers presented in Table 2 for the Pt-Ru/C catalysts are similar to those reported for other Pt-Ru/C catalysts [96]. For all samples, the total Pt-Pt and

Pt-Ru coordination numbers are considerably smaller than that in bulk Pt (N = 12), and this is consistent with the formation of highly dispersed and small particle Pt-based alloys. As shown in Table 3, when the heat treatment temperature of the Pt-Ru/C material is raised from 300 to 700 °C, the total $N_{Pt-Pt} + N_{Pt-Ru}$ increases from 9.0 to 9.3 in agreement with the enlargement of the catalyst particle size, as observed from XRD and TEM measurements. Based on the N_{Pt-Pt} and N_{Pt-Ru} values, it is estimated that the Pt:Ru atomic compositions are 3.9:1 for the Pt-Ru/C 200/300 and 3.4:1 for the Pt-Ru/C 200/700 catalysts. In agreement with the other evidences discussed above, this is indicative of a progressive increase in the conversion of ruthenium into the alloyed state with increasing temperature.

The EXAFS analysis is a more short-range order probe technique than XRD, meaning that the uncertainty in the interatomic distances in Table 1 is larger particularly for the materials with smaller particle sizes. In Table 2 it is seen that the Pt-Pt and Pt-Ru interatomic distances are similar. The values reveal a contraction of the Pt crystalline structure when the catalysts are submitted to high temperatures. This contraction of the reticular structure is one of the effects of the insertion of Ru atoms in the *fcc* Pt crystals, a fact related to the formation of a real alloy [101,102].

2.5. Electronic Structure of the Pt Atoms

X-ray absorption near edge structure (XANES) spectra have been used to characterize the electronic structure of the Pt atoms in the Pt-Os/C, Pt-Ru/C, and Pt-Mo/C catalysts. Figures 7 and 8 show XANES spectra at the Pt L_3 edge obtained at several electrode potentials for Pt-Os/C heat-treated at different temperatures and under oxidative/reductive environments and compared to Pt-foil [103]. The absorption at the Pt L_3 edge (11564 eV) corresponds to $2p_{3/2}$–5d electronic transitions and the magnitude of the hump (white line) located at about 5 eV from the edge energy is directly related to the occupancy of the 5d electronic states [96,104].

In Fig. 7 it is observed the XANES spectra for the as-prepared and heat treated materials at 500 °C in H_2. The Pt spectra are similar to that for the Pt-foil. However, for the material heat treated at 200° C in air, an increase of the white line of the Pt is observed, indicating that the heat treatment in oxidizing atmosphere affects the oxidation state of the Pt promoting the formation of PtO_2 in these conditions. The results in Fig. 8 show that above 0.8 V there is an increase of the white line magnitude in the XANES of Pt in the PtOs/C catalyst heat-treated at 500° C under H_2. The same effect of the electrode potential is observed for all other PtOs/C catalysts. In all cases the phenomenon can be attributed to the emptying of the Pt 5d band due to the well-known formation of a surface oxy-hydroxide layer on the catalyst particles at potentials above 0.85 V.

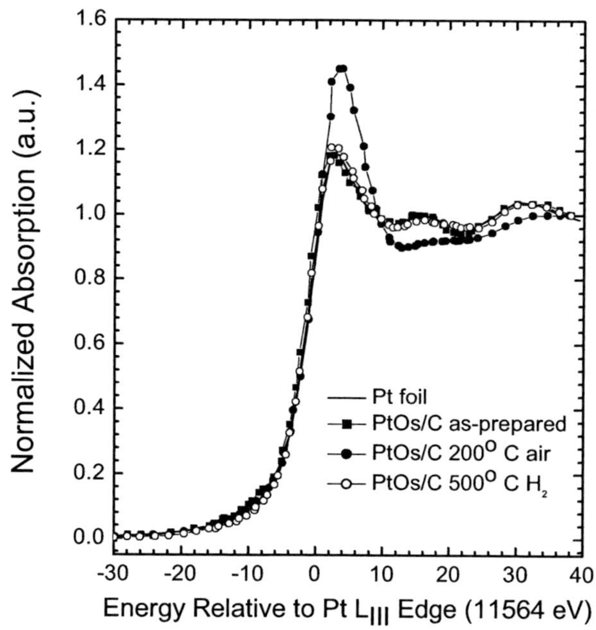

Figure 7. XANES spectra at the Pt L_3 edge at 0.1 V vs. HRE for PtOs/C heat-treated at different atmospheres and temperatures.

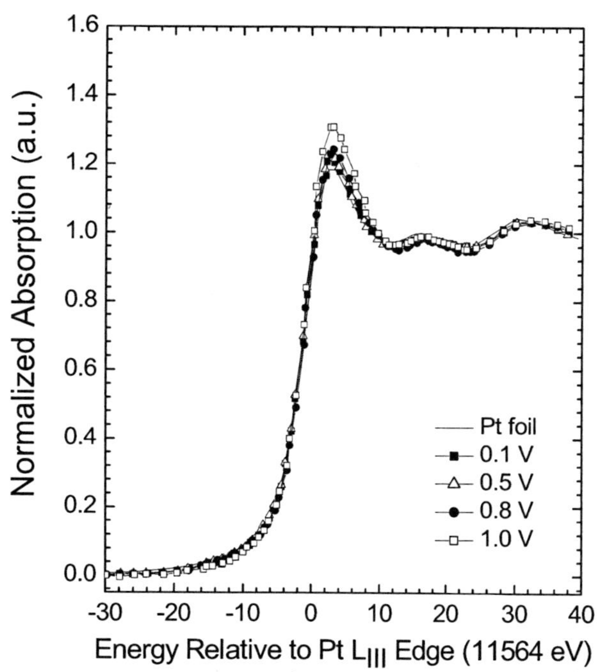

Figure 8. XANES spectra at the Pt L_3 edge at several electrode potentials for PtOs/C heat-treated at 500 °C under hydrogen.

The XANES spectra on the Os-edge for the Os-based materials compared to metallic Os are presented in Figure 9. It is observed that the as-prepared and treated at 500° C in H_2 materials are partially produced as an osmium oxide phase (OsO_2). A remarkable increase of the white line magnitude on the Os-edge is seen for Pt-Os/C treated at 200° C in air, indicating that the oxidative treatment promotes the formation of inactive oxides ~~of~~ for both Pt and Os. The XANES data on the Os-edge spectrum as a function of the potential for the as-prepared material (Fig. 9) show an increase of the white line for Os when the material is polarized at 500 mV, indicating a phase transition from Os to OsO_2 at potentials more positive than 500 mV. In the case of the Pt-Ru/C and Pt-Mo/C catalysts [105,106], the XANES spectra at the Pt L_{III} edge show that the Pt atoms present more vacant 5d electronic states as compared to Pt/C, evidencing the electron withdrawing effect of Ru.

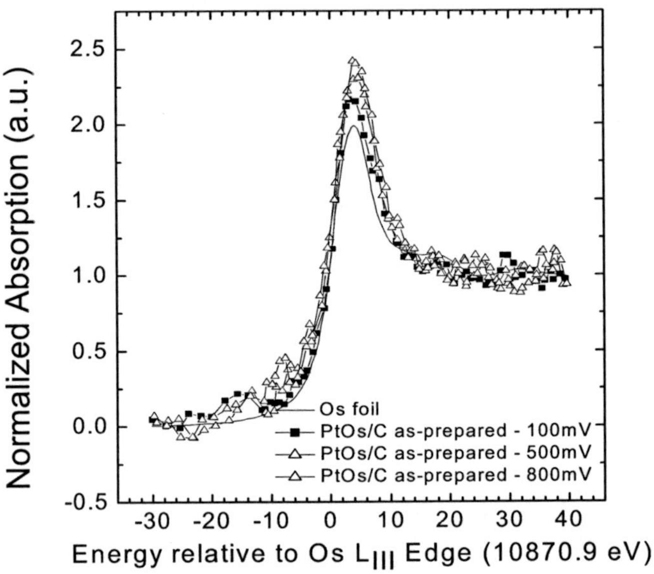

Figure 9. XANES spectra at the Os L_3 edge at several electrode potentials for as-prepared PtOs/C.

2.6. Analyses of the Particle Surfaces

The nature of surface species of Pt/C and Pt-Ru/C catalyst particles can be investigated by X-ray photoelectron spectroscopy (XPS). In general, XPS is considered a surface technique, which is true for non porous materials. For the case of nanoparticle materials, as supported catalysts, the XPS signal comes from the entire nanocrystal, because the crystal size is of the same order as the depth of escape of photoelectrons.

In general [23,67,85,107], the XPS analysis for the C 1s of Pt/C evidences the presence of graphitic carbon by a peak at 284.6 eV and –C=O species associated to the presence of hydroquinone at BE 286.4 eV [108,109]. Aricó et al. [93] also reported the presence of some oxygenated species. Other work [110] reports a peak around BE 281.05 eV due to the 3d Ru orbital, very close to the C 1s orbital, which may be attributed to the superposition of two peaks with very close binding energies, at 280,7 eV associated to the presence of RuO_2 and at 282.5 eV associated to RuO_3 (presents as amorphous species, not detected in XRD analysis).

Platinum is analysed using the 4f orbitals. The Pt 4f signal is derived from three pairs of Pt peaks. The most intense set of doublets with a first peak at 71.5 eV is attributed to metallic Pt [108]. The second set of doublets, where the first peak appearing at BE 72.86 eV can be assigned to a mixture of peaks due to the presence of the 2+ oxidation state in $Pt(OH)_2$ (72.6 eV [31]) and PtO (73.8 eV [108]) like species. An additional set of peaks of weaker intensity, at even higher BE (with a first peak at 74.71 eV), may be attributed to the presence of a small amount of PtO_2 species on the surface (74.6-75.0 eV [108]). For FAM (Pt/C, Pt-Ru/C 82:18) and E-TEK (Pt-Ru/C 50:50 analyzed for comparison) catalysts it was observed a shift of the 4f Pt orbital to higher values of energy, in agreement with the increase of Ru content, (71.72 eV for Pt-Ru/C 82:18 and 71.82 for Pt-Ru/C 50:50 E-TEK) but this shift may have contributions from effects associated to small particles or the carbon support as reported by Antolini et al. [111]. The analysis of 3p orbitals for Ru show for FAM catalysts the presence of two peaks at 462.14 eV and 465.84 eV attributed to RuO_2 and RuO_3 species. Yet, the width of ca 4 eV of the main peak suggests that this component could be derived from the contributions of more than one species with BEs very similar to the behavior reported by Arico et al. [93]

3. The Hydrogen Anode in the Presence of Co

3.1. The CO Tolerance Mechanism

The understanding of the CO tolerance for the hydrogen oxidation reaction is fundamental for the development of more efficient electrocatalysts. Among different possibilities, Pt-based alloys have been suggested as the most promising structure for the hydrogen oxidation reaction in the presence of CO [9,112]. Essentially, two mechanisms have been proposed for the effect of the second metal: (i) the so-called bifunctional mechanism [113-115] and (ii) a weakening of the Pt-CO interaction [116]. Watanabe and Motoo [113] have discussed the CO tolerance on Pt-Ru/C on the basis of the bifunctional mechanism. In this mechanism the Ru surface is activated with water and the CO electro-oxidation occurs on adjacent Pt sites, as follows:

$$Ru + H_2O \leftrightharpoons Ru\text{-}OH + H^+ + e^- \qquad (1)$$
$$Pt\text{-}CO + Ru\text{-}OH \leftrightharpoons Pt + Ru + CO_2 + H^+ + e^- \qquad (2)$$

The electro-oxidation of CO on Pt-Ru/C begins at 0.25V versus RHE, which corresponds to a shift to negative potentials of 0.2 – 0.35 V, compared to Pt/C, demonstrating the high capacity of adsorption of water on Ru [117].

More recently, Watanabe et al. [114,115] reported that the bifunctional model is only operative for Pt-precious metal alloys, such as those with Ru, Os, or Re due to stability of theses elements in the Pt structure. In contrast, the bifunctional mechanism could not be accepted for non-precious metal Pt alloys, such as Pt-Fe, Pt-Co, and Pt-Ni, and this was related to the formation of a Pt skin layer on the alloy particles.

Norskov et al. [116] conducted a systematic study by using density functional theory in order to evaluate the effects of the interaction between Pt and other metals in the CO tolerance process. The authors concluded that the bifunctional mechanism is operative at high overpotentials, but at low overpotentials, a weakening of the Pt-CO interaction predominates and favors the hydrogen adsorption on the Pt sites.

Koper et al. presented a detailed analysis of an electrocatalytic CO oxidation on Pt and Pt-Ru electrodes by using dynamic Monte Carlo simulations [118,119]. An important conclusion of these works is that the approach breaks down severely if the adsorbed species, which are involved in bimolecular surface reactions, do not diffuse, in contrast to the Pt/gas interface for which CO mobility is known to be high. The simulations for Pt-Ru electrodes, including the main bifunctional mechanism, show that CO mobility is vital for the Pt-Ru surface to be electrocatalytically more active than the pure elements Pt and Ru. In the absence of CO mobility the surface does not show any electrocatalytic enhancement with respect to pure Pt or Ru, even if Ru and Pt sites are well mixed. On the other hand, on a surface with large Ru islands, the overpotential for CO oxidation increases if the CO mobility is sufficiently low.

Rolison et al. [120-121] reported an enhancement in the electrocatalytic activity for methanol oxidation on mixed phases based on metallic Pt and Ru hydrated oxides (RuO_xH_y), which are electron and proton conductors able to dissociate –C≡O fragments from the Pt surface by a bifunctional mechanism (reactions (1) and (2)). The authors demonstrated that an improvement of the CO tolerance can be achieved for intimate phase neighbors and that for the bifunctional mechanism the surface conditions are more important than the presence of a true alloy itself. Schmidt et al. [55] proposed a method for the evaluation of catalysts for the oxidation of H_2 and CO by working with Pt and Pt-Ru particles supported on carbon and covered with a Nafion® film. A kinetic study of the oxidation of CO was presented later by the same authors [122].

As shown here, there are many controversies about the CO tolerance mechanisms around Pt-based, mainly Pt-Ru, catalysts. In this work the CO tolerance mechanism is analyzed for different catalysts and this is correlated with the material structure and electrochemical properties.

3.2. The CO Tolerance of Pt-Os/C and Pt-Ru/C Catalysts

Single cell polarization curves in the presence of H_2/100 ppm of CO for several Pt-Os/C catalysts prepared by different reduction processes and after several heat treatments are shown in Figure 10. These results show that the CO tolerance of Pt-Os/C (FAM) and Pt-Os/C (WM) is higher compared to that of Pt-Os/C (BM) catalysts, as seen from the magnitude of the plateau currents, which are determined by the number of Pt sites free from adsorbed CO. This is true even for Pt-Os/C (BM) with high Os contents, indicating that the amount of Os is not the most important factor in determining the CO tolerance. However, the increase of the Os content may also provide some enhancement of the CO tolerance. Table 1 shows that the Pt-Os/C (FAM) and Pt-Os/C (WM) materials present similar particle sizes (or surface area) and Os contents. In agreement to this, they also present similar CO tolerance. However, besides the surface area or particle size, the reasons for the different activities of the catalysts may be related to several other factors such as the material morphology and real composition and the chemical/electronic properties of the particle surfaces, particularly with respect to the Pt atoms.

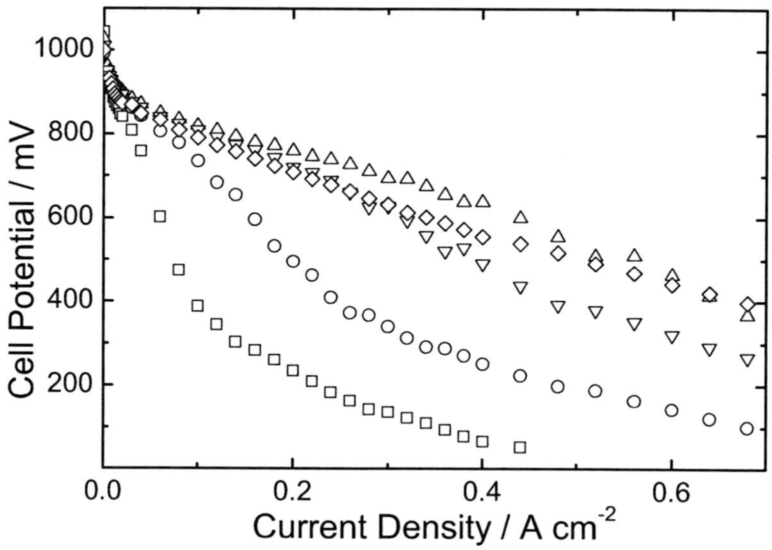

Figure 10. Polarization behavior of fuel cells based on different PtOs/C anodes, fed with O_2 in the cathode and H_2/CO (100 ppm) in the anode fed with O_2 in the cathode and pure H_2 or H_2/CO (100 ppm) in the anode. T_{cell} = 85 °C, T_{O2} = 90 °C, and T_{H2} = 100 °C; p_{H2} = 2.0 atm and p_{O2} = 1.7 atm. Anodes: (□) Pt/C E-tek; (○) PtOs/C – B2 treated at 200° C in air and 500° C in H_2 atmospheres; (Δ) PtOs/C – F treated at 500° C in H_2 atmosphere; (∇) PtOs/C – W treated at at 200° C in air and 500° C in H_2 atmospheres; (◇) PtRu/C (50:50) E-tek.

XANES spectra obtained on the Os-edge as a function the electrode potential indicate that a layer of inactive Os oxide (OsO_2) is formed when the material is submitted to an oxidative atmosphere, and this causes a loss of surface area for the CO oxidation reaction. This oxide is also produced at potentials more positive than 0.5 V limiting the electrocatalytic performance of Pt-Os/C. A high electrocatalytic performance is obtained for Pt-Os/C submitted to a reductive atmosphere (500 °C H_2), indicating a more homogeneous distribution of the metallic elements, Pt and Os (Fig. 10).

Figure 11 shows the polarization curves obtained for Pt-Ru/C catalysts in the presence of H_2/100 ppm of CO. The performance of the Pt-Ru/C 200/700 catalyst is clearly superior to all other investigated so far. Results also indicate that the commercial powder is less sensitive to the thermal treatment. Since the same total amount of Pt and Ru (1:1 bulk atomic composition) is present in both catalysts, as confirmed by EDX [78], the different behavior must be related to the extension of Ru incorporation into the *fcc* Pt structure and/or to the catalyst surface area. XRD results show that the average particle size is significantly different for these materials. While the average particle dimension is around 4.5 nm for the Pt-Ru/C 200/700 catalyst, the value for the commercial material is *ca.* 7.0 nm. This difference may be the main factor responsible for the poorer electrochemical performance of the Pt-Ru/C E-Tek catalyst when compared to Pt-Ru/C 200/700 [105].

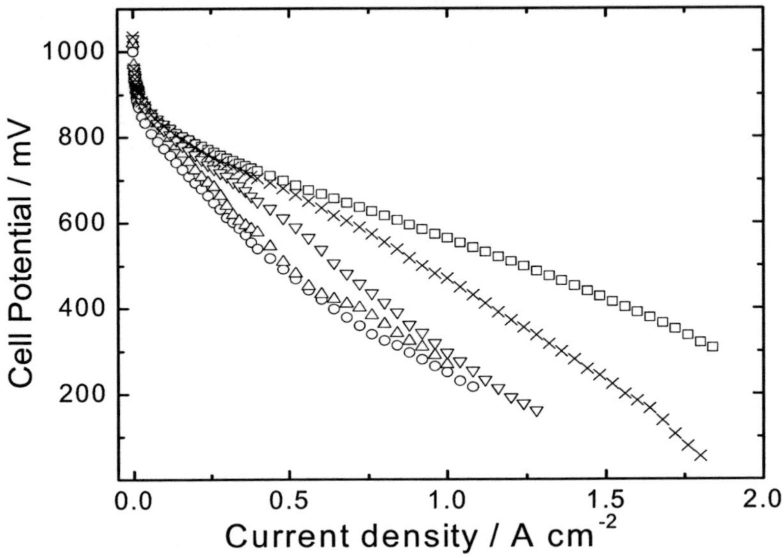

Figure 11. Potential-current density curves for PtRu/C catalysts: (□) PtRu/C E-Tek fed with pure H_2: (O) PtRu/C E-Tek as received; (Δ) PtRu/C E-Tek 200/300; (∇) PtRu/C E-Tek 200/700; (x) home-made PtRu/C 200/700 feed with H_2 + 100 ppm of CO.

It has been pointed out that not all available samples of unsupported and supported 1:1 PtRu (nominal overall atomic ratio) catalysts exhibit the same activity with respect to methanol oxidation [123,124] and this has been attributed to the formation of particles with different compositions and morphologies. Contrary to the current understanding of the oxidation of methanol on Pt-Ru, Long et al. [121] have proposed that the bimetallic alloy is not the most desired form of the catalyst. They proposed that the RuO_xH_y speciation of Ru in nanoscale Pt-Ru blacks affords a much more active catalyst for methanol oxidation than does Ru^0 as part of a bimetallic alloy. Dinh et al. [123] have investigated the surface composition of three 1:1 Pt-Ru catalyst unsupported powders by CO stripping voltammetry. They have concluded that the Pt-Ru catalysts that exhibit higher CO stripping charges have larger number of (exposed) metal alloy surface sites. This condition of the catalyst is achieved by a combination of smaller particle size and a more reduced state of the catalyst surface. In turn, a more reduced Pt-Ru catalyst surface seems to be obtained when the overall average state, bulk and surface, is deeper reduced.

In order to evaluate the effect of the presence of the RuO_xH_y speciation of Ru in nanoscale Pt-Ru blacks on the CO tolerance, Pt-Ru/C powders were synthesized by an impregnation method and submitted to a water vapor treatment [125]. The electrocatalytic performance for hydrogen oxidation in the presence of CO was compared to that of a commercial catalyst ($Pt_{50}Ru_{50}$/C, E-TEK). Figure 12 presents the polarization curves for the single cell with Pt-Ru/C anodes supplied with pure H_2 and with H_2/CO (100 ppm), after the heat treatments under H_2 at 550 °C for 3 h, followed or not by a treatment in a water vapor atmosphere at 100° C for 24 h. Some improvement in the polarization response is observed for the vapor treated commercial Pt-Ru/C, in contrast to Pt-Ru/C prepared by the impregnation method. In the case of the commercial material, the positive effect may be due to a decontamination of the surface of the metallic particles and/or the formation of RuO_xH_y

species promoted by the water vapor treatment. A higher CO tolerance is observed for the Pt-Ru/C catalyst prepared by the impregnation method without water vapor treatment and this may result from an increased catalytic active area promoted by an homogeneous distribution of the Ru and Pt particles, or by an electronic effect on the Pt atoms due to Pt-Ru interactions in a phase of true Pt-Ru alloy [105]. The decrease of the electrocatalytic performance observed after the water vapor treatment can be related the destabilization of the Pt-Ru interaction in the alloy phase.

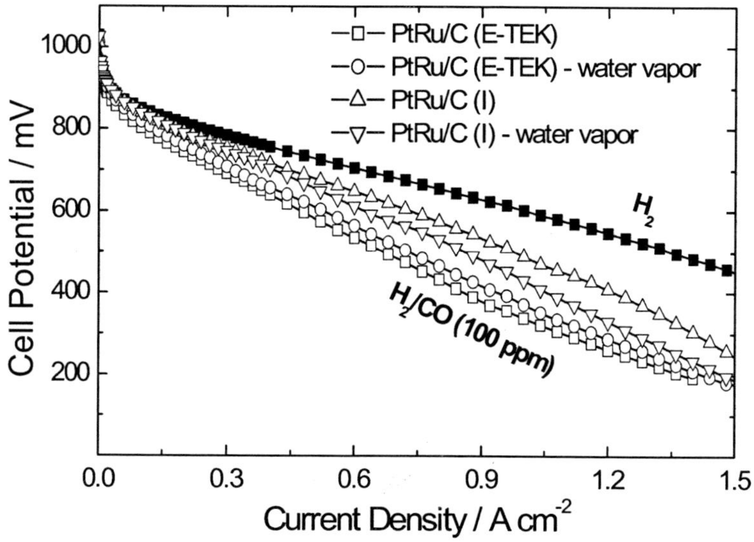

Figure 12. Polarization curves of pure H_2 and H_2/CO (100 ppm) for gas diffusion electrodes based on PtRu/C (20 wt.%, 0.4 mg cm^{-2}) supplied by E-TEK or prepared by the impregnation method. As-prepared and water vapor heat-treated materials. The cathode was fed with O_2. Oxygen humidification at 90°C and 1.7 atm and fuel humidification at 100°C and 2.0 atm. Cell temperature = 85°C.

3.3. The Particular Case of Pt-Mo/C

Attempts to elucidate the mechanism involved in the CO tolerance have been carried out on several Pt-Mo/C catalysts. Baltruschat et al. [126] analyzed the effect of Mo deposited onto different smooth Pt structures, such as Pt (111), Pt (332) single and porous polycrystals and attributed the CO tolerance to an oxygen spillover effect on weakly adsorbed CO and not to an electronic effect. Therefore, the effect of Mo is very small in the case of CO produced during the direct methanol oxidation.

In spite of the enhanced CO tolerance observed for Pt-Mo/C, good performance is not achieved for direct methanol and ethanol oxidations [47,127], demonstrating that the mechanism of CO tolerance differs from that of direct methanol oxidation. Cabrera et al. [128,129] have conducted experiments aiming to identify the active species for methanol oxidation. The authors associated the enhanced catalytic activity for methanol oxidation to the sub-stoichiometric lower valence molybdenum oxides (MoO_x – $2 < x < 3$) and the proton spillover effect from the hydrogen molybdenum bronze (H_xMoO_3).

Mukerjee and Urian [130] used electrochemical and X-ray absorption spectroscopy techniques to elucidate the CO and methanol oxidation mechanism on Pt-Mo/C. An important conclusion of this work is that the Mo surface presents no affinity to CO and then it is always available for CO oxidation even at low overpotentials, in contrast to the case of methanol oxidation where the active Mo species are not able to oxidize methanol intermediates. More recently, a low activity of Pt-Mo/C catalysts was found for the hydrogen oxidation in the presence of carbon dioxide (CO_2) and this has been associated to an effective electro-reduction of CO_2 into CO producing a higher CO coverage than those observed in the same experiments with Pt/C and Pt-Ru/C [131].

The electrocatalytic activity of highly active Pt-Mo/C dispersed catalysts prepared by chemically reducing the corresponding precursor ions with formic acid has been investigated. A comparison of the polarization responses of Pt-Mo/C, Pt/C, and Pt-Ru/C is shown in Figure 13 for pure H_2 and H_2/CO (100 ppm). For both Pt-Ru/C and Pt-Mo/C enhanced electrocatalytic performances compared to Pt/C are observed, but the overpotential loss at higher overpotentials is lower for Pt-Mo/C.

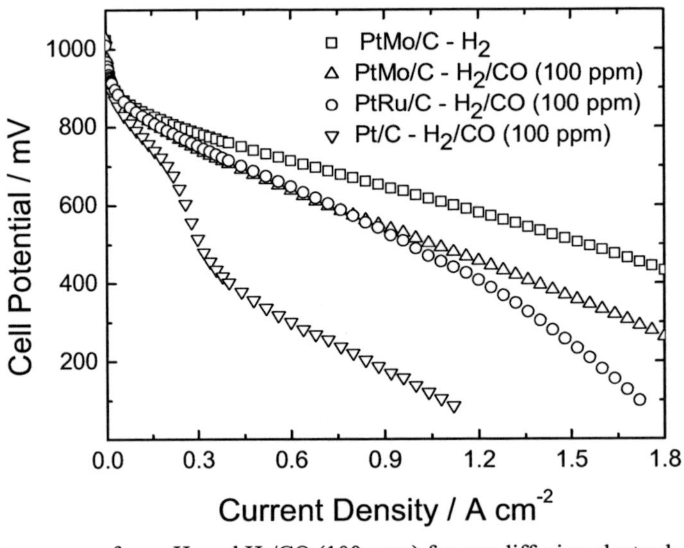

Figure 13. Polarization curves of pure H_2 and H_2/CO (100 ppm) for gas diffusion electrodes based on Pt/C, PtRu/C prepared by WM and heat treated at 200°C/700°C in H_2 atmosphere, and PtMo/C prepared by FAM (20 wt.%, 0.4 mg cm^{-2}). The cathode was fed with O_2. Oxygen humidification at 90°C and 1.7 atm and fuel humidification at 100°C and 2.0 atm. Cell temperature = 85°C.

Figure 14 shows single cell polarization curves for both, pure H_2 and H_2/CO (100 ppm), on standard electrodes with Pt/C, Mo/C and Pt-Mo/C catalyst layers and those having Mo-based filtering layers [18,132]: a) a CO filtering layer composed by Mo/C applied onto the gas diffusion layer facing a catalyst layer formed by Pt/C and b) a filter layer of Pt-Mo/C applied onto the gas diffusion layer facing a catalyst layer formed by Pt/C. The polarization curves presented in Fig. 14 show that for Mo/C there is no adsorption and oxidation of H_2. In the case of the Pt/C electrode with a Mo/C filter in the diffusion layer, a significant enhancement of the electrocatalytic performance is detected, compared to that of the electrodes with a single layer of Pt/C.

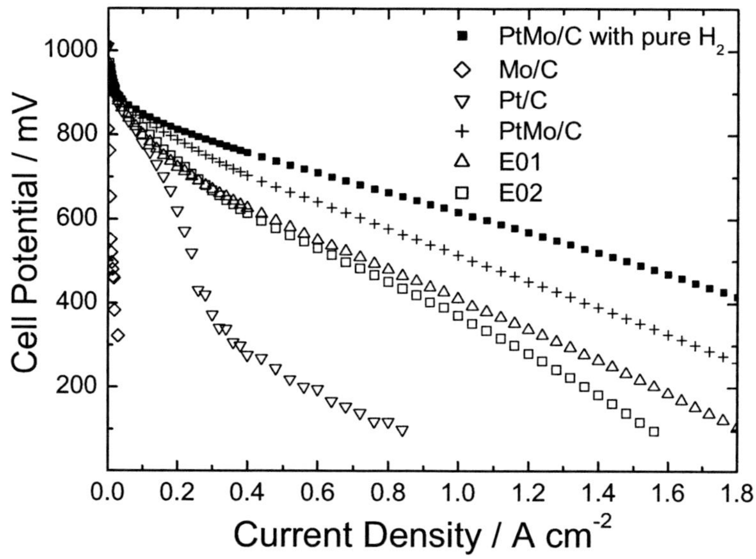

Figure 14. Polarization curves for Pt/C (0.4 mg cm^{-2}); Mo/C (0.4 mg cm^{-2}), PtMo/C (0.4 mg cm^{-2}); E01: Mo/C filter (0.064 mg cm^{-2}) plus a Pt/C (0.336 mg cm^{-2}) catalyst layer; E02: catalyst layer with Mo/C (0.064 mg cm^{-2}) plus Pt/C (0.336 mg cm^{-2}) physically mixed.

Recently, Ioroi et al. evaluated the CO-tolerance on several Pt-MoO_x/C electrodes. The authors analyzed the influence of the gas flow rate and the thickness of the Nafion membrane in the response of anodes based on Mo oxide and concluded that the CO tolerance is attained by an intrinsic factor given by the bifunctional mechanism in combination with a non-electrochemical CO oxidation involving O_2 permeated from the cathode [133]. However, this effect can not explain the present results because it would be essentially the same for all catalyst layers, independent of the composition of the diffusion layer. The behavior of the electrodes with the filtering layers are more consistent with the response of an electrode supplied with a lower CO content, as for Pt/C or Pt-Ru/C supplied with H_2/CO (20 ppm) [134]. This lowering of the CO content may occur inside the gas channels in the catalyst layer of the electrode through a chemical reaction of CO catalyzed by Mo sites. This may be the well-known water gas shift (WGS) reaction

$$CO + H_2O \rightarrow CO_2 + H_2 \qquad \Delta H = -41 \text{ kJ mol}^{-1}$$

which involves the conversion of CO to CO_2 on the Mo sites without direct participation of protons and electrons. Because of its exothermic character, this reaction is more favorable at lower temperatures (100 – 200 °C) and, generally, WGS catalysts are based on supported (SiO_2 and Al_2O_3) or unsupported iron, cooper, chromium, or zinc particles [135,136]. Compounds of molybdenum have also shown potentiality for application as WGS catalysts [137-139], mainly molybdenum carbide that has demonstrated an enhanced performance in comparison to a commercial Cu-Zn-Al catalyst [140].

To check if the Mo/C catalyst is really effective for lowering the CO concentration as claimed above, experiments of "on-line" gas chromatography were made to analyze the gas composition at the outlet of a WGS reactor fed with different $H_2O/H_2/He/CO$ mixtures. Figure 15 shows plots of the relative CO conversion factor as a function of time for the reactor at 100° C. In the absence of water (He/CO 1%), no CO conversion was observed. On the other hand, in the presence of water under the same conditions (H_2O/He/CO 1%), a lowering of the CO concentration was detected in the outlet of the reactor (*ca.* 5%, for t ≥ 40 min), indicating that oxidation of CO on Mo/C involves water, as required by the WGS reaction. Experiments with decreasing inlet concentrations of CO (1 to 0.1%) show decreasing outlet concentrations, reaching a CO conversion of *ca.* 15% for 0.1%, equivalent to a continuous CO consumption of about 150 ppm. These results clearly demonstrate the occurrence of the WGS reaction on the Mo/C material.

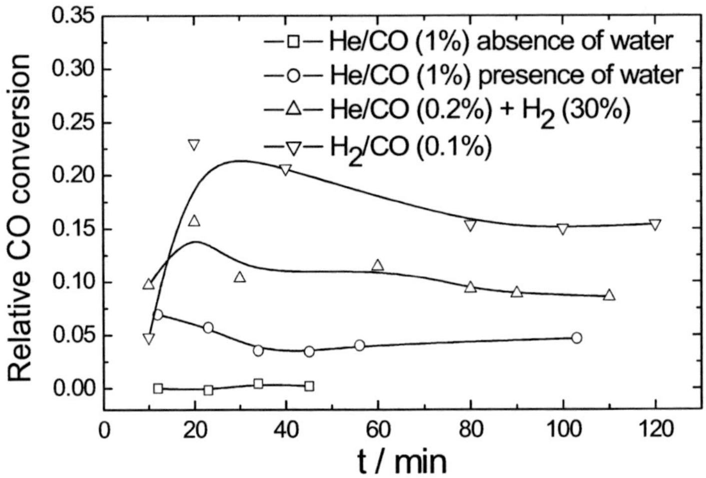

Figure 15. Plots of the relative CO conversion parameter as a function of time, measured by "on line" gas chromatographic analyses at the outlet/inlet of a Mo/C chemical reactor fed with different CO composition mixtures. The reactor was kept at 100 °C.

The quantity of converted CO (mol) per unit of catalyst mass and time (conversion rate) was evaluated for the experiment with (H_2O/He/CO 1%) and the average value resulted 8.5×10^{-8} mol g^{-1} s^{-1} (for t ≥ 60 min). This value is smaller than that presented for a conventional shift catalyst Cu/ZnO supported on Al_2O_3 for which the conversion rate was 5.5×10^{-7} mol g^{-1} s^{-1} for the catalyst showing the higher CO conversion factor (82%) [141]. The lower activity of Mo/C must be associated to the lower temperature of the reactor (100 °C) compared to that employed for Cu/ZnO/Al_2O_3 (200 °C), which presents no catalytic activity at 100 °C.

The effect of the presence of CO_2 in the performance of electrodes with the diffusion layer formed with Carbon or Mo/C and catalyst layers with Pt/C, Pt-Ru/C, and Pt-Mo/C, was also investigated. Figure 16 presents the polarization curves for Pt/C in pure H_2, H_2/CO_2 (75:25 v:v), and H_2/CO (100 ppm)/CO_2 (75:25 v:v). A similar performance is observed with pure H_2 on both electrodes. A small decrease in the cell potential is observed for the standard electrode when CO_2 is inserted within the H_2 medium. This decrease has been reported before for Pt/C and attributed to a CO poisoning effect caused by the CO formed from the reversing of the WGS reaction taking place in the catalyst layer [142,143]. Results in Fig. 16 show that the performance of the electrode based on the Mo/C diffusion layer supplied with H_2/CO_2

(75:25, v:v) is lower than that presented by the standard electrode. This effect must be associated to an additional CO production by the reverse WGS reaction, now also promoted by the Mo/C layer. In any case, it is noted that the total CO production is small so that the poisoning effect is well below that caused by 100 ppm of CO. Finally, it is noted that, compared to the standard Pt/C electrode, the Mo/C modified electrode shows a remarkable CO tolerance when supplied with H_2/CO(100 ppm)/CO_2(75:25 v:v). This result indicates that in this case there is a lowering the CO content, corresponding to the net effect of the direct/reverse WGS reaction.

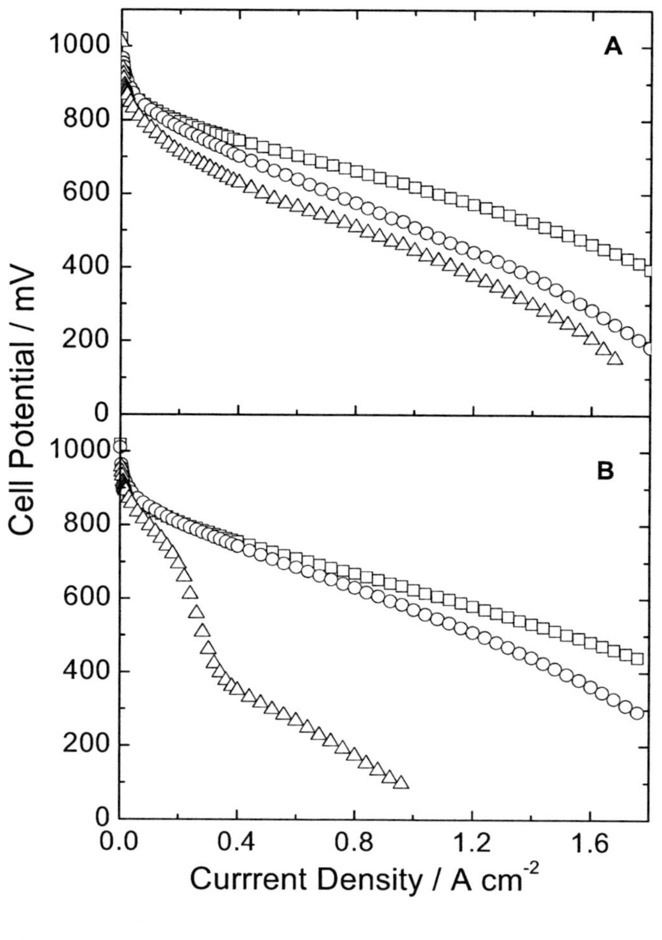

Figure 16. (A) Polarization curves for PEM single cells with gas diffusion anodes composed by a diffusion layer of Mo/C and catalyst layer of Pt/C: (□) pure hydrogen; (O) H_2/CO_2 (75:25 v:v); (Δ) H_2/CO(100 ppm)/CO_2 (75:25 v:v). (B) Results for an electrode with the same catalyst layer but with a standard carbon diffusion layer. The cathode was fed with O_2. T_{cell} = 85 °C, oxygen humidifier at 90 °C and hydrogen humidifier at 100° C; p_{H2}=2.0 atm; p_{O2}=1.7 atm. Catalyst loading = 0.4 mg cm^{-2}. Anode flow rate = 100 cm^3 min^{-1} and cathode flow rate = 670 mL min^{-1}.

4. The Methanol Oxidation Reaction

4.1. Electrochemical Response

The activity of several electrodes containing supported catalysts has been evaluated for the methanol electro-oxidation and some results obtained with the thin porous electrode[144], the thin-film electrode [55] and with gas diffusion electrodes [145] in a half and single fuel cell will be discussed here.

Figure 17 shows cyclic voltammograms obtained in a half-cell in the absence of methanol, with gas diffusion electrodes (prepared as described previously [145]) recorded at 20 mV s^{-1} for Pt-Ru/C 75:25 prepared by the FAM and submitted to thermal treatment. For comparison the voltammogram for Pt/C E-TEK 20 wt % is also shown. The main difference between the voltammetric profiles of the catalysts in Figure 17 is the region of adsorption and oxidation of hydrogen, between 0.08 and 0.35 V vs. RHE. When the catalysts are submitted to thermal treatment the current density decreases which is due to the increase in particle size (Table 1), as discussed above. On the other hand, the profile obtained with the heat treated catalyst is more similar to that of Pt/C and this can be explained by the segregation of Pt. The higher value of the lattice parameter and the voltammetric profile, which is affected by the larger energy of adsorption of H_2 on Pt than Ru, are evidences that the material behaves more like pure platinum. The current densities between 0.35 and 0.8 V vs. RHE should be higher for the Pt-Ru/C materials when compared to those of Pt/C E-TEK as a consequence of the formation of oxygenated species on Ru. However, the differences are not evident from the results in Figure 17, which may be due to the differences in particle size and, consequently, on active area.

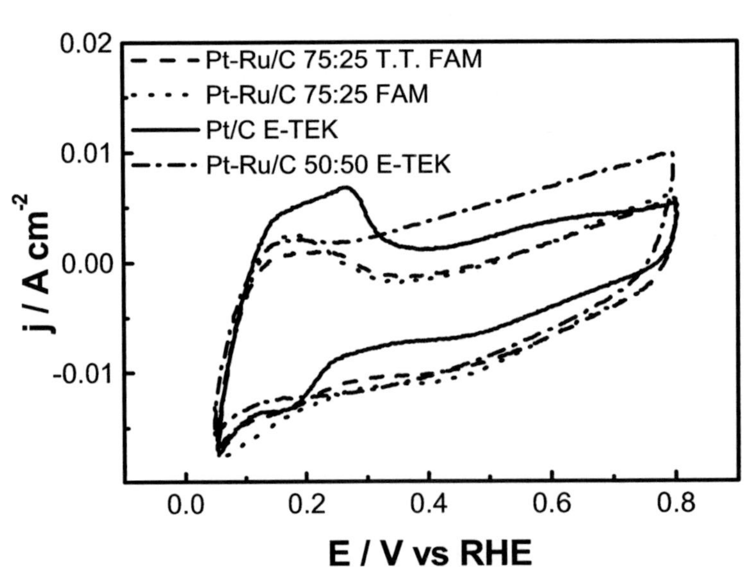

Figure 17. Cyclic voltammograms of Pt-Ru/C 75:25 and Pt-Ru/C 75:25 thermal treated, prepared by the FAM and of Pt/C E-TEK included for comparison.

Figure 18 presents the current densities at the 30th minute in chronoamperometric experiments at 0.5 V vs. RHE with different methanol concentrations for all the catalysts. The current densities are corrected for the respective cathodic current densities at the same potential in the electrolyte without methanol. It can be verified that the current increases very little for methanol concentrations above 2 mol L^{-1} of methanol. This is in agreement with results obtained with other Pt-Ru based catalysts [146]. Comparing the performance of the systems with just Pt and Pt-Ru prepared by the FAM, an improvement is verified with the increase of the Ru/Pt ratio in the catalysts up to 25:75. It is also interesting to see that the catalyst Pt-Ru/C 75:25, prepared in these laboratories by the FAM shows a better performance than the commercial Pt-Ru/C 50:50 E-TEK material. This may be the result of different preparation methods or the higher Ru contents of the E-TEK catalysts. One important difference between the two materials is the crystallinity. In Figure 1 it is observed that for the catalyst Pt-Ru/C prepared by the FAM the peaks corresponding to the plane (111) of Pt are better defined, i.e. they are much narrower, than for the E-TEK Pt-Ru/C 50:50 catalyst. This can be correlated with the fact that, according to the results of Chrzanowski et al. [147], the Pt(111) face is the most active for methanol oxidation and predominant in small supported particles [55,107].

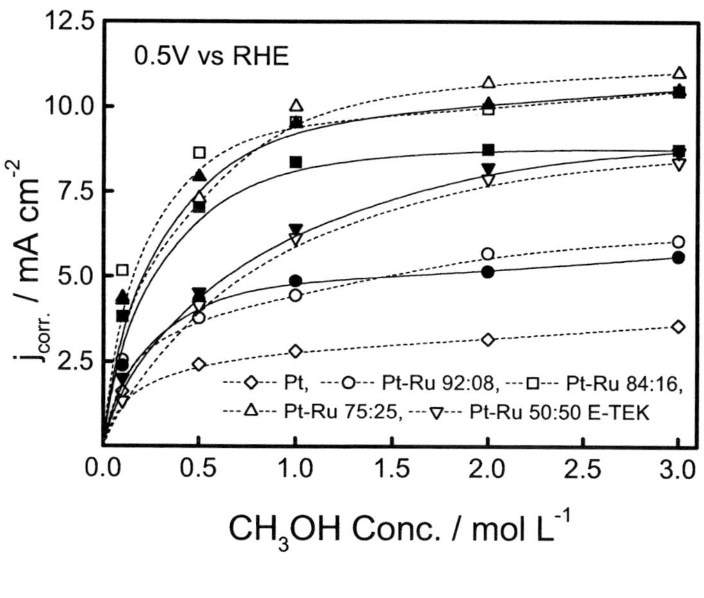

Figure 18. Current densities at the 30th minute of chronoamperometries at 0.5 V (RHE) in different methanol concentrations in 0.5 mol L^{-1} H_2SO_4, corrected for the respective current densities at the same potential in the absence of methanol. Electrodes with 1mg Pt cm^{-2}, 20 % wt/C. The corresponding total metal loads were 1.171, 1,098 and 1.045 mg cm^{-2} for Pt-Ru/C 75:25, 84:16 e 92:08, respectively. Catalysts as prepared (open symbols), and submitted to thermal treatment in H_2 (300 °C, 1h) (solid symbols).

It may be observed that the performance of the catalyst submitted to thermal treatment is not as good as that of the as prepared catalyst due the increase in particle size. Also, the as-prepared catalysts with Ru contents of 16 and 25 at. % show a similar behaviour for methanol electrooxidation, which shows no difference for concentrations between 0.5 and 3 mol L^{-1} in agreement with the results of Iwasita et al. [44]. Also, the performance of the heat treated catalyst with 25 at. % of Ru is similar to that of the Pt-Ru 84:16 material not submitted to thermal treatment. For methanol concentrations above 2.0 mol L^{-1} there is no improvement in the electrochemical response, so this concentration should be appropriate for the performance of the DMFC under the given conditions. These results are in accordance to those of Hogarth et al. [146]. They studied DMFC fed with different methanol concentrations and found that the best response was that for a 2 mol L^{-1} solution of methanol. The high activity of Pt-Ru/C 75:25 at 25 °C is due to the oxidation of CO while for Pt:Ru/C 50:50 there are nor enough available sites for the adsorption of methanol. For the case of Pt-Ru 90:10 and pure Pt/C, the number of Pt sites is high, but the bifunctional mechanism is not operative due to the small amounts of Ru in the catalyst.

The lower activity of Pt-Ru/C 50:50 E-TEK is related to the higher contents of Ru or Ru oxides which are not active for metanol adsorption. But an interesting fact is the effect of the thermal treatment, different from that for the FAM materials. In spite of the increase in particle size, the activity for methanol oxidation increases. The conclusion is that no general rules can be established on the effect of thermal treatment.

The evaluation of Pt-Ru/C catalysts prepared by using the modifications on the formic acid method was also done. Figure 19 shows the results of linear sweep voltammetry experiments, corrected for ohmic drop, for the electrooxidation of methanol on the Pt-Ru catalysts prepared by the P, F and D procedures, which are variations of the FAM. For comparison, linear sweep voltammetry curves for Pt-Ru/C 50:50 E-TEK were also made. The concentration of methanol was 2 mol L^{-1} and the best performance was observed with the type D catalysts. When the electrochemical performance of the catalysts, prepared by the same procedure but at different temperatures are compared, it is observed that the catalysts prepared at 80 °C generate larger currents. It is possible that, as the reduction rate of the metals at 80 °C is higher than at 53 °C, the final distribution of Ru on the surface is more homogeneous at the higher temperature favoring the oxidation of methanol. The DII catalyst has a good distribution of the metal on the carbon support, as shown in Figures 2 and 3. In these backscattering and TEM micrographs the presence of agglomerates of metal particles in micro and nanoscale is not observed. Also, the presence of carbon particles of the support without catalytic material is not observed. The good dispersion evidenced for the catalyst DII prepared at 80 °C may be related to the different conditions introduced in the preparation, such as the lower formic acid concentration, the Ru concentration and the temperature of the reaction. At the same time, it was observed that the catalysts prepared by the procedures P and F show similar performance, indicating that the order of addition of the solutions (the formic acid solution and the solution of the precursors) to the carbon is not relevant.

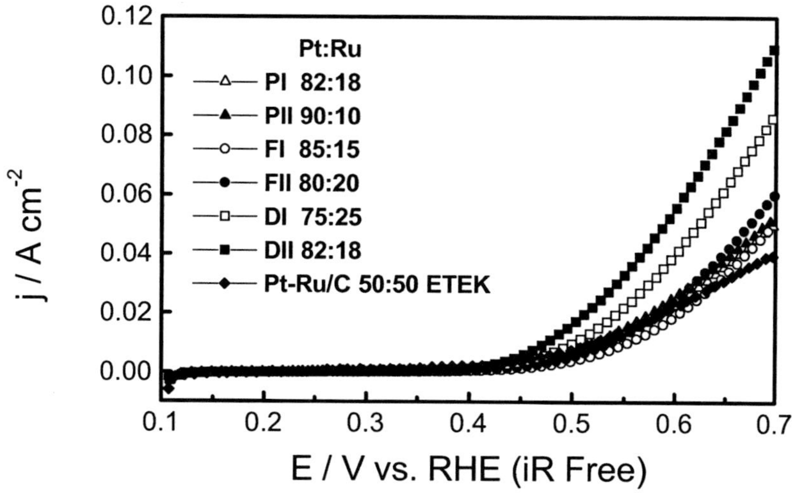

Figure 19. Methanol electrooxidation on diffusion electrodes with Pt-Ru/C. 0.5 mol L^{-1} H_2SO_4 as electrolyte support. Methanol concentration: 2.0 mol L^{-1}. v = 0.5 mVs^{-1}. For comparison Pt-Ru/C 50:50 E-TEK is included. Results corrected for ohmic drop.

The catalysts were also tested in a single DMFC and the potential-current plots are shown in Figure 20. It is apparent that there are differences in performance for the different catalysts. The best performance is obtained when the DII catalyst is used to prepare the anodes. For current densities of 0.3 A cm^{-2} a maximum power density of 93 mW cm^{-2} is reached. The performance of the different catalysts in the single DMFC follows the order of activity found for the oxidation of methanol in the half cell experiments so, in spite of the different conditions, the results of the two types of experiment are consistent. For comparison, Figure 20 shows the results for methanol oxidation using Pt-Ru/C 50:50 E-TEK in the anode. Under the conditions used here (90 °C and 3 atm pressure in the cathode) the current-potential response of the DII material is similar to that of the Pt-Ru/C 50:50 E-TEK catalysts. The comparison of the results in the DMFC with published data is not easy because the operational conditions are not always the same. However, the results are very similar to those published by Arico et al. [148] who used similar conditions to prepare the MEA and carry out the experiments in the DMFC. They obtained an electrochemical performance slightly better than in the present report.

The electroactive areas of the materials using the CO charge were determined subtracting from the CO stripping area the voltammetric area (recorded in a N_2 atmosphere) to eliminate the contributions of the double layer region and using the value of 420 μC cm^{-2} as the oxidation charge of one monolayer of CO on a smooth Pt surface. The results of electroactive areas are shown in Table 2 where the values are expressed both in m^2g^{-1} of Pt and m^2g^{-1} of Pt-Ru.

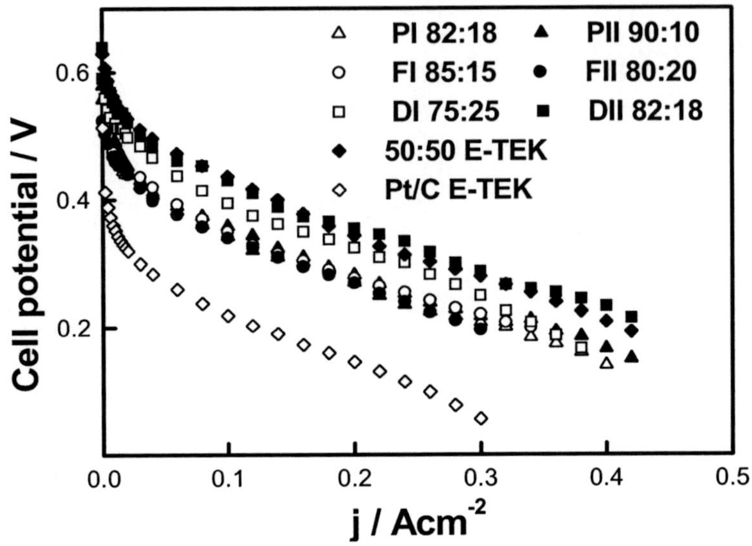

Figure 20. Cell potential vs. current density curves in a single DMFC. Cell temperature: 90 °C, PO_2 : 3 atm. Electrodes with 1 mgPt cm^{-2}, 35 % Nafion® and 20 % metal/C. 15 % PTFE in the diffusion layer. Anode: Pt-Ru/C FAM; cathode: Pt/C E-TEK. For comparison Pt-Ru/C 50:50 E-TEK is included.

It is observed that the material Pt-Ru/C 50:50 E-TEK presents a higher active area than the FAM materials as a consequence of the smaller particle size. The areas for FAM catalysts are similar to those calculated by Colmati et al. [86] with thin porous coating electrodes for catalysts prepared by chemical reduction with formic acid but using different conditions (the electroactive areas were 31 – 45 m^2g^{-1} Pt for the catalysts with a Pt:Ru ratio close to 80:20) . These results are in agreement with those of Takasu et al. [83] who found 34 m^2g^{-1} Pt for the specific surface area of Pt-Ru/C 50:50 using $RuCl_3$ as Ru precursor in the preparation. The D materials have a higher area than the P and F materials, which have similar areas, which is consistent with the electrochemical performances. It must be noted that the FAM materials are more electroactive than Pt-Ru/C 50:50 E-TEK for methanol electrooxidation at room temperature, probably because of the fact that methanol is not adsorbed on Ru at that temperature.

Within the range of values explored in this work, any influence of the particle size of the catalysts could not be considered because the different materials have particles of similar sizes. Also, the amounts of Ru in the prepared alloys are close 20% in almost all cases. Thus, it may be concluded that the main aspect that determines the observed differences in performance among the different materials is the morphology.

The voltammetric profile obtained with Pt/C and Pt-Ru/C 82:18 catalysts using the ultra-thin layer porous electrode is shown in Figure 21. The profile shows that the response of the supported electrode is very similar to that of smooth platinum, in particular with respect of the fine structure of the hydrogen peaks. Thus, it may be concluded that the method of preparation of the supported catalyst leads to a very clean active surface.

For Pt/C the characteristic profile is perfectly identified and the good definition of the hydrogen adsorption/desorption peaks attests on the cleanliness of the preparation method, which does not require any thermal treatment after the synthesis of the catalysts, although in some cases, and for certain reactions, the performance may be improved through this procedure [77]. When compared to platinum, the potentiodynamic profile for Pt-Ru shows an inhibition of the hydrogen peaks due to the presence of ruthenium on the material and the formation of RuOH species at low potentials

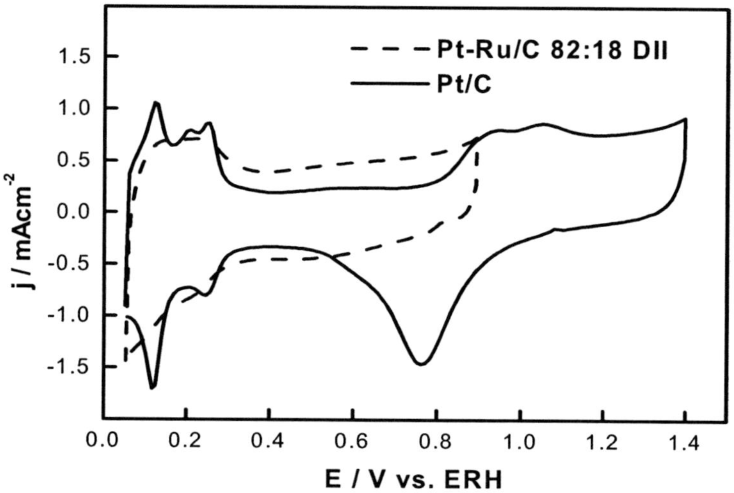

Figure 21. Cyclic voltamograms for Pt/C and Pt-Ru/C (82:18) electrodes prepared by chemical reduction with formic acid. 20 % wt. H_2SO_4 0.5 molL^{-1}. n= 20 mVs^{-1}. 10th scan.

4.2. Methanol Oxidation on PEMFC Anodes

In the presence of a solution 0.5 molL^{-1} of methanol, the current in the region from 0.05 to 0.4 V is almost completely inhibited as a result of the strong adsorption of products from the dehydrogenation of methanol on platinum sites, as shown in Figure 4. Comparison of the anodic portion of the voltamograms in Figure 21 indicates, after integration between 0.05 and 0.4 V, a reduction of 70% of the charge for Pt and 80% for Pt-Ru.

This can be taken as a rough estimative of the fraction of the surface covered by adsorbed species, revealing that a saturation coverage is attained under the experimental conditions and that at the beginning of the experiments the electrodes are approximately in the same conditions. The nature of the adsorbed species has been largelly investigated. Experiments using infrared spectroscopy techniques revealed that the calculated coverages obtained from the integrated absorbance bands for CO were insufficient to justify the high inhibition observed on the hydrogen peaks.

A comparison of the currents for methanol oxidation in a 0.5 mol L^{-1} solution on Pt/C and Pt-Ru/C is shown in Figure 22. In this figure only the 10th scan in the anodic direction is shown. In the range of potentials from 0.05 to 0.4 V the current is at the background level and there is no methanol oxidation. After 0.4 V an increase on the current is observed for Pt-Ru electrodes, which is compatible with the beginning of CO_2 production according to DEMS experiments [149]. For Pt electrodes the onset potential for methanol oxidation is about 0.2 V more positive, starting at about 0.6 V. An inversion on the apparent reactivity for Pt/C and Pt-Ru/C is observed at 0.67 V. Inversions of this type are not usually analysed in detail because they take place outside the potential range considered useful for the methanol fuel cell. However, an interpretation of this phenomenon may be of help in understanding the mechanism of the reaction on different catalysts. In order to identify the reasons for this inversion, electrochemical impedance spectroscopy experiments were carried out at potentials in the region from 0.4 to 0.7 V. The results plotted on the complex plane, are shown in Figure 23.

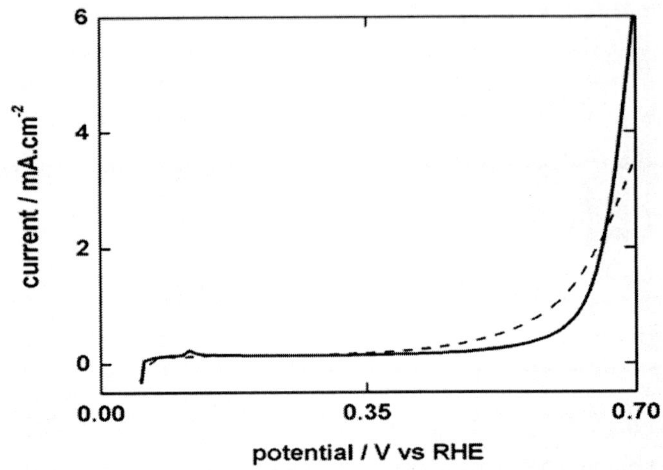

Figure 22. Linear sweep voltammetry for MeOH oxidation on thin film electrodes. Pt/C and Pt-Ru/C 82:18 FAM.

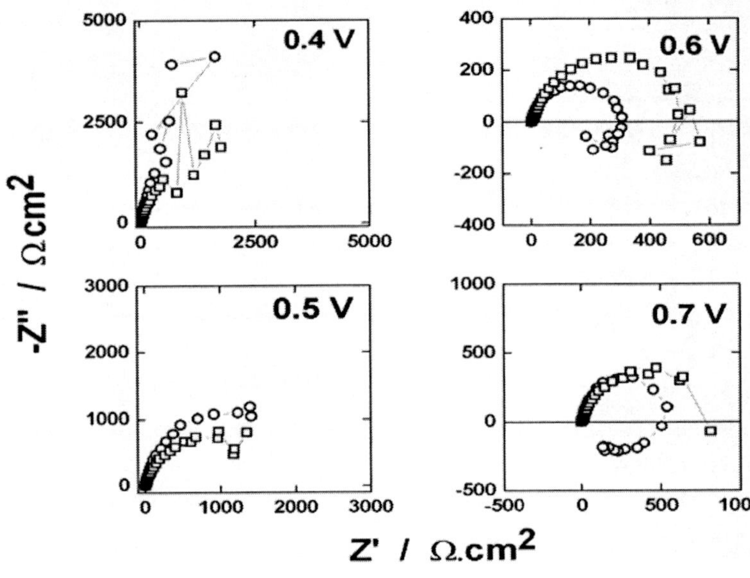

Figure 23. Impedance spectra for different potentials for MeOH oxidation on thin film electrodes with Pt/C and Pt-Ru/C 82:18 FAM.

The correlation between each relaxation time observed on the spectra and particular processes of the oxidation of methanol is being made in a related study still in progress, and in an impedance study of the oxidation of CO on Pt already published [150]. The diameter of the capacitive semi-circle shows, in general, an inverse dependence with the potential, which has been attributed to charge transfer processes. The inductive semi-circle is related to the beginning of the oxidation of the adsorbed intermediates of the reaction and has been attributed to the relaxation of the ad-layer of adsorbates. In the low potential region (under 0.6 V), where the formation of oxygenated species on Pt is not effective, Pt-Ru electrodes show a smaller value for the charge transfer resistance (R_{CT}) than Pt electrodes, as can be observed in Figure 23. This is an expected result, because Pt is not able to oxidize methanol at an appreciate rate at these low potentials.

At 0.7 V, similarly to what is observed with the oxidation currents in Figure 1, the performance of Pt electrodes improves. Impedance spectra in the region between 0.6 and 0.7 V show that R_{CT} for Pt electrodes is approximately two thirds of the value observed for Pt-Ru. By definition, R_{CT} is inversely proportional to the rate constant of the rate-determining step (rds) of the reaction. So it must be concluded that in this potential region the rate constant on Pt is larger than on Pt-Ru catalysts.

At this point it will be considered that, in the potential region from 0.4 to 0.7 V, the reaction of adsorbed CO and OH like species is the rds in agreement with other authors. Thus, the explanation for the inversion of the currents should lie on the intrinsic characteristics of both adsorbed reactants, CO and OH, on the surface.

Several studies have pointed out that the energy of adsorption of OH is larger on Ru that on Pt surfaces [151]. As the bond of CO with the surface has similar energies for both materials [53], the difference on the rate of the reaction between adsorbed CO and OH should be attributed to differences in the OH-metal bond, which is stronger on Ru, leading to a slower rate for the reaction. These results are in agreement with the energies of adsorption and the basic equation for the rate of CO oxidation discussed by Gasteiger et al. [99].

Other catalysts prepared by FAM methods and tested for methanol oxidation in a single DMFC have been Pt-Os/C [152], Pt-W/C [152] and Pt-Mo/C [47] in different compositions.

Figure 24 show cell potencial vs. current density curves for several anode materials at 70 °C and 1.0 atm of O_2 pressure. In all current density ranges it is observed a constant difference of potential, which must be due to the catalysts nature because the particle sizes are similar (2.5 – 3.4 nm). The best material in this case is Pt-Ru/C 75:25. The XRD patterns were similar to those of Pt-Ru/C materials, where only peaks associated to the Pt fcc structure are observed. In general, CO oxidation on Pt-Os/C takes place through the bifunctional mechanism as was reported by Santiago et al. [103], using the same method and conditions but a different Os precursor. They found values for the lattice parameter (obtained from XRD analysis) for Pt-Os/C similar to that for Pt/C. The complete discussion of the physical characterization and electrochemical evaluation is not published yet.

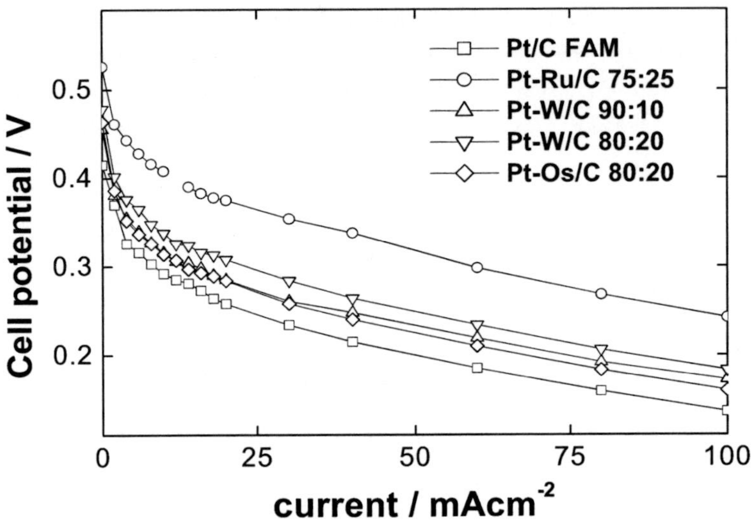

Figure 24. Cell potential vs. current density curves in a single DMFC for Pt/C, PtRu/C, PtOs/C, and PtW/C produced by FAM gas diffusion electrodes as anode. Cell temperature: 70 °C, PO_2 : 1.0 atm. Electrodes with 1 mgPt cm^{-2}, 35 % Nafion® and 20 % metal/C. 15 % PTFE in the diffusion layer. Cathode: Pt/C E-TEK.

4.3. Correlation between CO and Methanol Oxidation on Pt-Ru/C

Several published studies compare supported catalysts on the basis of results obtained in a half-cell and often those results are extrapolated to predict the performance of the catalysts in a fuel cell. One example of this is a study of CO oxidation with thin porous coating electrodes aiming at i) evaluating active areas and ii) comparing the electroactivity of Pt-Ru/C catalysts. The results obtained in a single cell show the risk of extrapolating the results [86]. At the same time an example of the influence of the preparation method of the catalyst on performance was found in the same work in the results for CO electrooxidation with Pt-Ru catalysts prepared by chemical reduction with formic acid and anchoring the metals successively and simultaneously. In these materials the Pt-Ru ratio was 80:20. It was found that the order and the procedure to anchor the metals affect the surface of the catalyst because the stripping curves for CO oxidation show non symmetric peaks, particularly in the case where the metals where anchored successively.

Differences are also found in the potential of ignition for the CO oxidation. For the as prepared catalysts a better response for CO oxidation was found when Pt was anchored first. These catalysts were submitted to thermal treatment in a reducing atmosphere (H_2, 300 °C, 1h) and some modifications in the response were observed. The CO oxidation curves were more symmetrical suggesting a modification of the distribution of the metals on the surface of the particles. Also, the electroactivity of the materials prepared differently and thermally treated showed some differences. For Ru-Pt/C (Ru anchored first) better current levels and lower peak potentials for the oxidation of CO were observed after the thermal treatment in comparison with the as prepared catalysts. Thus, for the thermally treated materials it was found that in spite of the particle size increase, the performance was better. When the experiments were made in a single fuel cell fed with either H_2 or H_2 + 100 ppm of CO the electrochemical response of the materials prepared and submitted to thermal treatment in H_2 (1h, 300 °C) with different anchoring procedures was also different. For catalysts with similar Ru contents the best response was found for catalysts in which the metals were anchored simultaneously, which may be attributed to a better Ru distribution on the particle surface and is confirmed by the values of the electroactive areas measured by the CO oxidation charge [86]. This is a good example of the difficulty of extrapolating results from half-cells or fundamental systems to practical fuel cell systems.

Catalysts with a Ru contents close to 20 % were tested for methanol electrooxidation at room temperature [67]. It was found that materials with several Ru contents thermally treated show lower activities than the as prepared catalysts, as a consequence of the increase in particle size. It is important to note the opposite response for CO oxidation and for methanol oxidation with the same catalysts when thermally treated, showing again the difficulties in the extrapolation of results.

In general, FAM supported bimetallic catalysts based on Pt with Ru [16,76,86,88], Mo [16], Os [152], W [152] show interesting performances for methanol electrooxidation but the particle sizes of these materials are still larger than it would be desirable.

4.4 Discussion about the Techniques - Physical Characterization of Pt-Ru/C Catalysts

The physical characterization of bimetallic alloys has been done in general by XRD and TEM analyses. Using Scherrer's equation the mean particle size can be calculated after adjusting the peaks in the diffractograms using one of several methods, such as Warren [153], or Warren-Averbach, using Rietveld's method [94] or by fitting the peaks with gaussian, lorentzian or pseudovoigt functions [79]. In most published work, the procedure to do the analysis of the XRD data is not reported. Approximations like considering the particles as spherical are usually done, while corrections for instrumental broadening (generally about 1 %) and the contributions due to strain lattice (5-10%) in small particles are not usually introduceded. In general, the results suggest the insertion of Ru on the Pt fcc structure because the XRD patterns do not show peaks associated to metallic Ru with hcp structure. However, the profile of the diffractograms may not show Ru peaks because the intensity of these is very low compared with those of Pt. In supported bimetallic catalysts the hcp structure is only observed for Ru contents above 70 % as Chu and Gilman have shown using electrodeposits [89]. Changes in the nature of the catalysts such as an increase in crystallinity after a thermal treatment can be observed as a consequence of the increment of the particle size and the diminution of the crystalline defects [67]. Is important to mention that the diffractograms have to be recorded with appropriate equipment. In some reports a conventional XRD diffractometer [67] is used and the fitting of XRD peaks is uncertain. A rotary anode diffractometer leads to a higher number of counts giving a higher resolution and as a consequence a better definition of the peaks. Data obtained with this equipment are appropriate to carry out a refinement by the Rietveld method.

An important parameter that contributes to the discussion of the structural information on Pt-Ru is the lattice parameter, obtained by XRD. For these materials with fcc structure, the lattice parameter can be calculated as shown by Radmilovic et al. [79]. Some bimetallic catalysts show values of the lattice parameter close to those expected by Vegard's law, but this is not the case with the FAM catalysts. The results of Takasu et al. [83] show a dependence of the lattice parameter of Pt-Ru catalysts on the Ru precursor used in the preparation of the catalyst. It was found that carbonylic or nitrate precursors favour the insertion of Ru in the Pt lattice. On the other hand, chloride precursors favour the formation of stable complexes at the temperature used in the reduction process. Antolini et al. [102,154] show that, for Pt/C materials, when the metal/C ratio increases from 20 % to 60 % the particle size increases from 2.2 to 9.0 nm. They found that the lattice parameter is smaller in Pt/C than in smooth Pt, which was attributed to the interaction metal/support. However, the XRD patterns for Pt/C obtained in this laboratory under different conditions show that the position of the maximum of the peaks was always the same giving a value for a_{fcc} of 0.39244 nm, similar to that of smooth Pt. In the case of Pt-Ru/C materials the particle size reported by Antolini et al. [154] is similar for different metal/C ratios. In this case it was found for Pt-Ru/C 50:50 that the lattice parameter decreases from 0.3885 to 0.3865 nm when the metal/carbon ratio increases from 20 to 60 %.

The results of Radmilovic et al. [79] suggest the existence of true alloys in the bimetallic nanoparticles but, on the other hand, the results of McBreen and Mukerjee [33] using XAS show that the percentage or true alloy is only 10%. Also using XAS, Camara et al. [105] found that in Pt-Ru materials with a ratio close to 50:50 the percentage of true alloy is only 30 %. At present, the relative activity of true alloys and simple aggregates is under debate. Friedrich et al. [155] show that agglomerates and alloys produce similar electrochemical responses for methanol electrooxidation and the results of Lizcano-Valbuena et al. [76] with Pt-Ru/C 82:18 catalysts with a lattice parameter around 0,391 nm show similar performance as that of materials with smaller lattice parameter values (more similar to true alloys), which is in agreement with the results of Friedrich et al. [155]. On the other hand, the formation of true alloys may be important for the stability of the material. Ru oxides present in Pt-Ru catalysts may show some solubility in water and this affects the stability and the performance of the materials in the fuel cell. The presence of metallic Ru or Ru oxides in the catalysts has been verified by XPS analysis [93,148,154], in particular the presence of RuO_2 and RuO_3 and of metallic Ru in the alloy, but not by XRD, suggesting that these species are present in the catalysts as amorphous materials, not detectable by X-ray diffraction analysis. The activity of the hydratated amorphous Ru oxides was reported by Garche et al. [156] showing that the existence of true alloys is not necessary for a good activity for methanol oxidation in agreement with the results of Long et al. [121]. TEM analysis is important in nanostructured materials because it gives the particle size distribution and the morphology of the catalyst, and shows whether all the carbon in the support was used to anchor the metal. As it was observed [76], the preparation procedure influences the particle size and its distribution. For example, considering the preparation of supported catalysts by reduction with formic acid, when the precursors are added to the carbon support impregnated in formic acid the form of the histograms is different than that observed when the formic acid is added to the carbon impregnated in the solutions of the precursors, with smaller particle sizes in the last case. However, this did not have an influence in the current-potential results obtained in the single cell because the current values were lower for the catalysts prepared by adding the precursors to the carbon impregnated in the formic acid. In all cases, for Pt-Ru/C catalysts prepared by the FAM the distribution of particle sizes is broad and in general the particle size measured by

TEM shows mean values around 4-5 nm [76]. In a high-resolution equipment (HRTEM) it is possible to examine the surface of the nanoparticles, determining crystalline structures, the lattice parameter and the composition of bimetallic catalysts by local EDX [79]. The compositional analysis by this technique shows different values for each particle indicating that overall results of the composition of the catalyst should be interpreted with care. HRSEM has also been used to determine particle sizes and to determine the morphology of the catalysts [76] but the resolution (around 3 nm) is lower than that obtained with TEM, HRTEM or XRD.

5. References

[1] F.J. Du Melle, *J. Power Sources*, 71 (1998) 7.

[2] K. Kordesch, G. Simader, Fuel Cells and their applications, VCH, Weinheim (1996).

[3] B. Emonts, H.J. Bogild, J.S. Loegsgaard, B. Hohlein, R. Peters, *J. Power Sources*, 71 (1998) 288.

[4] M.V. Twigg, Catalysts Handbook, Frome: Manson (1996).

[5] A. Hamnett, In: Interfacial Electrochemistry. Theory, Experimental, and Applications, ed. A. Wieckowski, Marcell Dekker, Inc.: New York (1998).

[6] S. Gottesfeld, J. Pafford, *J. Electrochem. Soc.*, 135 (1998) 2651.

[7] X. Gang, L. Qinfeng, H.A. Hjuler, N.J. Bjerrum, *J. Electrochem. Soc.*, 142 (1995) 2890.

[8] Y. Teng, H. Sakurai, A. Ueda, T. Kabayashi, *Int. J. Hydrogen Energy*, 24 (1999) 355.

[9] H.A. Gasteiger, N.M. Markovic, P.N. Ross, *J. Phys. Chem.*, 99 (1995) 8290.

[10] H.A. Gasteiger, N.M. Markovic, P.N. Ross, *J. Phys. Chem.*, 99 (1995) 16757.

[11] H.A. Gasteiger, N.M. Markovic, P.N. Ross, E.J. Cairns, *J. Phys. Chem.*, 98 (1994) 617.

[12] Y. Morimoto, E.B. Yeager, *J. Electroanal. Chem.*, 441 (1998) 77.

[13] H.-F. Oetjen, V.M. Schmidt, U. Stimming, F. Trila, *J. Electrochem. Soc.*, 143 (1996) 3838.

[14] M. Shibata, N. Furuya, *J. Electroanal. Chem.*, 274 (1989) 245.

[15] P.N. Ross, K. Kinoshita, A.;J. Scarpellino, P. Stonehart, *J. Electroanal. Chem.*, 59 (1975) 177.

[16] A. Oliveira-Neto, M.J. Giz, J. Perez, E.A. Ticianelli, E.R. Gonzalez, *J. Electrochem. Soc.*, 149 (2002) A272.

[17] S. Mukerjee, S.J. Lee, E.A. Ticianelli, J. McBreen, B.N. Grgur, N.M. Markovic, P.N. Ross, J.R. Giallombardo, E.S. De Castro, *Electrochem. Solid-St. Lett.*, 2 (1999) 12.

[18] E.I. Santiago, G.A. Camara, E.A. Ticianelli, *Electrochim. Acta*, 48 (2003) 3527.

[19] B.N. Grgur, N.M. Markovic, P.N. Ross, *Electrochim. Acta*, 43 (1998) 3631.

[20] T.J. Schmidt, H.A. Gasteiger, R.J. Behm, Abstract of the Workshop Electrocatalysis in Indirect and Direct Methanol PEM fuel cells. 3rd. International Symposium in Electrocatalysis, Portoroz, Slovenia (1999).

[21] V.M. Schmidt, H.F. Oetjen, J. Divisek, *J. Electrochem. Soc.*, 144 (1997) L237.

[22] J. Divisek, H.-F. Oetjen, V. Peinecke, V.M. Schmidt, U. Stimming, *Electrochim. Acta*, 43 (1998) 3811.

[23] M. Watanabe, M. Uchida, S. Motoo, *J. Electroanal. Chem.*, 229 (1987) 395.

[24] T. Iwasita, F.C. Nart, In: Advances in Electrochemical Science and Engineering, ed. H. Gerischer, C. Tobias, VCH: Weinheim (1995).
[25] Wolter, J. Heitbaum, *Ber Bunsenges. Phys. Chem.*, 88 (1984) 6.
[26] S.-C. Chang, M.J. WeaVer, *J. Chem. Phys.*, 92 (1990) 4582.
[27] M. Watanabe, S. Motoo, *J. Electroanal. Chem.*, 206 (1986) 197.
[28] Y. Morimoto, E.B. Yeager, *J. Electroanal. Chem.*, 441 (1998) 77.
[29] R. Ianniello, V.M. Schmidt, U. Stimming, J. Stumper, A. Wallau, *Electrochim. Acta*, 39 (1994) 1863.
[30] A.S. Arico, E. Modica, E. Passalacqua, V. Antonucci, P.L. Antonucci, *J. Appl. Electrochem.*, 27 (1997) 1275.
[31] H.A. Gasteiger, N.M. Markovic, P.N. Ross, *J. Phys. Chem.*, 99 (1995) 8290.
[32] D.C. Azevedo, W.H. Lizcano-Valbuena, E.R. Gonzalez, *J. New Mat. Electr. Syst.*, in press (2004).
[33] J. McBreen, S. Mukerjee, *J. Electrochem. Soc.*, 142 (1995) 3399.
[34] F. Buatier de Mongeot, M. Scherrer, B. Gleich, E. Kopatzki, R.J. Behm, *Surf. Sci.*, 411 (1998) 249.
[35] B.D. McNicol, R.T. Short , *J. Electroanal.Chem.*, 81 (1977) 249.
[36] S. Sriramulu, T.D. Jarvi, E.M. Stuve In: Interfacial Electrochemistry: Theory, Experiment, and applications, ed. Marcel Dekker, Inc.: New York (1998).
[37] E.A. Batista, G.R.P. Malpass, A.J. Motheo, T. Iwasita, *Electrochem. Commun.*, 5 (2003) 843.
[38] W. Vielstich, In: Enciclopédia of Electrochemistry, ed. A.J. Bard, M. Stratmann, Wiley-VCH, Weinheim (2003).
[39] S. Wasmus, A. Küver, *J. Electroanal. Chem.*, 461 (1999) 14.
[40] T. Iwasita, *J. Braz. Chem. Soc.,* 13 (2002) 401.
[41] T. Iwasita, *Electrochim. Acta*, 47 (2002) 3663.
[42] T. Iwasita In: *Handbook of Fuel Cells: Fundamental, Technology, Applications*, ed. W. Vielstich, A. Lamm, H.A. Gasteiger, chap. 41, volume 2, Wiley-VCH (2003).
[43] W. Chrzanowki, A. Wieckowski, *Langmuir*, 14 (1998) 1967.
[44] T. Iwasita, H. Hoster, A. John-Anacker, W.F. Lin, W. Vielstich, *Langmuir*, 16 (2000) 522.
[45] T. Iwasita, F.C. Nart *In: Advances in Electrochemical Science and Engineering*, ed. H. Gerischer, C. Tobias, VCH, Weinheim (1995)
[46] T. Iwasita, F.C. Nart, *J. Electroanal. Chem.*, 317 (1991) 291.
[47] A. Oliveira Neto, J. Perez, W.T. Napporn, E.A. Ticianelli, E.R. Gonzalez, *J. Braz. Chem. Soc.*, 11 (2000) 39.
[48] E.I. Santiago, G.A. Camara, E.A. Ticianelli, *Electrochim. Acta*, 48 (2003) 3527.
[49] T. Frelink, W. Visccher, J.A.R. van Veen, *Surf. Sci.*, 335 (1995) 353.
[50] M.T. Paffet, S.C. Gebhard, R.G. Windham, B.E. Koel, *J. Phys. Chem.*, 94 (1990) 6831.
[51] M. Götz, H. Wendt, *Electrochim. Acta*, 43 (1998) 3637.
[52] A. S. Aricó, Z. Poltarzewski, H. Kim, A. Morana, N. Giordano, V. Antonucci, *J. Power Sources*, 55 (1995) 159.
[53] K.L. Ley, R. Liu, C. Pu, Q. Fan, N. Leyarovska, C. Segre, E.S. Smotkin, *J. Electrochem. Soc.*, 144 (1997) 1543.

[54] B. Gurau, R. Viswanathan, R. Liu, T.J. Lafrenz, K.L. Ley, E.S. Smotkin, E. Riddington, A. Sapienza, B.C. Chan, T.E. Mallouk, S. Sarangapani, *J. Phys. Chem. B*, 102 (1998) 9997.
[55] T.J. Schmidt, H.A. Gasteiger, G.D. Stab, P.M. Urban, D.M. Kolb, R.J. Behm, *J. Eletrochem. Soc.*, 145 (1998) 2354.
[56] P. Stonehart, *J. Appl. Electrochem.*, 22 (1992) 995.
[57] H.A. Gasteiger, N.M. Markovic, P.N. Ross, E.J. Cairns, *J. Electrochem. Soc.*, 141 (1994) 1795.
[58] C. Coutanceau, A. Lima, E. Garnier, J.-M. Leger, C. Lamy, In: Joint International Meeting, Electrochemical Society – International Society of Electrochemistry ECS-ISE, San Francisco, USA (2001).
[59] P.A. Attwood, B.D. McNicol , R.T. Short, *J. Appl. Electrochem.*, 10 (1980) 213.
[60] T. Frelink, W. Visscher, J.A.R. van Veen, *J. Electroanal. Chem.*, 382 (1995) 65.
[61] A. Kabbabi, F. Gloagen, F. Andolfatto, R. Durand, *J. Electroanal. Chem.*, 373 (1994) 251.
[62] S. Mukerjee, J. McBreen, *J. Electroanal. Chem.*, 448 (1998) 163.
[63] W.H. Lizcano-Valbuena, D.C. Azevedo, E.R. Gonzalez, *Electrochim. Acta*, 49 (2004) 1289.
[64] X. Ren, M.S. Wilson, S.J. Gottesfeld, *J. Electrochem. Soc.*, 143 (1996) L12
[65] H. Dohle, T. Bewer, J. Mergel, R. Neitzel, D. Stolten, 2002 Fuel Cell Seminar, Oregon, USA, (2000), p. 130.
[66] H. Dohle, J. Divisek, J. Mergel, H.F. Oetjen, C. Zingler, D. Stolten, J. Power Sources, 105 (2002) 274.
[67] W.H. Lizcano-Valbuena, A. de Souza, C.A.P. Leite, F. Galembeck, E.R. Gonzalez, *Fuel Cells*, 3-4 (2003) 159.
[68] A. Heinzel, V.M. Barragán, *J. Power Sources*, 84 (1999) 70.
[69] J. Muller, G. Frank, K. Colbow, D. Wilkinson, "Transport/kinetic limitations and efficiency losses" In: *Handbook of Fuel Cells: Fundamentals, Technology, Applications*, ed. W. Vielstich, A. Lamm, H.A. Gasteiger, ch. 62, v. 4, Wiley-VCH (2003).
[70] M. Doyle, G. Rajendran, "Perfluorinated Membranes" In: Handbook of Fuel Cells – Fundamentals, Technology and Applications, Edited by W. Vielstich, H.A. Gasteiger, A. Lamm, John Wiley & Sons Ltd, Chichester, UK, chapter 30, v. 3, (2003) p. 351-395.
[71] P.A.Christensen, A. Hamnett, J. Munk, G.L. Troughton, *J. Electroanal. Chem.*, 370 (1994) 251.
[72] S.A.M. Silva, J. Perez, R.M. Torresi, C.A. Luengo, E.A. Ticianelli. *Electrochim. Acta*, 44 (1999) 3565.
[73] A.L.N. Pinheiro, A. Oliveira-Neto, E.C. de Souza, J. Perez, V.A. Paganin, E.A. Ticianelli, E.R. Gonzalez, *J. New Mat. Elect. Syst.* 6 (2003) 1.
[74] M. Carmo, V.A. Paganin, J.M. Rosolen, E.R. Gonzalez, *Electrochim. Acta* (2004) submitted.
[75] E.R. Gonzalez, E.A. Ticianelli, A.L.N. Pinheiro, J. Perez, BR INPI 00321 (1997)
[76] W.H. Lizcano-Valbuena, V.A. Paganin, C.A.P. Leite, F. Galembeck, E.R. Gonzalez, *Electrochim. Acta*, 48 (2003) 3869.
[77] M. Watanabe, M. Uchida, S. Motoo, *J. Electroanal. Chem.*, 229 (1987) 395.

[78] A.M. Castro-Luna, G.A. Camara, V.A. Paganin, E.A. Ticianelli, E.R. Gonzalez, *Electrochem. Commun*, 2 (2000) 2.
[79] V. Radmilovic, H.A. Gasteiger, P.N. Ross, *J. Catal.*, 154 (1995) 98.
[80] T. Frelink, W. Visscher, J.A.R. van Veen, *J. Electroanal. Chem.*, 382 (1995) 65.
[81] E. Auer, A. Freund, T. Lehmann, K.A. Starz, R. Schwarz, U. Stenke, EP 0800 188 A2.
[82] J.B. Goodenough, A. Hamnett, B.J. Kennedy, R. Manoharam, S.A. Weels, *Electrochim. Acta*, 35 (1990) 199.
[83] Y. Takasu, T. Fujiwara, Y. Murakami, K. Sasakim M. Oguri, T. Asaki, W. Sugimoto, *J. Electrochem. Soc.*, 147 (2000) 4421.
[84] K. Aika, L.L. Ban, I. Okura, S. Namba, J. Turkevich, *J. Res. Inst. Ctal. Hokkaido Univ.*, 24 (1976) 54.
[85] H. Bönnemann, W. Brijoux, R. Brinkmann, E. Dinjus, T. Joussen, B. Korall, *Angew. Chem. Int. Ed. Engl.*, 30 (1991) 1312.
[86] F. Colmati-Jr, W.H. Lizcano-Valbuena, G.A. Camara, E.A. Ticianelli, E.R. Gonzalez, *J. Braz. Chem. Soc.*, 13 (2002) 474.
[87] A.K. Shukla, M. Neegat, P. Bera, V. Jayaram, M.S. Hedge, *J. Electroanal. Chem.*, 504 (2001) 111.
[88] W.H. Lizcano-Valbuena, V.A. Paganin, E.R. Gonzalez, *Electrochim. Acta*, 47 (2002) 3715.
[89] D. Chu, S. Gilman, *J. Electrochem. Soc.*, 143 (1996) 1685.
[90] B.D. McNicol, R.T. Short, *J. Electroanal. Chem.*, 81 (1997) 249.
[91] C. Roth, N. Martz, H. Fuess, *Phys. Chem. Chem. Phys.*, 3 (2001) 315.
[92] B.D. Cullity. Elements of X-ray diffraction, 1st. ed. Reading, MA:Addison-Wesley (1967).
[93] A.S. Arico, P. Creti, H. Kim, R. Mantegna, N. Giordano, V. Antonucci, *J. Electrochem. Soc.*, 43 (1996) 3950.
[94] H.M. Rietveld, *J. Appl. Cryst.*, 10 (1969) 65.
[95] S. Mukerjee, S. Srinivasan, M.P. Soriaga, *J. Electrochem. Soc.*, 142 (1995) 1409.
[96] J. McBreen, S. Mukerjee, *J. Electrochem. Soc.,* 142 (1995) 3399.
[97] S.Mukerjee, S. Srinivasan, M.P. Soriaga, J. McBreen, *J. Chem. Phys.*, 99 (1995) 4577.
[98] S. Mukerjee, J. McBreen, *J. Electrochem. Soc.*, 143 (1996) 2285.
[99] T. Page, R. Johnson, J. Hormes, S. Noding, and B. Rambabu, *J. Electroanal. Chem.*, 485, 34 (2000).
[100] J.J. Rehr, R.C. Albers, *Phys. Rev. Sect. B*, 41 (1990) 139.
[101] E. Antolini, F. Cardellini, J. Alloys comp.., 315 (2001) 118.
[102] E. Antolini, F. Cardellini, L. Giorgi, E. Passalacqua, J. Mater. Sci. Lett., 19 (2000) 2099.
[103] E.I. Santiago, M.J. Giz, E.A. Ticianelli, J. Solid State Electrochem., 7 (2003) 607.
[104] M.K. Min, J. Cho, K. Cho, H. Kim, Electrochim. Acta, 45 (2000) 4211.
[105] G. A. Camara, M.J. Giz, V.A. Paganin, E.A. Ticianelli, *J. Electroanal. Chem.*, 537 (2002) 21.
[106] E.I. Santiago, M.J. Giz, E.A. Ticianelli, Structure and CO tolerance of a PtMo/C electrocatalysts In: XIV Annual Users Meeting of Brazilian Synchrotron Light Laboratory (LNLS), Campinas, Brazil, 2004.
[107] P. Stonehart, *J. Appl. Eletrochem.*, 22 (1992) 995.

[108] J.F. Moulder, W.F. Sticle, P.E. Sobol, K.D. Bomben, Handbook of X-ray photoelectron spectroscopy. A reference book of standard spectra for identification and interpretation of XPS data. Eden Prairie, Perkin Elmer Corporation (1992).

[109] E.A. Ticianelli, J.G. Berry, M.T. Paffet, S. Gottesfeld, *J. Electroanal. Chem.*, 258 (1989) 61.

[110] T.J. Schmidt, M. Noeske, H.A.Gasteiger, R.J. Behm, P. Britz, W. Brijoux, H. Bonnemann, *Langmuir*, 13 (1997) 2591.

[111] E. Antolini, L. Giorgi, F. Cardellini, E. Passalacqua, *J. Solid St. Electrochem.*, 5 (2001) 131.

[112] A. Kabbabi, R. Faure, R. Durand, B. Beden, F. Hahn, J.-M. Leger, C. Lamy, *J. Electroanal. Chem.*, 444 (1998) 41.

[113] M. Watanabe, S. Motoo, *Electroanal. Chem. And Interfacial Electrochem.*, 60 (1975) 267.

[114] M. Watanabe, Y. Zhu, H. Igarashi, H. Uchida, *Electrochemistry*, 68 (2000) 244.

[115] M. Watanabe, "New CO-tolerant catalyst concepts" In: Handbook of Fuel Cells – Fundamentals, Technology and Applications, Edited by W. Vielstich, H.A. Gasteiger, A. Lamm, Volume 2: Electrocatalysis, John Wiley & Sons Ltd, Chichester, UK, chapter 28 (2003) p. 408- 415.

[116] E. Christoffersen, P. Liu, A. Ruban, H.L. Skriver, and J.K. Norskov, *Journal of Catalysis*, 199, 123 (2001).

[117] S. Jayaraman and A.C. Hillier, *J. Phys. Chem. B*, 107, 5221 (2003).

[118] M.T.M. Koper, A.P.J. Jansen, R.A. van Santen, J.J. Lukkien, and P.A.J. Hibers, *J. Chem. Phys.*, 109 (1998) 6051.

[119] M.T.M. Koper, J.J. Lukkien, A.P.J. Jansen, and R.A van Santen, *J. Phys. Chem. B*, 103 (1999) 5522.

[120] D.R. Rolison, P.L. Hagans, K.E. Swider, J.W. Long, *Langmuir*, 15, 774 (1999).

[121] J.W. Long, E.M. Stroud, K.E. Swider-Lyons, D.R. Rolison, *J. Phys. Chem. B*, 104, 9772 (2000).

[122] T.J. Schmidt, H.A. Gasteiger, R.J. Behm, *J. Electrochem. Soc.*, 146 (1999) 1296.

[123] H.N. Dinh, X. Ren, F.H. Garzon, P. Zelenay, S. Gottesfeld, *J. Electroanal. Chem.*, 491 (2000) 222.

[124] C. Roth, N. Martz, H. Fuess, *Phys. Chem. Chem. Phys.*, 3 (2001) 315.

[125] E.I. Santiago, V.A. Paganin, M. do Carmo, E.R. Gonzalez, E.A. Ticianelli, J. Electroanal. Chem., submitted (2004).

[126] G. Samjeské, H. Wang, H. Löffler, H. Baltruschat, *Electrochim. Acta*, 47 (2002) 3681.

[127] A. Oliveira Neto, J. Perez, W.T. Napporn, E.A. Ticianelli, E.R. Gonzalez, *J. Braz. Chem. Soc.*, 1 (2000) 39.

[128] H. Zhang, Y. Wang , E.R. Fachini, and C.R. Cabrera, *Electrochem. Solid-Sate Lett.*, 2 (1999) 437.

[129] Y. Wang, E. R. Fachini, G. Cruz, Y. Zhu, Y. Ishikawa, J.A. Colluci, and C.R. Cabrera, *J. Electrochem. Soc.*, 148 (2001) C222.

[130] S. Mukerjee and R. C. Urian, *Electrochim. Acta*, 47 (2002) 3219.

[131] S. Ball, A. Hodgkinson, G. Hoogers, S. Maniguet, D. Thompsett, B. Wong, *Electrochem. Solid-State Lett.*, 5 (2002) A31.

[132] E.I. Santiago, M.S. Batista, E.M. Assaf, and E.A. Ticianelli, *J. Electrochem. Soc.*, 151 (2004) A944.

[133] T. Ioroi, K. Yasuda, Z. Siroma, N. Fujiwara, and Y. Miyazaki, *J. Electrochem. Soc.*, 150, A1225 (2003).

[134] G.A. Camara, E.A. Ticianelli, S. Mukerjee, S.J. Lee, J. McBreen, *J. Electrochem. Soc.*, 149 (2002) A748.

[135] J.R. Ladebeck and J.P. Wagner, in: Handbook of Fuel Cells – Fundamentals, Technology, and Applications, W. Vielstich, H.A. Gasteiger, and A. Lamm, Editors, vol. 3, p. 190, John Willey & Sons, Chichester (2003).

[136] C. Song, *Catal. Today*, 77, 17 (2002).

[137] A.A. Andreev, V.J. Kafedjiysky, and R.M. Edreva-Kardjieva, *Appl. Catal. A*, 179, 223 (1999).

[138] Y.M. Lim R.J. Wang, and L. Chang, *Catal. Today*, 51, 25 (1999).

[139] R.N. Nickolov, R.M. Edreva-Karjieva, V.J. Kafdjiysky, D.A. Nikolova, N.B. Stankova, D.R. Mehandjiev, *Appl. Catal. A*, 190,191 (2000).

[140] J. Patt, D.J. Moon, C. Phillips, and L. Thompson, *Catal. Lett.*, 65, 193 (2000).

[141] Y. Tanaka, T. Utaka, R. Kikuchi, K. Sasaki, and K. Eguchi, *Appl. Catal. A*, 238, 11 (2003).

[142] F.A. de Bruijn, D.C. Papageorgopoulos, E.F. Sitters, and G.J.M. Janssen, *J. Power Sources*, 110, 117 (2002).

[143] D. C. Papageorgopoulos and F. A. de Bruijn, *J. Electrochem. Soc.*, 149, A140 (2002).

[144] J. Perez, A.A. Tanaka, E.R. Gonzalez, E.A. Ticianelli, *J. Electrochem. Soc.*, 141 (1994) 431.

[145] V.A. Paganin, E.A. Ticianelli, E.R. Gonzalez, *J. Appl. Electrochem.*, 26 (2996) 297.

[146] M. Hogarth, P.A. Cristensen, A. Hamnett, A. Shukla, *J. Power Sources*, 69 (1997) 125.

[147] W. Chrzanowski, A. Wieckowski, *Langmuir*. 14 (1998) 1967.

[148] A.S. Arico, P. Creti, P. Antonucci, J. Cho, H. Kim, V. Antonucci, *Electrochim. Acta*, 43 (1998) 3719.

[149] Z. Jusys, J. Kaiser, R. Behm, *J. Electrochim. Acta*, 47 (2002) 3693.

[150] D.C. Azevedo, A.L.N. Pinheiro, R.M. Torresi, E.R. Gonzalez, *J. Electroanal. Chem.*, 523 (2002) 43.

[151] H.A. Gasteiger, N.M. Markovic, P.N. Ross, E. Cairns, *J. Phys. Chem.*, 98 (1994) 617.

[152] W.H. Lizcano-Valbuena, E.C. Bortholin, A. Oliveira Neto, V.A. Paganin, E.R. Gonzalez, ECS San Francisco, USA, 2001, Abs. 330.

[153] B.E. Warren, X-rays Diffraction, Dover Publications Inc., New York (1990).

[154] E. Antolini, L. Giorgi, F. Cardellini, E. Passalacqua, *J. Solid St. Electrochem.*, 5 (2001) 131.

[155] K.A. Friedrich, K.P. Geyzers, A.J. Dickinson, U. Stimming, J. Electroanal. Chem., 524-525 (2002) 261.

[156] K. Lasch, L. Jörrisen, K.A. Friedrich, J. Garche, *J. Solid State Electrochem.*, 7 (2003) 619.

ACKNOWLEDGEMENTS

The authors wish to thank Fundação de Amparo à Pesquisa do Estado de São Paulo (FAPESP), Coordenação de Aperfeiçoamento de Pessoal de Nível Superior (CAPES),

Conselho Nacional de Desenvolvimento Científico e Tecnológico (CNPq), and LNLS-National Synchrotron Light Laboratory, Brazil for financial and technical supports.

In: Focus on Electrochemistry Research
Editor: Magdalena Nunez, pp. 169-213

ISBN: 1-59454-545-6

INDEX

A

absorbance bands, 199
absorption, 40, 61, 62, 169, 176, 179, 182, 190
absorption spectrum, 40, 61
acetic acid, 159, 162, 164
acetone, 39, 85
acetonitrile, 68
acidity, 45
active site, 19, 27, 150
additives, 120, 124, 135, 155
adhesion, 60
adsorption, 146, 148, 149, 150, 151, 152, 153, 154, 156, 160, 161, 163, 165, 168, 169, 170, 185, 186, 190, 194, 196, 198, 199, 201
aggregates, 204
alcohols, 77, 78, 81, 84, 148, 150, 152, 153, 154, 155, 156, 157, 158, 159, 165, 168
alkaline hydrolysis, 80
alkylation, 85
alloys, 55, 168, 169, 170, 172, 174, 178, 179, 181, 182, 185, 198, 203, 204
aqueous solutions, 38, 151, 156, 165
argon, 151
aromatic compounds, 69
atmospheric pressure, 159
atomic force microscope (AFM) , 40, 133

B

backscattering, 175, 176, 196
batteries, vii
benzene, 79, 80, 150, 165
binding energies, 185
biochemistry, 67
bioelectricity, vii
biological activity, 158
biopolymers, 164
biosynthesis, 164
blocks, 19, 27
breakdown, 104

C

cables, 118
calibration, 39
calorimetry, 169
carbohydrate, 164
carbon, 1, 4, 7, 10, 11, 12, 13, 14, 15, 16, 33, 70, 147, 167, 169, 170, 171, 172, 173, 174, 176, 179, 181, 184, 185, 186, 190, 193, 196, 204
carbon nanotubes, 171
catalysts, 1, 2, 4, 6, 7, 14, 15, 16, 17, 18, 19, 23, 24, 26, 27, 28, 29, 30, 32, 33, 146, 149, 150, 159, 168, 17, 19, 154, 155, 165, 166, 167, 169, 170, 171, 172, 173, 174, 175, 176, 177, 178, 179, 180, 181, 182, 184, 185, 186, 187, 188, 189, 190, 191, 194, 195, 196, 197, 198, 199, 200, 201, 202, 203, 204
catalytic activity, 146, 149, 155, 156, 163, 189, 192
catalytic hydrogenation, 151, 159
cathodic process, 50
cation, 84
C-C, 148, 156
CH_2, 82, 86, 88
chemical bonds, 94
chemical deposition, 38
chemical etching, 108
chemical properties, 68, 92, 141
chemical reactions, vii, 6, 55, 94
chemical vapor deposition (CVD), 67, 140
chlorine, 71, 150

chromatography, 191
CO_2, 6, 168, 169, 185, 190, 191, 192, 193, 199
coatings, 11, 55, 155, 158, 198, 202
combined effect, 170
combustion, 6, 168
compatibility, 38
composition, 37, 38, 39, 40, 43, 44, 50, 52, 54, 56, 58, 60, 62, 145, 146, 148, 151, 152, 153, 154, 155, 156, 157, 160, 162, 163, 172, 173, 179, 187, 188, 191, 192, 205
compounds, 38, 40, 55, 67, 68, 70, 73, 74, 77, 78, 79, 82, 83, 84, 85, 86, 92, 147, 150, 153, 156, 158, 160, 161, 163, 164, 165, 170
condensation, 20, 153
conductivity, 2, 24, 26, 47, 94, 100
conversion, vii, 2, 169, 179, 182, 191, 192
copper, 68, 82, 83, 85, 91, 92, 93, 96, 97, 102, 103, 104, 107, 110, 112, 119, 120, 122, 124, 128, 129, 130, 131, 132, 133, 135, 136, 137, 141, 143, 148, 149, 150, 151, 154, 156, 158, 164, 165, 166
corrosion, vii, 15, 112, 113, 144
crystallinity, 174, 195, 203

D

data analysis, 119
decomposition, 63, 77
defects, 102, 149, 150, 203
deformation, 149
degradation, 168
density, 2, 4, 5, 6, 7, 8, 9, 10, 11, 12, 17, 18, 20, 23, 24, 25, 26, 27, 29, 60, 62, 92, 94, 96, 99, 101, 102, 108, 124, 125, 126, 135, 140, 141, 155, 167, 186, 188, 194, 197, 198, 201, 202
density functional theory, 186
density of states, 62
deposition, 37, 38, 41, 49, 50, 52, 53, 55, 56, 60, 61, 62, 63, 96
desorption, 151, 198
destruction, 147, 151, 152, 156, 159
dielectrics, 91, 92
diffusion, 2, 4, 6, 10, 11, 13, 21, 22, 23, 24, 27, 28, 29, 30, 31, 32, 33, 94, 99, 100, 101, 105, 106, 115, 116, 122, 123, 124, 131, 132, 138, 141, 142, 151, 170, 190, 191, 192, 193, 197, 198, 202
diffusivities, 21
dimethylformamide, 165
discharges, 113
dislocations, 92
dispersity, 154
dissolved oxygen, 149
distilled water, 39

E

electric field, 126, 127, 128
electrical properties, 113
electrocatalysis, 154, 163, 164, 165, 167
electrocatalyst, 154
electrochemical deposition, 97
electrochemical impedance, 106, 143, 169, 200
electrochemistry, vii, 55, 164
electrocrystallization, 155
electrodeposition, 37, 38, 39, 43, 45, 48, 50, 52, 54, 55, 62, 63, 156, 158, 159, 163
electrolysis, 38, 46, 146, 147, 148, 149, 151, 155, 156, 157, 158, 159, 160, 161, 162, 163
electrolytes, 1, 2, 3, 15, 38, 39, 44, 45, 46, 48, 49, 50, 51, 53, 54, 55, 62, 92, 93, 94, 96, 100, 103, 108, 110, 111, 112, 113, 115, 124, 125, 133, 135, 138, 150, 158, 159, 161, 162, 165, 167, 168, 170, 195, 197
electronic structure, 169, 170, 182
electrons, 2, 94, 96, 97, 99, 100, 113, 114, 125, 160, 191
electroplating, 166
electroreduction, 38, 159, 164, 165, 166
energy, vii, 2, 43, 62, 92, 94, 102, 146, 156, 168, 169, 172, 181, 182, 185, 194, 201
environment, 146, 156, 163, 168
etching, 92, 102, 108, 140
ethanol, 72, 83, 86, 147, 166, 189
ethylene, 86, 112, 119, 120, 122, 124, 135, 136, 150, 156, 166
evaporation, 20
extrapolation, 170, 203

F

fabrication, 150
fast Fourier transform infrared (FTIR), 168, 169
film formation, 63, 144
film thickness, 15
films, 37, 39, 40, 48, 50, 52, 54, 56, 60, 62, 91, 104, 133, 138, 143
fluorescence, 39, 46, 48, 50, 52, 53, 54
formaldehyde, 172, 179
free energy, 38, 43, 49
frequency, 113, 114, 118, 119, 120, 126, 128, 130
friction, 101
fuels, vii, 1, 2, 3, 4, 15, 17, 18, 19, 26, 27, 28, 31, 32, 33, 34, 167, 168, 170, 187, 189, 190, 194, 200, 202, 203, 204, 205

G

gas diffusion, 1, 4, 7, 12, 15, 19, 24, 26, 32, 170, 189, 190, 193, 194, 202
glycol, 112, 119, 120, 122, 124, 133, 135, 136
gold, 85
grains, 60, 149, 156
graphite, 7

H

halogens, 76
heat, 39, 155, 174, 175, 179, 180, 182, 183, 186, 188, 189, 190, 194, 196
heterocycle, 75, 159, 161, 162
heterogeneous catalysis, 164
hexafluoropropylene, 15
High Resolution Transmission Electron Microscopy (HRTEM), 205
holes, 138
hydrazine, 172, 179
hydrocarbons, 164
hydrogen, vii, 2, 6, 15, 17, 39, 43, 47, 55, 68, 70, 96, 104, 108, 132, 133, 134, 135, 138, 142, 146, 148, 149, 150, 152, 154, 155, 156, 157, 159, 163, 164, 165, 166, 167, 168, 170, 171, 179, 183, 185, 186, 188, 189, 190, 193, 194, 198, 199
hydrogen bonds, 152
hydrogen gas, 15, 17
hydrogen peroxide, 168, 171
hydrogenation, 145, 146, 147, 148, 149, 154, 155, 156, 157, 158, 159, 160, 161, 162, 163, 164, 165, 166
hydroquinone, 184
hydroxide, 182

I

impregnation, 7, 16, 188, 189
impurities, 171
infrared spectroscopy, 168, 169, 199
interface, 2, 4, 19, 22, 23, 24, 27, 28, 29, 30, 32, 33, 44, 92, 95, 98, 116, 165, 186
intermolecular interactions, 84
ionization, 94, 114, 166
ions, 38, 93, 94, 95, 96, 97, 99, 100, 101, 105, 111, 116, 123, 128, 129, 130, 132, 138, 165, 171, 172, 173, 179, 190
IR spectra, 83
isomerization, 84

K

kinetic parameters, 146
kinetics, 40, 55, 98, 104, 166

L

lattice parameters, 37, 48, 173
ligands, 85
liquid phase, 19
lithium, 71
low temperatures, 165
lubricants, 38

M

magnetic fields, 1, 5, 6, 8, 9, 10, 11, 13, 14, 15, 16, 18, 19, 23, 24, 30, 33
magnetic field effect, 10
magnetic moment, 5
magnetic particles, 16, 17
mass, 2, 4, 5, 7, 20, 21, 24, 28, 31, 44, 79, 82, 97, 98, 99, 100, 101, 102, 103, 104, 105, 106, 111, 113, 115, 116, 123, 124, 126, 129, 131, 132, 138, 140, 141, 168, 169, 192
medium composition, 147, 151
melt, 93
melting, 71, 73, 75
membrane, 2, 4, 15, 17, 23, 24, 25, 164, 168, 170, 191
metal salts, 85, 148
metals, 38, 92, 96, 146, 150, 154, 155, 156, 158, 163, 164, 165, 168, 169, 170, 172, 174, 186, 196, 202
methanol, 67, 82, 133, 165, 167, 168, 169, 170, 186, 188, 189, 190, 194, 195, 196, 197, 198, 199, 201, 203, 204
microelectronics, 91
microstructures, 60
mobility, 100, 186
molecular structure, 93
molecular weight, 20, 23, 102, 168
molybdenum, 38, 42, 48, 55, 63, 189, 191
monomolecular, 147
morphology, 37, 38, 40, 56, 60, 63, 176, 187, 198, 204
Mo–Se thin film, 37, 38, 40, 49, 50, 54, 55, 56, 58, 60, 61, 62, 63

N

NaCl, 94, 112, 133

nanoparticles, 170, 204
nickel, 55, 85, 106, 143, 144, 154, 155, 156, 158, 159, 166
nitrogen, 1, 20, 67, 84, 133, 161, 162
nitrogen gas, 1
nucleophiles, 67

O

olefins, 150, 156
oligomers, 155
optical density, 58, 60
optical micrographs, 56, 60
optical microscopy, 60
optical parameters, 40
optical properties, 37, 38
organic compounds, 5, 82, 145, 148, 150, 154, 156, 161, 163, 164, 165, 179
oscillation, 104, 113, 179
oxidation, 40, 49, 96, 98, 103, 125, 150, 167, 168, 169, 170, 179, 182, 185, 186, 187, 188, 189, 190, 191, 192, 194, 195, 196, 197, 199, 200, 201, 202, 203, 204
oxidation rate, 98
oxides, 55, 155, 168, 169, 171, 174, 184, 186, 189, 196, 204
oxygen, 1, 4, 6, 9, 10, 11, 13, 14, 15, 16, 17, 18, 19, 20, 21, 23, 27, 28, 31, 33, 67, 69, 84, 92, 108, 149, 150, 151, 168, 170, 189, 193

P

palladium, 85
particle morphology, 167
particles, 1, 4, 5, 15, 16, 17, 18, 19, 24, 26, 27, 28, 30, 33, 108, 169, 170, 172, 174, 175, 176, 178, 182, 184, 185, 186, 188, 191, 195, 196, 198, 202, 203
permeability, 5, 18, 19, 20, 25
permittivity, 123
pH, 37, 39, 40, 43, 44, 45, 46, 47, 48, 49, 51, 55, 172
photocorrosion, 38
photoelectron spectroscopy, 149, 209
photolysis, 84
photon, 62
physical properties, 151, 152
physicochemical methods, 149
physicochemical properties, 149, 153
plasticity, 92
platinum, 6, 11, 15, 39, 85, 149, 150, 165, 172, 174, 194, 198, 199
polarity, 147, 155
polarization, 2, 3, 4, 6, 12, 16, 17, 24, 103, 104, 106, 107, 110, 111, 113, 122, 133, 134, 140, 148, 186, 187, 188, 190, 192
pollutants, 168
Polyimide [PI], 39
polyimide film, 15
polymer, 1, 3, 7, 10, 14, 15, 167, 168, 170
polymeric membranes, 170
polymerization, 153
porosity, 4, 19, 20, 25, 142
precipitation, 112, 148, 155, 156, 157
prediction, 55
propagation, 91
protons, 4, 191
pure water, 5

Q

quality, 150

R

radiation, 2, 39, 172
radicals, 146
rate constant, 20, 25, 98, 201
rate constants, 20
reactants, 2, 4, 11, 28, 100, 201
reaction medium, 146
reaction rate, 4, 98
reaction time, 179
reaction zone, 6
reagents, 74
recrystallization, 73, 149, 150, 165
resistance, 2, 4, 30, 39, 91, 92, 104, 106, 112, 113, 115, 116, 117, 122, 123, 124, 130, 131, 138, 201
room temperature, 5, 47, 73, 74, 81, 82, 110, 120, 133, 142, 159, 168, 198, 203

S

scattering, 167, 176
selenium, 37, 40, 41, 46, 47, 50, 54, 55, 63, 67, 84, 86
semiconductors, 37, 62, 63
sensitivity, 76, 151
silicon, 40, 137
silver, 69, 92
SiO2, 191
sodium hydroxide, 77
solid phase, 43, 63
solid solutions, 179
solid surfaces, 101

solubility, 112, 151, 152, 204
solubility in water, 151, 204
solvents, 93, 145, 151, 152, 153, 156, 163, 165
solvothermal synthesis, 38
specific adsorption, 157
specific surface, 198
spectrophotometry, 40
spectroscopy, 106, 112, 145, 149, 169, 179, 190, 200
stoichiometry, 38, 43, 154
structural changes, 62, 151
substitution reaction, 69
substrates, 37, 40, 171
sulfuric acid, 106, 135, 136, 165
superconductor, 38
surface area, 2, 108, 120, 133, 142, 167, 171, 186, 187
surface layer, 133, 145, 149, 151
surface structure, 148, 155, 165
surface treatment, 148
synthesis, 38, 67, 68, 71, 72, 73, 82, 85, 86, 147, 163, 199

T

temperature, 4, 15, 17, 18, 19, 20, 21, 23, 37, 38, 40, 45, 47, 68, 94, 97, 99, 106, 111, 142, 149, 151, 155, 168, 169, 179, 182, 189, 190, 192, 196, 198, 202, 204
temperature dependence, 21
tetrahydrofuran, 81
tetrahydrofurane, 172
thermal treatment, 149, 156, 167, 171, 173, 174, 175, 178, 187, 194, 195, 196, 199, 202, 203
thermodynamical stability, 43
thermodynamics, 37, 38
thin films, 37, 38, 40, 45, 47, 48, 50, 52, 54, 55, 56, 60, 61, 63, 93, 200
thin film layers, 37, 48, 53, 56, 60, 63
tin oxide, 37, 39
transformation, 81
transition metal, 37, 38, 67, 68, 82, 85
transmission electron microscopy (TEM), 175
transport processes, 98
triphenylphosphine, 71, 81

V

vacuum, 5, 15, 18
viscosity, 20, 21, 25, 101, 106, 138, 141, 151, 153

W

water, 1, 4, 5, 6, 9, 14, 17, 18, 19, 20, 21, 23, 25, 26, 27, 28, 29, 30, 31, 32, 33, 39, 86, 92, 93, 94, 95, 96, 104, 111, 123, 124, 138, 147, 150, 151, 152, 153, 156, 159, 165, 166, 168, 172, 185, 188, 189, 191, 192
water clusters, 151
water permeability, 21, 25
water vapor, 19, 20, 21, 188, 189
wavelengths, 61, 62

X

X-ray analysis, 85
x-ray diffraction (XRD), 37 46, 48, 49, 50, 52, 53, 54, 55, 56, 61, 63, 85, 167, 173, 178, 182, 204, 208
X-ray photoelectron spectroscopy (XPS), 112, 144, 149, 150, 151, 184, 204, 209
X-rays, 167, 179, 210

Z

ZnO, 192